T0185373

ELECTRON CORRELATION IN METALS

The motion of electrons in metals is correlated so that they avoid each other in order to reduce the Coulomb repulsion energy. The purpose of this book is to describe the role of electron correlation in metals. Electron interaction as a result of Coulomb repulsion is very important to our understanding of magnetism and superconductivity in metals. The on-site electron interaction between strongly correlated electrons is the origin of magnetism and superconductivity. In this book the author describes the problem of magnetic moment in metals using the Anderson Hamiltonian, and the Kondo problem is explained on the basis of the Fermi liquid theory. Then, by using the Hubbard Hamiltonian, magnetism and the Mott-transition are discussed.

Superconductivity due to electron correlation is discussed. These systems are cuprates, organic superconductors, Sr_2RuO_4 and heavy fermions. The superconductivity in these systems arises universally from on-site Coulomb repulsion through the momentum dependence of electron interaction. The role of electron correlation in superconductivity is an important new issue in condensed matter physics since the discovery of high T_c superconductivity.

The theory of electron correlation is described on the basis of the Fermi liquid theory. In this book the Fermi liquid theory is described in detail and applied to various systems.

KOSAKU YAMADA has been a professor in the Department of Physics at Kyoto University since 1995.

ELECTRON CORRELATION IN METALS

K. YAMADA
Department of Physics, Kyoto University, Japan

CAMBRIDGE
UNIVERSITY PRESS

CAMBRIDGE UNIVERSITY PRESS
Cambridge, New York, Melbourne, Madrid, Cape Town, Singapore,
São Paulo, Delhi, Dubai, Tokyo, Mexico City

Cambridge University Press
The Edinburgh Building, Cambridge CB2 8RU, UK

Published in the United States of America by Cambridge University Press, New York

www.cambridge.org
Information on this title: www.cambridge.org/9780521147682

First published in Japanese 1993
English translation published 2004
First paperback printing 2010

A catalogue record for this publication is available from the British Library

Library of Congress Cataloguing in Publication data

Yamada, K. (Kosaku)–
[Denshi sokan. English]
p. cm.
Includes bibliographical references and index.
ISBN 0 521 57232 0 (hardback)
1. Free electron theory of metals. 2. Superconductivity. 3. Electron configuration. I. Title.

QC176.8.F74Y3613 2004
530.4′1 – dc22 2004045121

ISBN 978-0-521-57232-3 Hardback
ISBN 978-0-521-14768-2 Paperback

Contents

Preface

P. W. Anderson has achieved many brilliant theories in the wide field of condensed matter physics. His book titled *Basic Notions of the Condensed Matter Physics* was published in 1984. In this book Anderson stresses two basic principles of condensed matter physics. One of the principles is 'broken symmetry'. This means that condensed matter systems undergo phase transition to take a state possessing lower symmetry than that of the Hamiltonian. This statement corresponds to the appearance of a ferromagnetic state and a superconducting state, etc. at low temperatures. This principle manifests discontinuous change.

Another basic principle is the principle of 'adiabatic continuity'. This principle tells us that when we study a generally complicated physical system we can refer to a simple system that contains the essential nature of the real system and understand the complicated system on the basis of knowledge of the simple system. Anderson stresses that the most beautiful and appropriate example showing the importance of the continuity principle is Landau's Fermi liquid theory. Following the continuity principle, we start from a non-interacting Fermi gas and introduce interactions among particles gradually. There exists a one-to-one correspondence between the free particle system before the introduction of the interactions and the Fermi liquid after the introduction. It is the basic character of the Fermi liquid at low temperatures that we can introduce interactions as slowly as possible owing to the long lifetime of quasi-particles. Even though many-body interactions exist among particles, by considering quasi-particles renormalized by the interactions we can treat them as if they are free particles. By this procedure strongly interacting Fermi systems are much simplified. Strictly speaking, however, the systems cannot be completely free particle systems even after renormalization; there remain damping effects giving a finite lifetime and weak renormalized interactions among quasi-particles. In particular, since attractive forces make the Fermi surface unstable, it is only the repulsive force that can be continuously renormalized on the basis of the Fermi liquid theory. This fact plays an important role in many-body problems.

After reducing systems to weakly interacting quasi-particle systems following the Fermi liquid theory, we can then discuss phase transitions such as superconductivity, induced by the renormalized interactions among quasi-particles.

Fermi liquid theory, which was introduced by Landau for liquid ^{3}He, has been applied to electrons in metals and developed. Along with progress in the electron theory of solids, strongly interacting electron systems have been studied actively for various metals. By this development, the Fermi liquid theory has been made profound and based on microscopic foundations. After completion of the many-body theory of electron gas and the BCS theory of superconductivity, in the 1960s magnetism of transition metals and Mott transitions of their oxides were studied as main issues of electron correlation. Then, the discovery of the Kondo effect and the study of the Anderson Hamiltonian relating to the appearance and disappearance of localized moments in metals followed. Recently, we have studied heavy electrons whose masses are thousands of times as large as the free electron mass. These electrons are nothing but the quasi-particles in the Fermi liquid theory. Our present subject in electron correlation is the study of copper-oxide superconductors as metallic systems near the Mott insulator. These are called strongly correlated electron systems. The normal state of strongly correlated electron systems can be described as the Fermi liquid. In recent years it has been made clear that the Fermi liquid theory contains various fruitful physical properties. At present the Fermi liquid theory is in the course of development.

The Hubbard Hamiltonian contains a wealth of physics, such as metal–insulator transitions and magnetic transitions. Although the theory of the Hubbard Hamiltonian has been greatly developed, as seen in the study of the Mott transition using an infinite-dimensional model, we are still far from a complete understanding. Recently, the new problem of whether the Hubbard system can exhibit superconductivity has been added, as an important problem to be solved. In this book we show that superconductivity is actually realized owing to electron correlation in the Hubbard Hamiltonian.

The purpose of this book is to introduce the Fermi liquid theory and to describe the physics of strongly correlated electron systems. I intend to introduce unique and rich physical phenomena in each system mentioned above; I would like to describe the universality of the Fermi liquid and the basic principles common to various systems. At this point, I believe there exists a powerful theory that is not changed by details of experimental data. Nature is much more complicated than we can imagine, but it is not capricious.

To accomplish the purpose of this book, I intend to introduce the microscopic Fermi liquid theory developed by Luttinger *et al.* and describe the concepts confirmed through its development, although I am not sure whether I have succeeded in this task or not.

I am grateful to Professor Kei Yosida for valuable discussions over the course of 30 years. I should like to thank Drs Hiroshi Kontani and Hiroaki Ikeda for valuable discussions and comments during the preparation of this book. Professor Veljko Zlatić also gave me valuable advice.

This manuscript was typed by Kumiko Kishii. The author would like to express his sincere thanks to her for her continuous collaboration and encouragement.

1

Fermi gas

The basic properties of free electron systems are introduced. Then, the many-body effects of electron gas are discussed. The ground state energy is obtained by taking the screening effect into account.

1.1 Metals

Metals are composed of positive ions and conduction electrons making itinerant motion all over the crystal. A positive ion is composed of a nucleus and core electrons bounded around it. Conduction electrons lower the kinetic energy by making itinerant motion compared with the state bounded to a positive ion. This point is important in metallic cohesion.

For simple metals such as Na and Al, the Hamiltonian is given by

$$\mathcal{H} = \mathcal{H}_{\rm i} + \mathcal{H}_{\rm e} + \mathcal{H}_{\rm e-i}, \tag{1.1}$$

$$\mathcal{H}_{\rm i} = \sum_i \frac{\boldsymbol{P}_i^2}{2M} + \frac{1}{2}\sum_{i \neq j} V(\boldsymbol{R}_i - \boldsymbol{R}_j), \tag{1.2}$$

$$\mathcal{H}_{\rm e} = \sum_i \frac{\boldsymbol{p}_i^2}{2m} + \frac{1}{2}\sum_{i \neq j} \frac{e^2}{|\boldsymbol{r}_i - \boldsymbol{r}_j|}, \tag{1.3}$$

$$\mathcal{H}_{\rm e-i} = \sum_{ij} v(\boldsymbol{r}_i - \boldsymbol{R}_j). \tag{1.4}$$

Here, $\mathcal{H}_{\rm i}$ in (1.2) represents the positive ion system; we assume one kind of positive ion with mass M. $\boldsymbol{P}_i$ is the momentum of ion i and $V(\boldsymbol{R}_i - \boldsymbol{R}_j)$ is the potential between ions, which depends only on their distance. $\mathcal{H}_{\rm e}$ in (1.3) is the Hamiltonian for the electron system with electron mass m; the first and second terms denote the kinetic energy and the coulomb interaction between electrons, respectively. $\mathcal{H}_{\mathrm{e-i}}$ in (1.4) represents the potential between electrons and positive ions. The Hamiltonian combined with these terms can describe various properties including magnetism and superconductivity. In this book we discuss mainly the second term of (1.3), electron interaction. Owing to the coulomb interaction, electrons move so as to avoid each other. This kind of correlated electron motion is called 'electron correlation'.

When we are mainly interested in electron interaction, we simplify the positive ions by replacing them with a uniform positive charge distributed over the crystal and discuss only the term $\mathcal{H}_{\mathrm{e}}$. By this replacement, we can avoid the difficulty arising from a periodic potential and lattice vibrations.

1.2 Free Fermi gas

An electron is a Fermi particle with a spin of 1/2. There exist 10^{22}–10^{23} conduction electrons per cubic centimetre in most metals. For simplicity, let us ignore electron interactions among these, and also the periodic potential due to the positive ions, and consider the system composed of free electrons. We assume a system with N electrons in a cube of side $L = \Omega^{1/3}$. The wave-function $\varphi_{\boldsymbol{k}}(\boldsymbol{r})$ for a free electron with wave-vector $\boldsymbol{k}$ is given by

$$\varphi_{\boldsymbol{k}} = \frac{1}{\sqrt{\Omega}} e^{i\boldsymbol{k}\cdot\boldsymbol{r}}. \tag{1.5}$$

Here we take the periodic boundary condition

$$\varphi_{\boldsymbol{k}}(x+L, y, z) = \varphi_{\boldsymbol{k}}(x, y+L, z) = \varphi_{\boldsymbol{k}}(x, y, z+L) = \varphi_{\boldsymbol{k}}(x, y, z). \tag{1.6}$$

By substituting (1.5) into (1.6), we obtain

$$e^{ik_xL} = e^{ik_yL} = e^{ik_zL} = 1. \tag{1.7}$$

By this condition, values of $\boldsymbol{k}$ are given by integers n_1, n_2 and n_3 as

$$k_x = 2\pi n_1/L, \qquad k_y = 2\pi n_2/L, \qquad k_z = 2\pi n_3/L. \tag{1.8}$$

Thus, wave-vector $\boldsymbol{k}$ corresponds to a lattice point with unit of $2\pi/L$ in the wave-vector space.

The energy $\varepsilon_{\boldsymbol{k}}$ of a free electron with wave-vector $\boldsymbol{k}$ is given by

$$\varepsilon_{\boldsymbol{k}} = \frac{\hbar^2 k^2}{2m} = \frac{\hbar^2}{2m}\left(\frac{2\pi}{L}\right)^2 (n_1^2 + n_2^2 + n_3^2). \tag{1.9}$$

Let us construct the ground state composed of N free electrons. Each one-electron state specified by wave-vector $\boldsymbol{k}$ and spin quantum number σ can be occupied by only one electron because of the Pauli exclusion principle. In the ground state electrons occupy the N states from the lowest energy state to the Nth lowest state.

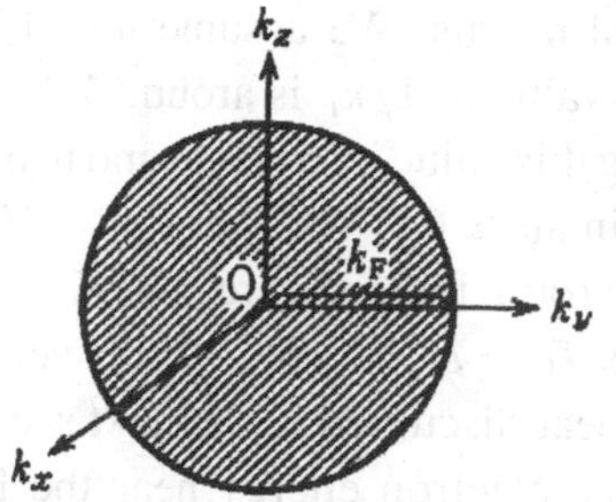

Fig. 1.1 Fermi sphere. Each $\boldsymbol{k}$ point inside the sphere with radius k_F is occupied by one up-spin electron and one down-spin electron.

The highest occupied energy ε_F and wave-number k_F are called the Fermi energy and Fermi wave-number, respectively; ε_F and k_F satisfy the relation

$$\varepsilon_F = \frac{\hbar^2}{2m} k_F{}^2. \tag{1.10}$$

Thus, we have the ground state in which two electrons with up- and down-spins occupy the state $\boldsymbol{k}$ within the sphere with radius k_F. Since the wave-vector $\boldsymbol{k}$ is situated at a lattice point spaced $2\pi/L$ apart, wave-vectors distribute in uniform density $(L/2\pi)^3 = \Omega/(2\pi)^3$. As a result, the electron number N is related to k_F by

$$N = \frac{2\Omega}{(2\pi)^3} \frac{4\pi}{3} k_F{}^3, \tag{1.11}$$

where factor 2 represents the degree of freedom arising from electron spin. From (1.11), k_F is given by electron density $n = N/\Omega$ as

$$k_F = (3\pi^2 n)^{1/3}. \tag{1.12}$$

Let us calculate the number of electron states between energy E and $E + \Delta E$, ΔE being an infinitesimal energy. Assuming the total number of states below energy E as $N(E)$, we obtain

$$N(E + \Delta E) - N(E) = \frac{dN(E)}{dE} \Delta E. \tag{1.13}$$

Here,

$$dN(E)/dE = \rho(E) \tag{1.14}$$

is the energy density of states. Using (1.10) and (1.11), we obtain

$$\rho(E) = \frac{dN(k)}{dk} \frac{dk}{dE} = \frac{2\Omega}{(2\pi)^3} 4\pi k^2 \frac{m}{\hbar^2 k} = \frac{\Omega k m}{\pi^2 \hbar^2} = \frac{\Omega m}{\pi^2 \hbar^3} \sqrt{2mE}. \tag{1.15}$$

The density of states for a free electron system is proportional to $\sqrt{E}$.

Now let us consider real metals. We assume $n = 10^{22}\,\text{cm}^{-3}$ and obtain $k_\text{F} \simeq 10^8\,\text{cm}^{-1}$ from (1.12). The value of $1/k_\text{F}$ is around $1\,\text{Å} = 10^{-8}\,\text{cm}$, corresponding to atomic distance. Inserting this value into (1.10) and using $\hbar = 1 \times 10^{-27}$ erg s and $m = 9 \times 10^{-28}$ g, we obtain $\varepsilon_\text{F} \simeq 6 \times 10^{-12}\,\text{erg} \simeq 4\,\text{eV}$. Since 1 eV is the energy corresponding to 10^4 K, the room temperature, 300 K, is sufficiently low compared with the Fermi temperature, $T_\text{F} = \varepsilon_\text{F}/k_\text{B}$. This fact is very important in understanding the electronic specific heat discussed in the next section.

The level splitting of one-electron energy near the Fermi surface is given by $\Delta\varepsilon = \varepsilon_\text{F}/N \simeq 10^{-22}\,\text{eV} \simeq 10^{-18} k_\text{B}$ K. As a result, electron–hole pair excitations near the Fermi energy can be created with vanishingly small excitation energy. These electron–hole pair excitations exist in an infinite number. Thus, the conduction electron system in the Fermi degeneracy is degenerate in the infinite number of states. In general, the degenerated states suffer a strong effect even under small perturbations. The special nature of the Fermi surface, which can be called 'fragility', plays an important role in the orthogonality theorem and the theory of superconductivity.

1.3 Electronic specific heat and Pauli susceptibility

If we apply the principle of equipartition to the free electron system to calculate the specific heat, we obtain the following result. The internal energy W is given by

$$W = \frac{3}{2} N k_\text{B} T, \tag{1.16}$$

k_B being the Boltzmann constant. The specific heat C_V takes a constant value,

$$C_\text{V} = \frac{dW}{dT} = \frac{3}{2} N k_\text{B}. \tag{1.17}$$

This value is expected to give the same order of contributions as the lattice specific heat around room temperature. However, in actual metals we cannot observe such a large electronic specific heat at room temperature. This is because the room temperature is too low compared with the Fermi temperature for the principle of equipartition to be applicable. Only the electrons limited within the narrow width of temperature in the vicinity of the Fermi energy contribute to the specific heat.

Now, let us calculate correctly the specific heat due to free electrons. Using the chemical potential μ, the distribution of electrons is given by the Fermi distribution function

$$f(\varepsilon_k) = \left[1 + \exp\left(\frac{\varepsilon_k - \mu}{k_\text{B} T}\right)\right]^{-1}. \tag{1.18}$$

In this case the internal energy of the electron system W is given by

$$W = 2\sum_{k} \varepsilon_k f(\varepsilon_k), \tag{1.19}$$

where the factor 2 arises from spin degeneracy. Using the density of states $\rho(\varepsilon_k)$ including both spins, we obtain

$$W = \int_0^\infty \varepsilon\rho(\varepsilon)f(\varepsilon)d\varepsilon. \tag{1.20}$$

The total number of electrons N is given by

$$N = \sum_{k\sigma} f(\varepsilon_k) = \int_0^\infty \rho(\varepsilon)f(\varepsilon)d\varepsilon. \tag{1.21}$$

This equation at $T = 0$ becomes

$$N = \int_0^{\varepsilon_F} \rho(\varepsilon)d\varepsilon. \tag{1.22}$$

Now let us consider the following integral I to accomplish the calculation at low temperatures:

$$I = \int_0^\infty g(\varepsilon)f(\varepsilon)d\varepsilon. \tag{1.23}$$

Here, $g(\varepsilon)$ is a smooth function of energy ε. Using a partial integration, we obtain

$$I = [G(\varepsilon)f(\varepsilon)]_0^\infty - \int_0^\infty G(\varepsilon)\frac{\partial f}{\partial \varepsilon}d\varepsilon, \tag{1.24}$$

$$G(\varepsilon) = \int_0^\varepsilon g(\varepsilon)d\varepsilon. \tag{1.25}$$

The first term of (1.24) vanishes because $f(\infty) = 0$. To calculate the second term, we expand $G(\varepsilon)$ around $\varepsilon = \mu$. Writing the nth derivative of G as $G^{(n)}$, we obtain

$$G(\varepsilon) = G(\mu) + (\varepsilon - \mu)G'(\mu) + \frac{1}{2}(\varepsilon - \mu)^2 G''(\mu) + \cdots \tag{1.26}$$

Inserting this into (1.24), we get

$$I = G(\mu)\int_0^\infty \left(-\frac{\partial f}{\partial \varepsilon}\right) d\varepsilon + G'(\mu)\int_0^\infty (\varepsilon - \mu)\left(-\frac{\partial f}{\partial \varepsilon}\right) d\varepsilon + \cdots$$
$$+ \frac{G^{(n)}(\mu)}{n!}\int_0^\infty (\varepsilon - \mu)^n \left(-\frac{\partial f}{\partial \varepsilon}\right) d\varepsilon + \cdots \tag{1.27}$$

The first term gives $G(\mu)$. The general terms are given by

$$\frac{1}{n!}\int_0^{\infty}(\varepsilon-\mu)^n\left(-\frac{\partial f}{\partial\varepsilon}\right)d\varepsilon = \frac{(k_\mathrm{B}T)^n}{n!}\int_{-\infty}^{\infty}\frac{z^n}{(e^z+1)(1+e^{-z})}dz$$
$$=\begin{cases}2c_n(k_\mathrm{B}T)^n & (n\ \text{even})\\ 0 & (n\ \text{odd}).\end{cases} \tag{1.28}$$

As an example, for $n=2$,

$$2c_2=\frac{1}{2}\int_{-\infty}^{\infty}\frac{z^2dz}{(e^z+1)(1+e^{-z})}=\frac{\pi^2}{6}. \tag{1.29}$$

As a result, I is given by

$$I=\int_0^{\mu}g(\varepsilon)d\varepsilon+\frac{\pi^2}{6}(k_\mathrm{B}T)^2\left[\frac{\partial g(\varepsilon)}{\partial\varepsilon}\right]_{\varepsilon=\mu}+\cdots \tag{1.30}$$

Applying this result to (1.20) and (1.21), we obtain

$$W=\int_0^{\mu}\varepsilon\rho(\varepsilon)d\varepsilon+\frac{\pi^2}{6}(k_\mathrm{B}T)^2\left[\frac{\partial}{\partial\varepsilon}(\varepsilon\rho(\varepsilon))\right]_{\varepsilon=\mu}+\cdots \tag{1.31}$$

$$\begin{aligned}C_\mathrm{V}=\frac{dW}{dT}&=\mu\rho(\mu)\frac{d\mu}{dT}+\frac{\pi^2}{3}{k_\mathrm{B}}^2T\left[\rho(\varepsilon)+\mu\frac{\partial\rho}{\partial\varepsilon}\right]_{\varepsilon=\mu}+O(T^2)\\ &=\frac{\pi^2}{3}{k_\mathrm{B}}^2\rho(\varepsilon_\mathrm{F})T+\mu\rho(\mu)\left[\frac{d\mu}{dT}+\frac{\pi^2{k_\mathrm{B}}^2T}{3\rho(\mu)}\frac{\partial\rho}{\partial\varepsilon}\right]_{\varepsilon=\mu}\\ &=\frac{\pi^2}{3}{k_\mathrm{B}}^2\rho(\varepsilon_\mathrm{F})T.\end{aligned} \tag{1.32}$$

In (1.32) we have used the shift of chemical potential

$$\mu=\varepsilon_\mathrm{F}-\frac{\pi^2}{6}(k_\mathrm{B}T)^2\left[\frac{\partial}{\partial\varepsilon}\log\rho(\varepsilon)\right]_{\varepsilon=\mu}, \tag{1.33}$$

which is obtained by substituting $g(\varepsilon)=\rho(\varepsilon)$ in (1.30) and shifting μ so as to conserve the total electron number N.

Thus, the electronic specific heat at low temperatures is proportional to the density of states $\rho(\varepsilon_\mathrm{F})$ on the Fermi surface and given by the T-linear term as shown in (1.32). Equation (1.32) can be written as

$$\begin{aligned}C_\mathrm{V}&=\gamma T,\\ \gamma&=\frac{\pi^2}{3}{k_\mathrm{B}}^2\rho(\varepsilon_\mathrm{F}).\end{aligned} \tag{1.34}$$

Here, the coefficient of the specific heat γ is called the Sommerfeld constant.

Let us apply a weak magnetic field H to the free electron system at low temperature and obtain the expression for the Pauli susceptibility. Using a g-value of $g = 2$ and the Bohr magneton μ_B, the Zeeman energy with spin σ is given by $g\sigma\mu_B H/2 = \sigma\mu_B H$. The Zeeman energy induces the magnetization given by $\Delta M = \mu_B(\delta n_\downarrow - \delta n_\uparrow) = {\mu_B}^2\rho(\varepsilon_F)H$. Thus, the magnetic susceptibility χ is given by

$$\chi = \Delta M/H = {\mu_B}^2\rho(\varepsilon_F). \tag{1.35}$$

The Pauli susceptibility is proportional to the density of states $\rho(\varepsilon_F)$ on the Fermi surface; this is common to the coefficient of specific heat γ.

1.4 Many-body effect of electron gas

The effects of coulomb interaction on electron gas had been made clear in the 1950s by the efforts of many people, such as Bohm, Pines, Nozières, Gell-Mann, Brueckner and Sawada. As shown below, initially, by a straight perturbation calculation, the essential points of the problem were made clear. Then, the difficult key problems were solved by physical considerations, creating new concepts and developing the methods of the many-body problem. Since the history of the investigations on electron gas is instructive, we describe it here in detail as a starting point for the many body problem [1–6].

In order to discuss mainly the effect of electron interaction, we assume the electron gas model in which the positive charge due to ions is replaced by a uniform one. As a result, in this model the uniform distribution of electrons in space cancels with the positive background to give no effect. The deviation from the uniform distribution gives rise to the coulomb interaction among electrons. The Hamiltonian of this system is given by $\mathcal{H}_e$ in (1.3) and is written as

$$\mathcal{H}_e = \sum_i \frac{{\boldsymbol{p}_i}^2}{2m} + \frac{1}{2}\sum_{i\neq j} \frac{e^2}{|\boldsymbol{r}_i - \boldsymbol{r}_j|}, \tag{1.36}$$

where $\boldsymbol{r}_i$ and $\boldsymbol{p}_i$ are the position and momentum of electron i, respectively. Here we define electron density $\rho(\boldsymbol{r})$ and its Fourier transform $\rho_{\boldsymbol{q}}$:

$$\rho(\boldsymbol{r}) = \sum_i \delta(\boldsymbol{r} - \boldsymbol{r}_i) = \sum_{\boldsymbol{q}} \rho_{\boldsymbol{q}} e^{i\boldsymbol{q}\cdot\boldsymbol{r}}, \tag{1.37}$$

$$\rho_{\boldsymbol{q}} = \frac{1}{\Omega}\sum_i e^{-i\boldsymbol{q}\cdot\boldsymbol{r}_i} = {\rho_{-\boldsymbol{q}}}^\dagger. \tag{1.38}$$

The $\boldsymbol{q} = 0$ component of $\rho_{\boldsymbol{q}}$, $\rho_0 = N_e/\Omega = n$, is the average electron density and cancels with the uniform positive charge. Using $\rho_{\boldsymbol{q}}$, we can write (1.36) as

$$\mathcal{H}_e = \sum_i \frac{\boldsymbol{p}_i^{\,2}}{2m} + \frac{1}{2}\sum_{\boldsymbol{q}} V_{\boldsymbol{q}}(\Omega \rho_{\boldsymbol{q}}{}^{\dagger}\rho_{\boldsymbol{q}} - n), \tag{1.39}$$

where $V_{\boldsymbol{q}}$ is the Fourier transform of the coulomb interaction e^2/r:

$$V_{\boldsymbol{q}} = \frac{4\pi e^2}{q^2}. \tag{1.40}$$

Here it is noted that since

$$\rho_{\boldsymbol{q}}{}^{\dagger}\rho_{\boldsymbol{q}} = \frac{1}{\Omega^2}\sum_{ij} e^{i\boldsymbol{q}\cdot(\boldsymbol{r}_i - \boldsymbol{r}_j)}, \tag{1.41}$$

the term $i = j$ gives n/Ω and (1.39) excludes the coulomb interaction with $\boldsymbol{r}_i = \boldsymbol{r}_j$, in agreement with (1.36).

To treat the coulomb interaction as a perturbation, let us represent (1.36) in the second quantization form (see Appendix B). The coulomb integral is written by the wave-functions $\varphi_{\sigma_1}(\boldsymbol{r}_1)$ and $\varphi_{\sigma_2}(\boldsymbol{r}_2)$, as

$$\iint d\boldsymbol{r}_1 d\boldsymbol{r}_2 \varphi_{\sigma_1}{}^*(\boldsymbol{r}_1)\varphi_{\sigma_2}{}^*(\boldsymbol{r}_2)\frac{e^2}{|\boldsymbol{r}_1 - \boldsymbol{r}_2|}\varphi_{\sigma_2}(\boldsymbol{r}_2)\varphi_{\sigma_1}(\boldsymbol{r}_1). \tag{1.42}$$

By expanding the wave-function $\varphi_\sigma(\boldsymbol{r})$ with the plane waves as

$$\varphi_\sigma(\boldsymbol{r}) = \sum_{\boldsymbol{k}} a_{\boldsymbol{k}\sigma}\frac{1}{\sqrt{\Omega}}e^{i\boldsymbol{k}\cdot\boldsymbol{r}}, \tag{1.43}$$

we write the coulomb interaction as

$$\mathcal{H}_C = \frac{1}{\Omega}\sum_{\boldsymbol{q}\neq 0}\frac{2\pi e^2}{q^2}\sum_{\substack{\boldsymbol{k}_1\boldsymbol{k}_2\\ \sigma_1\sigma_2}} a_{\boldsymbol{k}_1+\boldsymbol{q}\sigma_1}{}^{\dagger}a_{\boldsymbol{k}_2-\boldsymbol{q}\sigma_2}{}^{\dagger}a_{\boldsymbol{k}_2\sigma_2}a_{\boldsymbol{k}_1\sigma_1}. \tag{1.44}$$

Here $a_{\boldsymbol{k}\sigma}$ ($a_{\boldsymbol{k}\sigma}{}^{\dagger}$) is the annihilation (creation) operator of the electron with wave-vector $\boldsymbol{k}$ and spin σ. The summation over $\boldsymbol{q}$ in (1.44) excludes the part of $\boldsymbol{q} = 0$ because it cancels with the positive charge.

The density fluctuation $\rho_{\boldsymbol{q}}$ is given by

$$\rho_{\boldsymbol{q}} = \frac{1}{\Omega}\sum_{\boldsymbol{k}\sigma} a_{\boldsymbol{k}-\boldsymbol{q}\sigma}{}^{\dagger}a_{\boldsymbol{k}\sigma}. \tag{1.45}$$

Thus, $\mathcal{H}_e$ is given in the second quantization as

$$\mathcal{H}_e = \sum_{k\sigma} \varepsilon_k a_{k\sigma}{}^\dagger a_{k\sigma} + \frac{1}{2\Omega} \sum_{\substack{kk',q\neq 0\\ \sigma\sigma'}} V_q a_{k+q\sigma}{}^\dagger a_{k'-q\sigma'}{}^\dagger a_{k'\sigma'} a_{k\sigma}, \tag{1.46}$$

where ε_k is given by

$$\varepsilon_k = \frac{\hbar^2 k^2}{2m}. \tag{1.47}$$

Operators $a_{k\sigma}{}^\dagger$ and $a_{k\sigma}$ satisfy the commutation rule for Fermi particles (see Appendix B):

$$\begin{aligned} \left[a_{k\sigma}, a_{k'\sigma'}\right]_+ &= \left[a_{k\sigma}{}^\dagger, a_{k'\sigma'}{}^\dagger\right]_+ = 0, \\ \left[a_{k\sigma}, a_{k'\sigma'}{}^\dagger\right]_+ &= \delta_{k,k'}\delta_{\sigma,\sigma'}. \end{aligned} \tag{1.48}$$

Now we study the effect of the coulomb interaction using the above results. First of all, let us calculate the ground state energy by regarding the second term, the coulomb interaction, as a perturbation on the first term, the kinetic energy. The unperturbed state given by the first term is the Fermi sphere occupied up to k_F by two electrons with up and down spin. We write it as $|0\rangle$ and obtain

$$n_{k\sigma} = \langle 0|a_{k\sigma}{}^\dagger a_{k\sigma}|0\rangle = \begin{cases} 1 & (k < k_F) \\ 0 & (k > k_F). \end{cases} \tag{1.49}$$

The average kinetic energy per electron in the unperturbed state is given by

$$\varepsilon_{kin} = \frac{1}{N_e}\langle 0|\sum_{k\sigma} \varepsilon_k a_{k\sigma}{}^\dagger a_{k\sigma}|0\rangle = \frac{3}{5}\varepsilon_F, \tag{1.50}$$

where ε_F is the Fermi energy given by (1.10). The result of (1.50) is obtained from

$$\varepsilon_{kin} = \int_0^{k_F} \frac{\hbar^2 k^2}{2m} 4\pi k^2 dk \bigg/ \int_0^{k_F} 4\pi k^2 dk. \tag{1.51}$$

Now let us introduce r_0 representing electron density as

$$\frac{\Omega}{N_e} = \frac{1}{n} = \frac{4\pi}{3} r_0{}^3. \tag{1.52}$$

The radius r_0 of the sphere for one electron is replaced by the dimensionless parameter r_s, which is given in units of Bohr radius as

$$r_s = r_0/a_B. \tag{1.53}$$

The Fermi energy ε_F is given by (1.12) as

$$\varepsilon_F = \frac{\hbar^2 k_F{}^2}{2m} = \frac{\hbar^2}{2m}(3\pi^2 n)^{2/3} = \left(\frac{9\pi}{4}\right)^{2/3} \frac{1}{r_s{}^2}\,\text{Ry}. \tag{1.54}$$

The unit of energy Ry corresponds to the ionization energy of atomic hydrogen:

$$1\,\text{Ry} = \frac{me^4}{2\hbar^2} = \frac{e^2}{2a_B} \simeq 13.5\,\text{eV} = 2.17 \times 10^{-11}\,\text{erg}. \tag{1.55}$$

As a final result, the average kinetic energy of (1.50) is given by

$$\varepsilon_{\text{kin}} = \frac{3}{5}\varepsilon_F \simeq \frac{2.21}{r_s{}^2}\,\text{Ry}. \tag{1.56}$$

1.5 Exchange energy

The first-order perturbation term with respect to the coulomb interaction in the ground state energy is given by

$$\begin{aligned} E_1 &= \frac{1}{\Omega}\sum_{\substack{kk'\\ q\sigma}} \frac{V_q}{2}\langle 0|a_{k+q\sigma}{}^\dagger a_{k'-q\sigma'}{}^\dagger a_{k'\sigma'} a_{k\sigma}|0\rangle \\ &= \frac{1}{\Omega}\left\{\sum_{\substack{kk'\\ \sigma\sigma'}} \frac{V_{q=0}}{2} n_{k\sigma} n_{k'\sigma'} + \sum_{\substack{kq\\ \sigma}} -\frac{V_q}{2} n_{k+q\sigma} n_{k\sigma}\right\}. \end{aligned} \tag{1.57}$$

The first term corresponding to $\boldsymbol{q} = 0$ in (1.57) cancels with the positive charge of the background. The second term with $\boldsymbol{q} = \boldsymbol{k}' - \boldsymbol{k}$, to which only the electrons possessing parallel spins contribute, arises from the exchange integral of the coulomb interaction and possesses a negative sign. The term denoted as E_{ex} is given by

$$\begin{aligned} E_{\text{ex}} &= -\frac{1}{\Omega}\sum_{\substack{kq\\ \sigma}} \frac{V_q}{2} n_{k+q\sigma} n_{k\sigma} = \frac{2}{\Omega}\sum_{\substack{k_1<k_F\\ k_2<k_F}} \frac{-2\pi e^2}{|\boldsymbol{k}_1 - \boldsymbol{k}_2|^2} \\ &= -\sum_{\substack{k_1<k_F\\ \sigma}} \frac{e^2}{2\pi}\left\{\left(\frac{k_F{}^2 - k_1{}^2}{2k_1}\right)\log\left|\frac{k_F + k_1}{k_F - k_1}\right| + k_F\right\}. \end{aligned} \tag{1.58}$$

Then, by integrating this over k_1,

$$E_{\text{ex}} = -2\frac{\Omega}{(2\pi)^3} e^2 k_F{}^4. \tag{1.59}$$

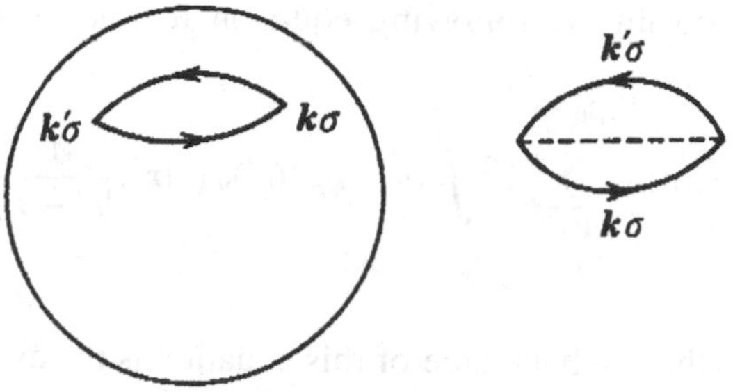

Fig. 1.2 Diagram of the exchange term.

Thus, the exchange energy per electron is given by

$$\varepsilon_{\text{ex}} = \frac{E_{\text{ex}}}{N_{\text{e}}} = -\frac{3}{2\pi}\left(\frac{9\pi}{4}\right)^{1/3}\frac{1}{r_{\text{s}}}\,\text{Ry} = -\frac{0.916}{r_{\text{s}}}\,\text{Ry}. \tag{1.60}$$

Compared with (1.56), this term is higher by one order in r_{s}. As a result, for small r_{s}, namely the high density case, this correction becomes relatively small. Before proceeding to higher-order terms, we study in detail the properties of the exchange term. By defining the first-order term of the self-energy $\Sigma^{(1)}(\boldsymbol{k})$ as

$$\Sigma^{(1)}(\boldsymbol{k}) = -\frac{1}{2\Omega}\sum_{q} V_q n_{\boldsymbol{k}+\boldsymbol{q}} = -\frac{e^2 k_{\text{F}}}{4\pi} F\left(\frac{k}{k_{\text{F}}}\right), \tag{1.61}$$

$$F(x) = 2 + \frac{1-x^2}{x}\log\left|\frac{1+x}{1-x}\right|, \tag{1.62}$$

we can express (1.58) as

$$E_{\text{ex}} = \sum_{k\sigma} \Sigma^{(1)}(\boldsymbol{k}) n_{\boldsymbol{k}\sigma}. \tag{1.63}$$

Here $\Sigma(\boldsymbol{k})$ represents the shift of one-electron energy due to the electron interaction. By adding the first-order term of the self-energy, the one-electron energy becomes

$$\tilde{\varepsilon}_{\boldsymbol{k}} = \frac{\hbar^2 \boldsymbol{k}^2}{2m} + \Sigma^{(1)}(\boldsymbol{k}).$$

However, from (1.61) the derivative $d\tilde{\varepsilon}_{\boldsymbol{k}}/d\boldsymbol{k}$ diverges on the Fermi surface and the density of states at the Fermi energy vanishes. This result gives rise to the difficulty that the electron system does not give the T-linear specific heat. This difficulty is solved by cutting the long-range part of the coulomb interaction, as will be shown in the following sections.

Now we consider the physical meaning of the exchange term using the Hartree–Fock equation [6]. By treating the coulomb interaction with the Hartree–Fock approximation, we obtain the following equation for the plane wave φ_k given by (1.5):

$$-\frac{\hbar^2\nabla^2}{2m}\varphi_{k\sigma}(\boldsymbol{r}) - \sum_{k'<k_F} e^2 \int d\boldsymbol{r}' \varphi_{k'\sigma}{}^*(\boldsymbol{r}')\varphi_{k\sigma}(\boldsymbol{r}')\frac{1}{|\boldsymbol{r}'-\boldsymbol{r}|}\varphi_{k'\sigma}(\boldsymbol{r}) = E_k\varphi_{k\sigma}(\boldsymbol{r}). \quad (1.64)$$

The second term on the left-hand side of this equation is the exchange term. Here we consider a nonlocal operator $\hat{A}$ defined as

$$\hat{A}\varphi_k(\boldsymbol{r}) = \int d\boldsymbol{r}' A(\boldsymbol{r},\boldsymbol{r}')\varphi_k(\boldsymbol{r}'), \quad (1.65)$$

$$A(\boldsymbol{r},\boldsymbol{r}') = \sum_{k'<k_F} \frac{e^2}{|\boldsymbol{r}-\boldsymbol{r}'|}\varphi_{k'}{}^*(\boldsymbol{r}')\varphi_{k'}(\boldsymbol{r}) = \frac{e^2}{\Omega}\sum_{k'<k_F}\frac{e^{i\boldsymbol{k}'\cdot(\boldsymbol{r}-\boldsymbol{r}')}}{|\boldsymbol{r}-\boldsymbol{r}'|}$$
$$= \frac{e^2}{|\boldsymbol{r}-\boldsymbol{r}'|}\rho(\boldsymbol{r}-\boldsymbol{r}'), \quad (1.66)$$

where $\rho(\boldsymbol{r}-\boldsymbol{r}')$ is given by

$$\rho(\boldsymbol{r}-\boldsymbol{r}') = \frac{1}{\Omega}\sum_{k'<k_F} e^{i\boldsymbol{k}'\cdot(\boldsymbol{r}-\boldsymbol{r}')}$$
$$= \frac{k_F{}^3}{2\pi^2}\left\{\frac{\sin(k_F|\boldsymbol{r}-\boldsymbol{r}'|)}{(k_F|\boldsymbol{r}-\boldsymbol{r}'|)^3} - \frac{\cos(k_F|\boldsymbol{r}-\boldsymbol{r}'|)}{(k_F|\boldsymbol{r}-\boldsymbol{r}'|)^2}\right\}. \quad (1.67)$$

The second term on the left-hand side of (1.64) gives a negative potential to the electron situated at $\boldsymbol{r}$. The result shows that electrons at $\boldsymbol{r}'$, giving the potential, distribute around $\boldsymbol{r}$ as shown in (1.67). Figure 1.3 shows this situation; electrons

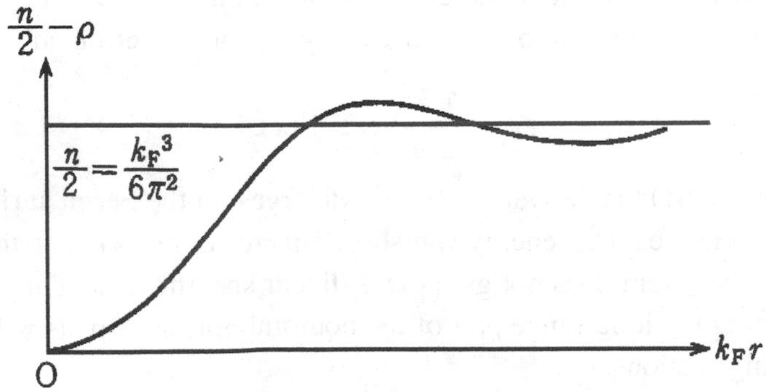

Fig. 1.3 Distribution of electrons with spin parallel to that of the electron situated at $\boldsymbol{r} = 0$. Owing to the exchange term, each electron accompanies one positive hole with parallel spin, which is called the exchange hole.

with the same spin as that of the electron at $\boldsymbol{r}$ distribute as if there exists one hole in the uniform distribution $n/2$. The one hole is called the exchange hole.

The exchange term included in the Hartree–Fock approximation is not directly related to the electron correlation due to the coulomb interaction between electrons. The electrons possessing the same spins cannot take the same momentum $\boldsymbol{k}$ owing to the restriction of the Pauli principle, and cannot approach each other. As a result, the coulomb interaction between electrons with parallel spins is reduced; the exchange term represents the reduction of the coulomb interaction. As stated below, electrons with antiparallel spins are also expected to move so as not to approach each other, owing to the repulsive coulomb interaction. As a result, the wave-functions of electrons are transformed so as to reduce the coulomb energy.

In this case the electron correlation arises not because of the statistical nature of the Pauli principle, but because of the interaction. We call this effect 'electron correlation'. This is the main subject of this book.

1.6 Screening effect

To consider the interaction between electrons, we assume that an electron stays at a position and consider it as an impurity in the electron gas [4]. For this purpose we assume a perturbed potential given by

$$\delta u(\boldsymbol{r}, t) = u e^{i\boldsymbol{q}\cdot\boldsymbol{r}} e^{i\omega t + \alpha t}. \tag{1.68}$$

Here the potential is assumed to oscillate in space and time with Fourier components $\boldsymbol{q}$ and ω, respectively. The time constant α is a positive infinitesimal and $\delta u = 0$ at $t = -\infty$. The unperturbed wave-function of the electron is written as the plane wave

$$\varphi_{\boldsymbol{k}} = \frac{1}{\sqrt{\Omega}} \exp\left[i \left(\boldsymbol{k}\cdot\boldsymbol{r} - \frac{\varepsilon_{\boldsymbol{k}}}{\hbar} t \right) \right]. \tag{1.69}$$

By the perturbation (1.68), $\varphi_{\boldsymbol{k}}$ changes into

$$\Psi_{\boldsymbol{k}} = \varphi_{\boldsymbol{k}} + b_{\boldsymbol{k}+\boldsymbol{q}}(t)\varphi_{\boldsymbol{k}+\boldsymbol{q}} e^{i(\varepsilon_{\boldsymbol{k}+\boldsymbol{q}} - \varepsilon_{\boldsymbol{k}})t/\hbar}. \tag{1.70}$$

The coefficient $b_{\boldsymbol{k}+\boldsymbol{q}}(t)$ is given by

$$b_{\boldsymbol{k}+\boldsymbol{q}}(t) = \frac{u e^{i\omega t + \alpha t}}{\varepsilon_{\boldsymbol{k}} - \varepsilon_{\boldsymbol{k}+\boldsymbol{q}} \pm \hbar\omega \mp i\hbar\alpha}. \tag{1.71}$$

The change of charge distribution due to (1.70) becomes

$$\begin{aligned} \delta\rho(\boldsymbol{r}, t) &= e \sum_{\boldsymbol{k}\sigma} \left\{ |\Psi_{\boldsymbol{k}}(\boldsymbol{r}, t)|^2 - 1/\Omega \right\} f(\boldsymbol{k}) \\ &\simeq \frac{e}{\Omega} \sum_{\boldsymbol{k}\sigma} \left\{ b_{\boldsymbol{k}+\boldsymbol{q}}(t) e^{i\boldsymbol{q}\cdot\boldsymbol{r}} + b_{\boldsymbol{k}+\boldsymbol{q}}{}^*(t) e^{-i\boldsymbol{q}\cdot\boldsymbol{r}} \right\} f(\boldsymbol{k}), \end{aligned} \tag{1.72}$$

where $f(\boldsymbol{k}) = f(\varepsilon_{\boldsymbol{k}})$ is the Fermi distribution function. To make $\delta\rho(\boldsymbol{r}, t)$ a real number, we add δu^*, and obtain

$$\delta\rho = \frac{e}{\Omega}\sum_{k\sigma} f(\boldsymbol{k})\left\{\frac{u}{\varepsilon_{\boldsymbol{k}} - \varepsilon_{\boldsymbol{k}+\boldsymbol{q}} \pm \hbar\omega \mp i\hbar\alpha} + \frac{u}{\varepsilon_{\boldsymbol{k}} - \varepsilon_{\boldsymbol{k}-\boldsymbol{q}} \mp \hbar\omega \pm i\hbar\alpha}\right\} e^{i\boldsymbol{q}\cdot\boldsymbol{r}+i\omega t+\alpha t} + \text{C.C.} \tag{1.73}$$

where C.C. means the complex conjugate of the first term. Equation (1.73) is rewritten as

$$\delta\rho = \frac{eu}{\Omega}\sum_{k\sigma}\left\{\frac{f(\boldsymbol{k}) - f(\boldsymbol{k} \mp \boldsymbol{q})}{\varepsilon_{\boldsymbol{k}} - \varepsilon_{\boldsymbol{k}\pm\boldsymbol{q}} + \hbar\omega - i\hbar\alpha}\right\} e^{i\boldsymbol{q}\cdot\boldsymbol{r}+i\omega t+\alpha t} + \text{C.C.} \tag{1.74}$$

The induced charge distribution $\delta\rho$ gives rise to the potential $\delta\Phi(\boldsymbol{r}, t)$ following the Poisson equation:

$$\nabla^2(\delta\Phi) = -4\pi e\delta\rho. \tag{1.75}$$

The $\delta\Phi(\boldsymbol{r}, t)$ term is assumed to show the same space and time dependence as δu:

$$\delta\Phi(\boldsymbol{r}, t) = \Phi e^{i\boldsymbol{q}\cdot\boldsymbol{r}+i\omega t+\alpha t} + \text{C.C.} \tag{1.76}$$

From (1.75), Φ is given by

$$\Phi = \frac{4\pi e^2}{q^2\Omega}\sum_{k\sigma}\frac{f(\boldsymbol{k}) - f(\boldsymbol{k} \mp \boldsymbol{q})}{\varepsilon_{\boldsymbol{k}} - \varepsilon_{\boldsymbol{k}\pm\boldsymbol{q}} + \hbar\omega - i\hbar\alpha}u. \tag{1.77}$$

This is the potential created by the redistribution of charge due to δu assumed first. This induced potential $\delta\Phi$ is added to the external potential $\delta V(\boldsymbol{r}, t)$ to give rise to δu determined self-consistently:

$$\delta u(\boldsymbol{r}, t) = \delta V(\boldsymbol{r}, t) + \delta\Phi(\boldsymbol{r}, t) \tag{1.78}$$

$$\delta V(\boldsymbol{r}, t) = V e^{i\boldsymbol{q}\cdot\boldsymbol{r}+i\omega t+\alpha t} + \text{C.C.} \tag{1.79}$$

As a result, the potential u induced by the external potential V is given by

$$u = V + \left\{\frac{4\pi e^2}{q^2\Omega}\sum_{k\sigma}\frac{f(\boldsymbol{k}) - f(\boldsymbol{k} \mp \boldsymbol{q})}{\varepsilon_{\boldsymbol{k}} - \varepsilon_{\boldsymbol{k}\pm\boldsymbol{q}} + \hbar\omega - i\hbar\alpha}\right\}u. \tag{1.80}$$

By introducing the dielectric function $\varepsilon(\boldsymbol{q}, \omega)$ as

$$\varepsilon(\boldsymbol{q}, \omega) = 1 + \frac{4\pi e^2}{q^2\Omega}\sum_{k\sigma}\frac{f(\boldsymbol{k}) - f(\boldsymbol{k} \mp \boldsymbol{q})}{\varepsilon_{\boldsymbol{k}\pm\boldsymbol{q}} - \varepsilon_{\boldsymbol{k}} - \hbar\omega + i\hbar\alpha}, \tag{1.81}$$

we obtain

$$u = \frac{V}{\varepsilon(\boldsymbol{q}, \omega)}. \tag{1.82}$$

Thus, we obtain finally the following result. When the external potential $\delta V(\boldsymbol{r}, t)$,

$$\delta V(\boldsymbol{r}, t) = \iint V(\boldsymbol{q}, \omega) e^{i\boldsymbol{q}\cdot\boldsymbol{r}+i\omega t} d\boldsymbol{q} d\omega \tag{1.83}$$

is applied to the electron system, the electrons screen it and $\delta u(\boldsymbol{r}, t)$ is realized as

$$\delta u(\boldsymbol{r}, t) = \iint \frac{V(\boldsymbol{q}, \omega)}{\varepsilon(\boldsymbol{q}, \omega)} e^{i\boldsymbol{q}\cdot\boldsymbol{r}+i\omega t} d\boldsymbol{q} d\omega. \tag{1.84}$$

To make clear the physical meaning of the screening effect, we consider the case of the static and smooth spatial change, by putting $\omega = 0$ and $\boldsymbol{q} \simeq 0$ in (1.81):

$$\varepsilon_{\boldsymbol{k}+\boldsymbol{q}} - \varepsilon_{\boldsymbol{k}} \simeq \boldsymbol{q} \cdot \nabla_{\boldsymbol{k}} \varepsilon_{\boldsymbol{k}}. \tag{1.85}$$

Here $f(\boldsymbol{k})$ is a function of $\varepsilon_{\boldsymbol{k}}$:

$$f(\boldsymbol{k}) - f(\boldsymbol{k}+\boldsymbol{q}) \simeq -\boldsymbol{q} \cdot \frac{\partial f}{\partial \varepsilon_{\boldsymbol{k}}} \nabla_{\boldsymbol{k}} \varepsilon_{\boldsymbol{k}}. \tag{1.86}$$

Using the above approximation, we obtain

$$\varepsilon(\boldsymbol{q}, 0) \simeq 1 + \frac{\lambda^2}{q^2}, \qquad (\boldsymbol{q} \to 0) \tag{1.87}$$

$$\lambda^2 = 4\pi e^2 \rho(\varepsilon_{\mathrm{F}})/\Omega, \tag{1.88}$$

where $\rho(\varepsilon_{\mathrm{F}})$ is the density of states at the Fermi energy. According to (1.87), the dielectric constant ε becomes infinite for the long wavelength component, $\boldsymbol{q} \to 0$. This means that in metals the screening is complete by changing the electron distribution. By applying (1.87) to the coulomb interaction, $V(\boldsymbol{q}) = 4\pi e^2/q^2$, we obtain

$$\frac{1}{(2\pi)^3} \int d\boldsymbol{q} \frac{4\pi e^2}{q^2} \left(\frac{q^2}{q^2 + \lambda^2} \right) e^{i\boldsymbol{q}\cdot\boldsymbol{r}} = \frac{e^2}{r} e^{-\lambda r}. \tag{1.89}$$

The coulomb interaction is screened at a distance of λ^{-1}. By substituting (1.15) into $\rho(\varepsilon_{\mathrm{F}})$ in (1.88), we obtain λ as

$$\lambda = \left(\frac{4me^2}{\pi\hbar^2} k_{\mathrm{F}} \right)^{1/2}. \tag{1.90}$$

Using the Bohr radius $a_{\mathrm{B}} = \hbar^2/me^2$, we rewrite λ as

$$\lambda \simeq \left(\frac{4}{\pi} \frac{k_{\mathrm{F}}}{a_{\mathrm{B}}} \right)^{1/2}. \tag{1.91}$$

The screening length $1/\lambda$ is nearly equal to the atomic distance. By considering the electron itself as the origin of the local charge, we can see that the electron charge is screened by the other electrons within the atomic distance. That is, electrons avoid each other and make a uniform charge distribution in a long-range scale.

Here it should be noted that, as described in Chapter 3, the short wavelength oscillation $\cos 2k_{\mathrm{F}}r$ corresponding to $q = 2k_{\mathrm{F}}$, called the Friedel oscillation, decreases with amplitude $1/r^3$ and extends over longer distance than the exponential decay near $\boldsymbol{q} = 0$.

1.7 Plasma oscillation

To study the motion of electron gas we will consider the Fourier transform of electron density $\rho_{\boldsymbol{q}}$ in (1.38). By putting $\dot{\rho}_{\boldsymbol{q}} = d\rho_{\boldsymbol{q}}/dt$, we obtain

$$\dot{\rho}_{\boldsymbol{q}} = [\rho_{\boldsymbol{q}}, \mathcal{H}]/i\hbar. \tag{1.92}$$

Here we use $\mathcal{H}_{\mathrm{e}}$ in (1.39) as our Hamiltonian:

$$\mathcal{H} = \sum_i \frac{\boldsymbol{p}_i{}^2}{2m} + \sum_{\boldsymbol{q}} \frac{2\pi e^2}{q^2}(\Omega \rho_{\boldsymbol{q}}{}^\dagger \rho_{\boldsymbol{q}} - n)$$

and obtain

$$\dot{\rho}_{\boldsymbol{q}} = -i\frac{1}{\Omega}\sum_i \left(\frac{\boldsymbol{q}\cdot\boldsymbol{p}_i}{m} + \frac{\hbar \boldsymbol{q}^2}{2m}\right) e^{-i\boldsymbol{q}\cdot\boldsymbol{r}_i}. \tag{1.93}$$

Further, calculating $\ddot{\rho}_{\boldsymbol{q}}$ from the commutation relation between $\dot{\rho}_{\boldsymbol{q}}$ and $\mathcal{H}$, we obtain

$$\ddot{\rho}_{\boldsymbol{q}} = -\frac{1}{\Omega}\sum_i \left(\frac{\boldsymbol{q}\cdot\boldsymbol{p}_i}{m} + \frac{\hbar \boldsymbol{q}^2}{2m}\right)^2 e^{-i\boldsymbol{q}\cdot\boldsymbol{r}_i} - \sum_{\boldsymbol{q}'} \frac{4\pi e^2}{m q'^2} \boldsymbol{q}\cdot\boldsymbol{q}' \rho_{\boldsymbol{q}-\boldsymbol{q}'} \rho_{\boldsymbol{q}'}, \tag{1.94}$$

where the second term on the right-hand side is derived by operating the first term in (1.93) to the coulomb interaction term in $\mathcal{H}$. By extracting the term with $\boldsymbol{q}' = \boldsymbol{q}$ from the second term of (1.94), we obtain

$$\ddot{\rho}_{\boldsymbol{q}} + \omega_{\mathrm{p}}{}^2 \rho_{\boldsymbol{q}} = -\frac{1}{\Omega}\sum_i \left(\frac{\boldsymbol{q}\cdot\boldsymbol{p}_i}{m} + \frac{\hbar \boldsymbol{q}^2}{2m}\right)^2 e^{-i\boldsymbol{q}\cdot\boldsymbol{r}_i} - \sum_{\boldsymbol{q}'\neq\boldsymbol{q}} \frac{4\pi e^2}{m \boldsymbol{q}'^2} \boldsymbol{q}\cdot\boldsymbol{q}' \rho_{\boldsymbol{q}-\boldsymbol{q}'} \rho_{\boldsymbol{q}'}. \tag{1.95}$$

Here, using $\rho_{\boldsymbol{q}=0} = n$, we have put

$$\omega_{\mathrm{p}} = (4\pi n e^2/m)^{1/2}. \tag{1.96}$$

When the right-hand side is small, (1.95) is written as

$$\ddot{\rho}_{\boldsymbol{q}} + \omega_{\mathrm{p}}{}^2 \rho_{\boldsymbol{q}} = 0. \tag{1.97}$$

Thus, ρ_q oscillates with frequency ω_p. This oscillation of electron density is called the plasma oscillation. The frequency is estimated as $\omega_p \simeq 2 \times 10^{16}\,\mathrm{s}^{-1}$ when $n \sim 10^{23}\,\mathrm{cm}^{-3}$, and the energy $\hbar\omega_p$ is about 12 eV. This plasma oscillation is not excited at room temperature. There exists only the zero point oscillation. Thus, in most metals, there is no electron density fluctuation and electrons move so as to keep the electron density constant.

In (1.97) we have assumed that the right-hand side of (1.95) is small. Now let us discuss the condition of the assumption. The $\rho_{q-q'}$ in the second term becomes, for $\boldsymbol{q}' \neq \boldsymbol{q}$,

$$\rho_{q-q'} = \frac{1}{\Omega} \sum_i e^{i(\boldsymbol{q}-\boldsymbol{q}')\cdot \boldsymbol{r}_i}. \tag{1.98}$$

This contribution becomes small by taking the sum over terms with random phases. We call it the random phase approximation, in which the sum over terms possessing the random phases is ignored. The first term becomes of order $\boldsymbol{q}^2 {v_F}^2 \rho_q$ for small $\boldsymbol{q}$, by approximating $\boldsymbol{p}_i/m \simeq v_F$, v_F being the Fermi velocity. Thus, the condition for (1.97) to hold is given by

$$q < \frac{\omega_p}{v_F} \simeq \left(\frac{4\pi n e^2}{m {v_F}^2}\right)^{1/2} \simeq \left(\frac{2\pi n e^2}{\varepsilon_F}\right)^{1/2} \simeq \left(\frac{2\pi \rho(\varepsilon_F) e^2}{\Omega}\right)^{1/2} = \frac{\lambda}{\sqrt{2}}, \tag{1.99}$$

where we have used λ given by (1.88). From (1.99), we can see that (1.97) holds for $\boldsymbol{q}$ corresponding to wavelengths longer than the screening length $1/\lambda$. As a result, the freedom of electron motion with small $\boldsymbol{q}$, wavelength longer than the screening length $1/\lambda$, can be described as the plasma oscillation. The rest freedoms appear as the electron system interacting via the screened coulomb interaction. From (1.52), we obtain

$$n = \left(\frac{4\pi}{3} {r_0}^3\right)^{-1} = \frac{2}{(2\pi)^3} \frac{4\pi}{3} {k_F}^3, \tag{1.100}$$

$$r_s = \left(\frac{9\pi}{4}\right)^{1/3} \Big/ (a_B k_F). \tag{1.101}$$

Thus, λ given by (1.91) is given by

$$\lambda \simeq \left(\frac{12}{\pi}\right)^{1/3} \frac{1}{a_B r_s^{1/2}}. \tag{1.102}$$

From the above result, the coulomb interaction given by the second term of (1.46) is cut in a small $\boldsymbol{q}$ region by λ proportional to $1/\sqrt{r_s}$ to give no divergence in the $\boldsymbol{q} = 0$ limit.

1.8 Ground state energy

We have calculated, in (1.60) of Section 1.5, up to the first-order term of the ground state energy with respect to the coulomb interaction. In Section 1.6, we have shown that the long-range part of the coulomb interaction is cut by the screening effect. As a result, we have only to treat the short-range part with $|\boldsymbol{q}| > q_c = \lambda$ in our perturbation expansion.

Now we proceed to the second-order term. It is given by two processes, the direct process $E_2{}^{(a)}$ and the exchange process $E_2{}^{(b)}$, as shown in Figs. 1.4 and 1.5, respectively. The energy denominator of the direct process is given by

$$E_{kk'}(\boldsymbol{q}) = \frac{\hbar^2}{2m}\{(\boldsymbol{k}-\boldsymbol{q})^2 + (\boldsymbol{k}'+\boldsymbol{q})^2 - k^2 - k'^2\} = \frac{\hbar^2}{m}\boldsymbol{q}\cdot(\boldsymbol{k}'-\boldsymbol{k}+\boldsymbol{q}). \quad (1.103)$$

The energy of the direct process $E_2{}^{(a)}$ is given by

$$E_2{}^{(a)} = -8\sum_{kk'q}\left(\frac{2\pi e^2}{q^2}\right)^2 \frac{m}{\hbar^2\boldsymbol{q}\cdot(\boldsymbol{k}'-\boldsymbol{k}+\boldsymbol{q})} f_k(1-f_{k-q})f_{k'}(1-f_{k'+q}), \quad (1.104)$$

where we have included the factor 4 due to the spin sum and 2 due to the same two terms given by the contraction. The contribution from the exchange term shown in

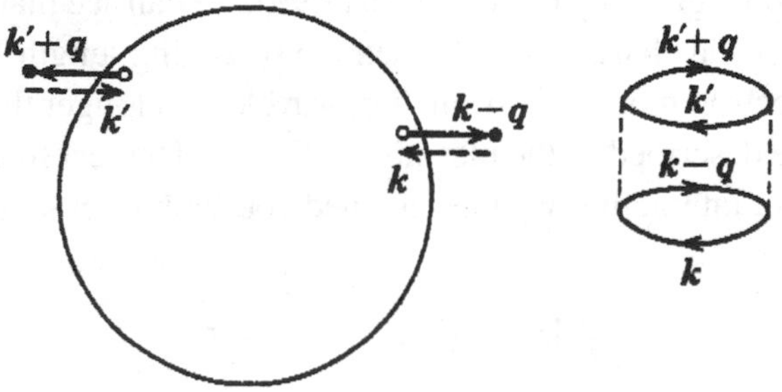

Fig. 1.4 Direct process in second order.

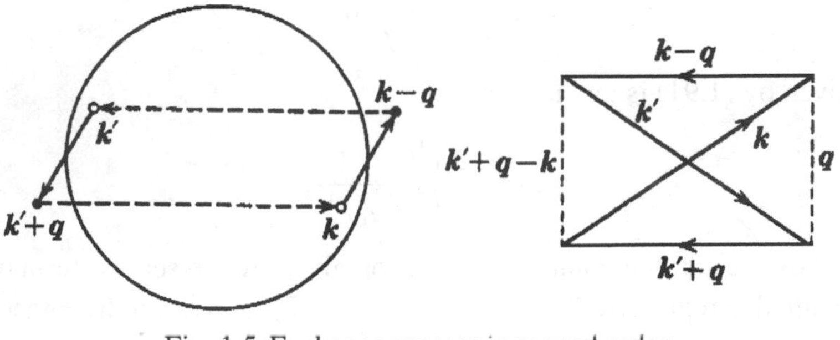

Fig. 1.5 Exchange process in second order.

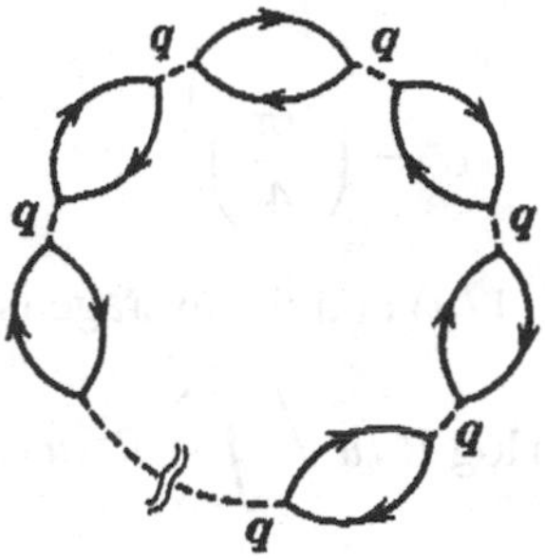

Fig. 1.6 Diagram gives the most dominant term, in which the coulomb interactions transferring the small momentum $\boldsymbol{q}$ are repeated.

Fig. 1.5 is written, with the factor 2 from the spin sum and 2 from the contraction process, as

$$E_2^{(b)} = 4\sum_{kk'q} \frac{2\pi e^2}{q^2} \frac{2\pi e^2}{(\boldsymbol{k}'-\boldsymbol{k}+\boldsymbol{q})^2} \frac{m}{\hbar^2 \boldsymbol{q}\cdot(\boldsymbol{k}'-\boldsymbol{k}+\boldsymbol{q})} f_k(1-f_{k-q}) f_{k'}(1-f_{k'+q}). \tag{1.105}$$

As for the first-order term, evaluating the energy $\varepsilon_2{}^a$ per electron in units of Ry and renormalizing the momentum as $\boldsymbol{k}_1 = \boldsymbol{k}/k_{\mathrm{F}}$, $\boldsymbol{k}_2 = \boldsymbol{k}'/k_{\mathrm{F}}$ and $\boldsymbol{q} = \boldsymbol{q}/k_{\mathrm{F}}$, we obtain

$$\begin{aligned} \varepsilon_2{}^a &= \frac{E_2{}^{(a)}}{N_{\mathrm{e}}} \Big/ \frac{me^4}{2\hbar^2} = -\frac{3}{8}\frac{1}{\pi^5} \int \frac{d\boldsymbol{q}}{q^4} \int_{\substack{k_1<1\\|\boldsymbol{k}_1+\boldsymbol{q}|>1}} d\boldsymbol{k}_1 \int_{\substack{k_2<1\\|\boldsymbol{k}_2+\boldsymbol{q}|>1}} \frac{d\boldsymbol{k}_2}{q^2+\boldsymbol{q}\cdot(\boldsymbol{k}_1+\boldsymbol{k}_2)} \\ &= \frac{4}{\pi^2}(1-\log 2)\log q_{\mathrm{c}}, \end{aligned} \tag{1.106}$$

where q_{c} is a cutoff momentum and proportional to $r_{\mathrm{s}}{}^{1/2}$, from (1.101) and (1.102):

$$q_{\mathrm{c}} = \lambda/k_{\mathrm{F}} \propto r_{\mathrm{s}}{}^{1/2}. \tag{1.107}$$

Here, if we put $q_{\mathrm{c}} = 0$, (1.106) diverges logarithmically. On the other hand, the second-order exchange term $\varepsilon_2{}^b = E_2{}^{(b)}/N_{\mathrm{e}}$ is given, in units of Ry, as

$$\varepsilon_2{}^b \simeq 0.046\,\mathrm{Ry}. \tag{1.108}$$

This value is a constant independent of r_{s}.

From the example of second-order terms, we can see that, as shown in Fig. 1.6, the interaction terms transferring the small momentum $\boldsymbol{q}$ make dominant contributions. According to the calculation done by Gell-Mann and Brueckner [7], the contributions from the higher-order terms than second order are given by

$$\varepsilon_{\mathrm{c}} = \frac{2}{\pi^2}(1-\log 2)\left[\log\frac{4\alpha r_{\mathrm{s}}}{\pi} + \langle\log R\rangle_{\mathrm{av}} - \frac{1}{2}\right], \tag{1.109}$$

where

$$\alpha = \left(\frac{9\pi}{4}\right)^{1/3}. \tag{1.110}$$

The function $R = 1 - u\tan^{-1}(1/u)$ and the average is given by

$$\langle \log R \rangle_{\rm av} = \int_{-\infty}^{\infty} R^2(u) \log R du \bigg/ \int_{-\infty}^{\infty} R^2(u) du = -0.551. \tag{1.111}$$

Using this value, we obtain the correlation energy $\varepsilon_{\rm c}$;

$$\varepsilon_{\rm c} = 0.0622 \log r_{\rm s} - 0.096. \tag{1.112}$$

Combining (1.56), (1.60) and (1.112), we obtain the total energy $\varepsilon_{\rm t}$ as

$$\varepsilon_{\rm t} = \frac{2.21}{{r_{\rm s}}^2} - \frac{0.916}{r_{\rm s}} + 0.0622 \log r_{\rm s} - 0.096. \tag{1.113}$$

This result is the most reliable in the limit of small $r_{\rm s}$, namely high electron density.

It is noted that the ground state energy given by the diagram in Fig. 1.6 can be calculated from the dielectric function $\varepsilon(\boldsymbol{q}, \omega)$ in (1.81) obtained by the random phase approximation (RPA), by using the Feynman relation in Appendix A.

1.9 Wigner crystal

The kinetic energy of an electron is proportional to ${r_{\rm s}}^{-2}$, as seen in (1.56). On the other hand, the coulomb interaction is proportional to ${r_{\rm s}}^{-1}$. As a result, if we decrease the electron density to increase $r_{\rm s}$, the potential energy due to the coulomb interaction should dominate over the kinetic energy. This fact shows that, in the limit of low electron density, electrons localize at positions far from each other so as to lower the energy due to the coulomb interaction. That is, electron crystallization occurs. This crystal, proposed by Wigner [8], is called the Wigner crystal.

Let us assume that one electron is localized in a unit cell, whose volume is equal to the volume of a sphere with radius r_0, and calculate the energy of the system. When the electron is situated at a position $\boldsymbol{r}$ from the centre of the sphere ($r < r_0$), the potential $V(\boldsymbol{r})$ due to the uniform positive charge is given by

$$V(\boldsymbol{r}) = \left\{ e \int_{r'>r}^{r_0} \frac{d\boldsymbol{r}'}{|\boldsymbol{r}' - \boldsymbol{r}|} + e \int_{r'<r} \frac{d\boldsymbol{r}'}{|\boldsymbol{r}' - \boldsymbol{r}|} \right\} \frac{3}{4\pi {r_0}^3} = \frac{3}{2}\frac{e}{r_0} - \frac{er^2}{2{r_0}^3}. \tag{1.114}$$

Thus, the Hamiltonian for the electron at $\boldsymbol{r}$ from the centre is written as

$$\mathcal{H} = \frac{\boldsymbol{p}^2}{2m} + \frac{e^2 r^2}{2{r_0}^3} - \frac{3}{2}\frac{e^2}{r_0}. \tag{1.115}$$

The third term on the right-hand side is constant and the motion of the electron is determined by the first two terms. Since the potential is proportional to r^2, the electron makes a harmonic oscillation. Its frequency is given by

$$\omega^2 = \frac{e^2}{m{r_0}^3} = \frac{{\omega_{\mathrm{p}}}^2}{3}. \tag{1.116}$$

Here, we have used ${\omega_{\mathrm{p}}}^2 = 4\pi ne^2/m$ in (1.96) and $n = 3/4\pi {r_0}^3$. The energy of the crystallized electron is given by

$$\varepsilon_{\mathrm{sol}} = -\frac{3}{2}\frac{e^2}{r_0} + \frac{\hbar\omega_{\mathrm{p}}}{2}\sqrt{3}. \tag{1.117}$$

The second term on the right-hand side arises from the energy of the zero point motion; the factor 3 due to the number of vibration modes is included there. By representing $\varepsilon_{\mathrm{sol}}$ in r_{s}, we obtain

$$\varepsilon_{\mathrm{sol}} = -\frac{3}{r_{\mathrm{s}}} + \frac{3}{{r_{\mathrm{s}}}^{3/2}}\,\mathrm{Ry}. \tag{1.118}$$

On the other hand, in the Hartree–Fock approximation, electrons distribute uniformly all over the crystal. Adding the exchange energy to the interaction energy among electrons and that with the positive charge, we obtain the energy $\varepsilon_{\mathrm{HF}}$:

$$\varepsilon_{\mathrm{HF}} = \frac{1.2}{r_{\mathrm{s}}} - \frac{2.4}{r_{\mathrm{s}}} - \frac{0.92}{r_{\mathrm{s}}} = -\frac{2.12}{r_{\mathrm{s}}}\,\mathrm{Ry}. \tag{1.119}$$

Actually, we can confirm that (1.118) is lower than (1.119) in energy, when r_{s} is large.

References

[1] D. Pines, *Elementary Excitations in Solids* (Benjamin, 1964).
[2] P. Nozières, *Theory of Interacting Fermi Systems* (Benjamin, 1964).
[3] D. Pines, *The Many Body Problem* (Benjamin, 1961).
[4] J. M. Ziman, *Principle of the Theory of Solids* (Cambridge University Press, 1971).
[5] P. W. Anderson, *Basic Notions of Condensed Matter Physics* (Benjamin, 1984).
[6] P. W. Anderson, *Concepts in Solids* (Benjamin, 1963).
[7] M. Gell-Mann and K. A. Brueckner, *Phys. Rev.* **106** (1957) 364.
[8] E. P. Wigner, *Phys. Rev.* **46** (1934) 1002.

2

Fermi liquid theory

We will consider the normal state of interacting Fermi particles without any long-range order. The normal state at low temperatures is called the Fermi liquid and is considered to be the system of free quasi-particles that is continuously connected with free Fermi gas. The concept of the Fermi liquid was introduced and developed by L. D. Landau [1]. Landau's Fermi liquid theory, which concentrates rich contents into a simple theory, is a good example to solve many-body problems. In this chapter, we introduce the basis and main contents of the Fermi liquid.

2.1 Principle of continuity

The basis of the Fermi liquid theory is the principle of adiabatic continuity, which connects free Fermi gas with the Fermi liquid by introducing gradually an interaction among particles. There is a one-to-one correspondence between two states before and after the introduction of the interaction. States belonging to the same symmetry do not cross each other, and new states with the interaction can be represented by the old quantum number. Since the distribution of free Fermi gas is given by the Fermi distribution function $n(\boldsymbol{k})$, that of the Fermi liquid is also written $n(\boldsymbol{k})$. The state denoted by $\boldsymbol{k}, \sigma$ in the Fermi liquid is called the quasi-particle. For the system to be described by quasi-particles, the following condition is necessary.

Let us consider the basis of the Fermi liquid theory following the explanation of the Landau theory given by Nozières [2] and Anderson [3]. Here, we define R as the rate introducing the electron interaction. The time necessary to introduce the interaction τ_0 is given by $R = \hbar/\tau_0$. Small R means the slow introduction of interaction and corresponds to the high resolution of energy. Since we discuss the excitations corresponding to the temperature T, it is necessary for our resolution to be finer than T:

$$R < k_{\mathrm{B}}T. \tag{2.1}$$

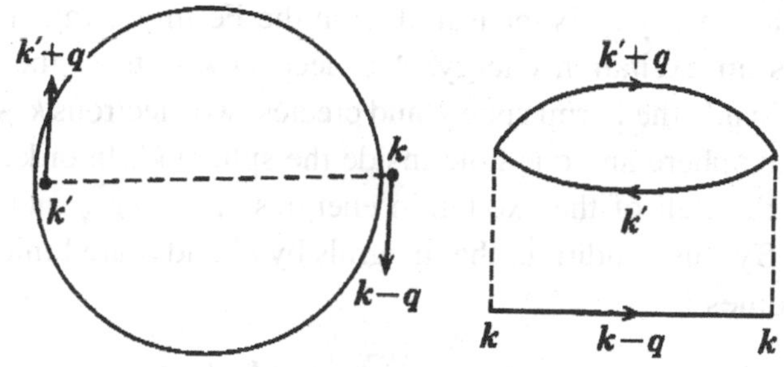

Fig. 2.1 Second-order term of self-energy. The broken lines and full lines denote interaction $V(\boldsymbol{q})$ and electron or hole, respectively.

On the other hand, we cannot consider the quasi-particles constructed by the interaction, unless the interaction is introduced more quickly than the damping rate of quasi-particles $1/\tau$, τ being the lifetime of quasi-particles:

$$R > \hbar/\tau. \tag{2.2}$$

For the Fermi liquid theory to hold, it is necessary for both (2.1) and (2.2) to be satisfied. That is, R should satisfy

$$k_B T > R > \hbar/\tau. \tag{2.3}$$

For this condition to be satisfied, the damping rate of quasi-particles h/τ must be smaller than $k_B T$. This condition is satisfied at sufficiently low temperatures below the Fermi temperature $T_F = \varepsilon_F/k_B$. This fact can be shown as follows. We introduce the Fourier transform of the electron interaction $V(\boldsymbol{r})$:

$$V(\boldsymbol{q}) = \int d\boldsymbol{r} e^{i\boldsymbol{q}\cdot\boldsymbol{r}} V(\boldsymbol{r}), \tag{2.4}$$

and consider an electron system interacting by this interaction. As shown in Fig. 2.1, we consider the second-order process, in which an electron $\boldsymbol{k}$ excited outside the Fermi sphere interacts with $\boldsymbol{k}'$ inside the sphere, and they are scattered into $\boldsymbol{k}-\boldsymbol{q}$ and $\boldsymbol{k}'+\boldsymbol{q}$, respectively. Then they return to $\boldsymbol{k}$ and $\boldsymbol{k}'$, respectively via the interaction $V(\boldsymbol{q})$. The self-energy of the electron $\boldsymbol{k}$ by this process is written

$$\Sigma^{(2)}(\boldsymbol{k}, \varepsilon_k) = \int d\boldsymbol{q} \int d\boldsymbol{k}' \frac{|V(\boldsymbol{q})|^2}{D - i\eta}, \tag{2.5}$$

where the denominator $D = \varepsilon_{\boldsymbol{k}} + \varepsilon_{\boldsymbol{k}'} - \varepsilon_{\boldsymbol{k}'+\boldsymbol{q}} - \varepsilon_{\boldsymbol{k}-\boldsymbol{q}}$ and η is a positive infinitesimal. The damping rate $1/\tau$ is given by

$$\begin{aligned} \frac{1}{\tau} &= \frac{1}{\hbar} \lim_{\eta\to 0} \int d\boldsymbol{q} \int d\boldsymbol{k}' |V(\boldsymbol{q})|^2 \,\mathrm{Im}\left(\frac{1}{D - i\eta}\right) \\ &= \frac{\pi}{\hbar} \int d\boldsymbol{q} \int d\boldsymbol{k}' \delta(D) |V(\boldsymbol{q})|^2. \end{aligned} \tag{2.6}$$

For simplicity, the energy ε_k is measured from the Fermi energy ε_F, that is, $\varepsilon_k = \hbar^2 k^2/2m - \varepsilon_F$ is an excitation energy. An electron $\boldsymbol{k}$ outside the Fermi sphere interacts with $\boldsymbol{k}'$ inside the Fermi sphere and creates two electrons $\boldsymbol{k}' + \boldsymbol{q}$ and $\boldsymbol{k} - \boldsymbol{q}$ outside the Fermi sphere and one hole inside the sphere $\boldsymbol{k}'$. In order to satisfy the energy conservation, all of the excitation energies $\varepsilon_{k'+q}$, ε_{k-q} and $-\varepsilon_{k'}$ must be smaller than ε_k. By this condition, the integrals by $\boldsymbol{k}'$ and $\boldsymbol{q}$ are limited by ε_k. As a result, (2.6) becomes

$$\frac{\hbar}{\tau} = \pi V^2 {\varepsilon_k}^2 \rho^3 = \pi \rho V^2 \left(\frac{k_B T}{\varepsilon_F} \right)^2. \tag{2.7}$$

Here we have assumed the density of states $\rho = 1/\varepsilon_F$ and the excitation energy ε_k is of order $k_B T$. The ratio of the damping rate to $k_B T$ is given by

$$\frac{\hbar}{\tau} \bigg/ k_B T = \pi \left(\frac{V}{\varepsilon_F} \right)^2 \frac{k_B T}{\varepsilon_F}. \tag{2.8}$$

As far as $k_B T \ll \varepsilon_F$, (2.3) holds. The reason that the damping rate $\hbar/\tau$ becomes small is due to the marked reduction of the scattering rate of degenerate electrons originating from the energy conservation and the Pauli principle. Thus the quasi-particles in the vicinity of the Fermi surface possess a long lifetime.

On the other hand, the real part of the self-energy is given by

$$\Delta\varepsilon_k = \mathrm{Re}\Sigma^{(2)}(\boldsymbol{k}, \varepsilon_k) = \int d\boldsymbol{q} \int d\boldsymbol{k}' \frac{|V(\boldsymbol{q})|^2}{D}. \tag{2.9}$$

This part is not always small. As discussed later, this energy shift $\Delta\varepsilon_k$ changes the mass and the velocity of quasi-particles.

The important point of Landau's Fermi liquid theory is that there exists a one-to-one correspondence between states of the free Fermi gas and those of the interacting Fermi liquid. Let us write the state of quasi-particle $\boldsymbol{k}$ added to the Fermi sphere $|0\rangle$ as ${Q_k}^\dagger|0\rangle$. When this state is represented in terms of the original bare particles, it is given by a linear combination of the states possessing many electron–hole pairs. With the coefficients, $\Gamma_1, \Gamma_2, \ldots$

$$\begin{aligned} {Q_k}^\dagger|0\rangle = \sqrt{z_k}\Big\{ {C_k}^\dagger &+ \sum_{k_1 k_2 k_3} \Gamma_1 {C_{k_1}}^\dagger {C_{k_2}}^\dagger C_{k_3} \\ &+ \sum_{k_1 k_2 \cdots k_5} \Gamma_2 {C_{k_1}}^\dagger {C_{k_2}}^\dagger {C_{k_3}}^\dagger C_{k_4} C_{k_5} + \cdots \Big\} |0\rangle. \end{aligned} \tag{2.10}$$

As shown in (2.10), the quasi-particle ${Q_k}^\dagger$ corresponds to the bare particle ${C_k}^\dagger$. ${Q_k}^\dagger|0\rangle$ contains the states accompanying electron–hole pair excitations in addition to ${C_k}^\dagger|0\rangle$. Each collision process conserves the charge, particle number, momentum and spin. As a result, every term on the right-hand side of (2.10) possesses one charge, one electron and the same momentum $\boldsymbol{k}$ as a result of the sum over all electrons and holes. For example, $\boldsymbol{k}_1 + \boldsymbol{k}_2 - \boldsymbol{k}_3 = \boldsymbol{k}$ and $\boldsymbol{k}_1 + \boldsymbol{k}_2 + \boldsymbol{k}_3 - \boldsymbol{k}_4 - \boldsymbol{k}_5 = \boldsymbol{k}$ hold in the second and third terms, respectively. Thus the renormalized state on the right-hand side can be considered as the quasi-particle $Q_{\boldsymbol{k}}{}^{\dagger}$. The value $z_{\boldsymbol{k}}$ is called the wave-function renormalization factor and expresses the weight of the bare particle $C_{\boldsymbol{k}}{}^{\dagger}$ contained in the quasi-particle $Q_{\boldsymbol{k}}{}^{\dagger}$. Although we have ignored any spin suffix, the same conservation law also holds for spin.

Starting with free particles and increasing gradually the strength of interaction among them, we can introduce the quasi-particles corresponding to the free particles. The important thing here is that the transformation is continuous with respect to the strength of interaction.

The continuity in the transformation from a free Fermi gas to an interacting Fermi liquid guarantees that physical quantities are analytic with respect to the strength of interaction, and the perturbation expansion is possible. We can see an actual example of continuity in the Anderson Hamiltonian discussed in Chapter 5.

Moreover, as a result of the one-to-one correspondence between $Q_{\boldsymbol{k}}{}^{\dagger}$ and $C_{\boldsymbol{k}}{}^{\dagger}$, the number of occupied states does not change and the volume of the Fermi sphere is invariant in the transformation from the free Fermi gas to the interacting Fermi liquid. The Fermi surface in general form can be transformed by the interaction, but the volume surrounded by the surface is invariant. Since this property was proved on the basis of the microscopic theory by Luttinger [4], this is called the Luttinger theorem.

Finally, we add a comment; (2.10) expresses the transformation from a bare particle with wave-vector $\boldsymbol{k}$ to a quasi-particle accompanying electron–hole pair excitations. On the other hand, (2.6) gives the transition probability with which the electron with wave-vector $\boldsymbol{k}$ decays into an electron–hole pair and an electron. By repeating this process the quasi-particle $\boldsymbol{k}$ decays into incoherent states. As a result, we can consider τ as the lifetime of the quasi-particle.

2.2 Landau's Fermi liquid theory

Hereafter, we introduce the Landau theory following Nozières [2].

Let us consider a free Fermi gas with N particles in volume Ω. At $T = 0$, particles occupy up to the Fermi wave-number k_{F} as shown in Fig. 2.2. The wave-number k_{F} is determined by

$$\frac{N}{\Omega} = \frac{k_{\mathrm{F}}{}^3}{3\pi^2}. \tag{2.11}$$

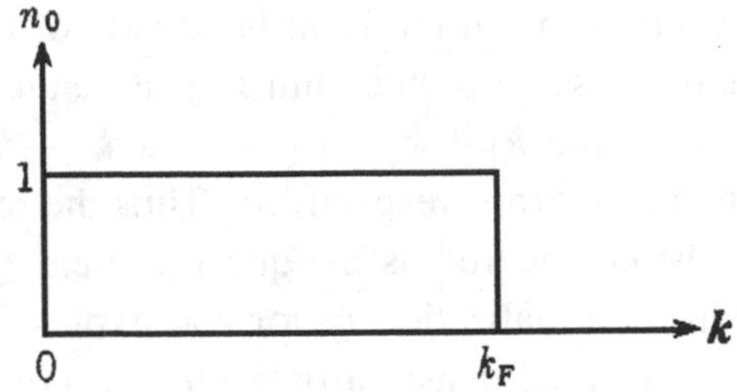

Fig. 2.2 Momentum distribution in the free Fermi gas at $T = 0$.

When a small deviation $\delta n(\boldsymbol{k})$ arises in the distribution function, the change in total energy becomes

$$\delta E = \sum_{\boldsymbol{k}} \frac{\hbar^2 k^2}{2m} \delta n(\boldsymbol{k}). \tag{2.12}$$

The energy $\hbar^2 k^2/2m$ of the particle with wave-number $\boldsymbol{k}$ and mass m can be defined as the functional derivative $\delta E/\delta n(\boldsymbol{k})$. The deviation $\delta n(\boldsymbol{k})$ is positive for $k > k_F$ and negative for $k < k_F$.

Now, let us introduce adiabatically the interaction in the free Fermi gas to transfer the system into the Fermi liquid. In this case, if the interaction between particles is attractive, it might create bound states and induce superconductivity. To avoid this complexity we confine ourselves to the repulsive force. Similarly to the Fermi gas, we consider a particle as an elementary excitation for $k > k_F$ and a hole for $k < k_F$. The distribution function for this particle, $n_0(\boldsymbol{k})$, is the same as that in Fig. 2.2. The excitation in the system is determined by

$$\delta n(\boldsymbol{k}) = n(\boldsymbol{k}) - n_0(\boldsymbol{k}). \tag{2.13}$$

For the quasi-particles to be well defined the deviation of distribution δn should be limited in the vicinity of the Fermi surface. As stated in the previous section, the quasi-particles are the elementary excitations near $k = k_F$ and do not give any information about the ground state. The total energy E is a functional of $n(\boldsymbol{k})$. When $n_0(\boldsymbol{k})$ changes by $\delta n(\boldsymbol{k})$, the linear term of E with respect to $\delta n(\boldsymbol{k})$ is expressed as (2.14) and (2.15):

$$\delta E = \sum_{\boldsymbol{k}} \varepsilon_{\boldsymbol{k}} \delta n(\boldsymbol{k}), \tag{2.14}$$

$$\varepsilon_{\boldsymbol{k}} = \delta E/\delta n(\boldsymbol{k}). \tag{2.15}$$

$\varepsilon_{\boldsymbol{k}}$ is a functional derivative of E with respect to $n(\boldsymbol{k})$; for $k > k_F$, $\varepsilon_{\boldsymbol{k}}$ means the energy change when a quasi-particle with wave-vector $\boldsymbol{k}$ is added to the Fermi system. It should be noted that ε_k defined by a functional derivative is generally a complicated function of $n(\boldsymbol{k})$ and the energy of the total system E cannot be given by summing up the energy of the quasi-particles. For $k = k_F$, ε_k is the energy needed to add one particle to the Fermi surface. Since the state thus obtained is the ground state for the $N+1$ particle system, the chemical potential is given at $T = 0$ by

$$\varepsilon_{k_F} = E_0(N+1) - E_0(N) = \mu, \tag{2.16}$$

where $\mu = \partial E_0(N)/\partial N$, E_0 being the ground state energy.

Equation (2.14) is correct when the second-order terms in $\delta n(\boldsymbol{k})$ can be ignored; it holds when the number of excited quasi-particles is sufficiently small compared with N. When the change of ε_k arising from the distribution function $n(\boldsymbol{k})$ is included, (2.14) becomes

$$\delta E = \sum_{\boldsymbol{k}} \varepsilon_k{}^0 \delta n(\boldsymbol{k}) + \frac{1}{2}\sum_{\boldsymbol{k}}\sum_{\boldsymbol{k}'} f(\boldsymbol{k},\boldsymbol{k}')\delta n(\boldsymbol{k})\delta n(\boldsymbol{k}'). \tag{2.17}$$

Here, $f(\boldsymbol{k},\boldsymbol{k}')$ is called the Landau parameter and is given by the second functional derivative of E by δn. From the symmetry of (2.17):

$$f(\boldsymbol{k},\boldsymbol{k}') = f(\boldsymbol{k}',\boldsymbol{k}). \tag{2.18}$$

Energy $\varepsilon_k{}^0$ is the energy of the isolated quasi-particle $\boldsymbol{k}$. Owing to the interaction with quasi-particles distributed with density $\delta n(\boldsymbol{k}')$, the energy of quasi-particle $\boldsymbol{k}$ becomes

$$\varepsilon_k = \varepsilon_k{}^0 + \sum_{\boldsymbol{k}'} f(\boldsymbol{k},\boldsymbol{k}')\delta n(\boldsymbol{k}'). \tag{2.19}$$

Since the sum over $\boldsymbol{k}'$ gives volume Ω, $f(\boldsymbol{k},\boldsymbol{k}')$ is a quantity of order $1/\Omega$. If we include the spin of the Fermi particle,

$$E = E_0 + \sum_{\boldsymbol{k}\sigma} \varepsilon_k{}^0 \delta n(\boldsymbol{k},\sigma) + \frac{1}{2}\sum_{\substack{\boldsymbol{k}\boldsymbol{k}'\\ \sigma\sigma'}} f(\boldsymbol{k}\sigma,\boldsymbol{k}'\sigma')\delta n(\boldsymbol{k}\sigma)\delta n(\boldsymbol{k}'\sigma'). \tag{2.20}$$

Since the interaction between two quasi-particles depends only on the relative direction between their spins σ and σ', the Landau parameters can be written as

$$f^{s,a}(\boldsymbol{k},\boldsymbol{k}') = \frac{1}{2}\{f(\boldsymbol{k}\sigma,\boldsymbol{k}'\sigma) \pm f(\boldsymbol{k}\sigma,\boldsymbol{k}'-\sigma)\}, \tag{2.21}$$

where s and a correspond to + and −, respectively in (2.21). In the isotropic system, we can define the Landau parameters $f_l{}^{s,a}$ as

$$f^{s,a}(\boldsymbol{k},\boldsymbol{k}') = \sum_l f_l{}^{s,a} P_l(\cos\theta), \tag{2.22}$$

where P_l and θ are the Legendre function and the angle between $\boldsymbol{k}$ and $\boldsymbol{k}'$, respectively.

Today, when we apply the Fermi liquid theory to electron systems, the systems possess symmetry of crystals, which generally violates the Galilei invariance and the spherical symmetry. Therefore, we cannot use the Landau parameters appropriate to the liquid, but should use the parameters appropriate to the solid. To avoid confusion, in this book we discuss the microscopic theory directly on the basis of the field theory.

References

[1] L. D. Landau, *Sov. Phys. JETP* **3** (1957) 920; **5** (1957) 101; **8** (1959) 70.
[2] P. Nozières, *Theory of Interacting Fermi Systems* (Benjamin, 1964).
[3] P. W. Anderson, *Basic Notions of Condensed Matter Physics* (Benjamin, 1984).
[4] J. M. Luttinger, *Phys. Rev.* **119** (1960) 1153.

3

Anderson's orthogonality theorem

The Fermi liquid possesses an infinite number of electron–hole pair excitations in the vicinity of the Fermi surface and can be considered an infinitely degenerate system. Because of this nature the Fermi surface is largely transformed even by a small perturbation. The fragility of the Fermi surface gives rise to interesting physics, such as superconductivity and the various many-body problems related to the Fermi surface. A typical example is the Kondo effect. In this chapter, we consider the system of dilute impurity atoms in metals.

First, in Section 3.1, we consider the screening effect of an impurity charge in metals, following the Friedel theory. In Section 3.2, we explain Anderson's orthogonality theorem as a singularity originating from the infinite number of low-energy excitations near the Fermi energy in metals. In Sections 3.3 and 3.4, we introduce the photoemission due to soft X-rays and the quantum diffusion of charged particles in metals, respectively, as typical examples in which the orthogonality theorem plays an important role.

3.1 Friedel sum rule

Let us assume that there exists an impurity atom in a normal metal such as Cu, Ag or Al. Owing to the local impurity potential $V(\boldsymbol{r})$, the electron distribution around the impurity changes. The local change of electron number was calculated by Friedel in 1958 [1]. We introduce it here.

Approximating a normal metal by a free electron system, we can write the Schrödinger equation for the wave-function $\psi(\boldsymbol{r})$ of an electron moving in the local potential $V(\boldsymbol{r})$,

$$-\frac{\hbar^2}{2m}\nabla^2\psi(\boldsymbol{r}) + V(\boldsymbol{r})\psi(\boldsymbol{r}) = E\psi(\boldsymbol{r}). \tag{3.1}$$

The asymptotic form of the wave-function in the region far from the impurity can be written as

$$\psi(\boldsymbol{r}) \overset{r\to\infty}{\longrightarrow} e^{i\boldsymbol{k}\cdot\boldsymbol{r}} + A(\theta, \varphi)e^{ikr}/r. \tag{3.2}$$

The first term on the right-hand side of (3.2) represents the ingoing wave and the second term represents the scattering wave. The coefficient A is the scattering amplitude and θ is the angle between $\boldsymbol{k}$ and $\boldsymbol{r}$. Here, we assume that the potential $V(\boldsymbol{r})$ has spherical symmetry, and we expand $\psi(\boldsymbol{r})$ in spherical waves using the Legendre polynomials:

$$\psi(\boldsymbol{r}) = \sum_l c_l \psi_l(r) P_l(\cos\theta). \tag{3.3}$$

The asymptotic form of $\psi_l(r)$ in the region $r \to \infty$, where $V(\boldsymbol{r}) = 0$, is given by

$$\psi_l(r) \sim \frac{1}{r}\sin\left(kr - \frac{l\pi}{2} + \delta_l\right), \tag{3.4}$$

where δ_l is the phase shift. Representing (3.2) in the spherical waves and equating it with (3.3), we obtain from the coefficient of $\exp(-ikr)$,

$$kc_l = (2l+1)i^l e^{i\delta_l}. \tag{3.5}$$

From that of $\exp(ikr)$, we obtain

$$A(\theta) = \frac{1}{2ik}\sum_{l=0}^{\infty}(2l+1)(e^{2i\delta_l} - 1)P_l(\cos\theta). \tag{3.6}$$

The differential scattering cross-section $\sigma(\theta)$ is given by

$$\sigma(\theta) = |A(\theta)|^2 = \frac{1}{k^2}\left|\sum_{l=0}^{\infty}(2l+1)e^{i\delta_l}\sin\delta_l P_l(\cos\theta)\right|^2. \tag{3.7}$$

Substituting (3.5) into (3.3), we obtain

$$\psi(\boldsymbol{r}) = \sum_l (2l+1)i^l e^{i\delta_l}k^{-1}P_l(\cos\theta)\psi_l(r). \tag{3.8}$$

Putting $\psi_l(r) = \varphi_l/r$ and taking the spin degeneracy into account, we obtain the total electron number within the sphere with radius R as

$$\begin{aligned}\int_0^R \rho(r)4\pi r^2 dr &= \int_0^R 4\pi r^2 dr \int_0^{k_\mathrm{F}} 4\pi k^2 dk\, 2\sum_l (2l+1)\frac{1}{(2\pi)^3}\frac{{\psi_l}^2(r)}{k^2}\\ &= \frac{4}{\pi}\sum_l (2l+1)\int_0^{k_\mathrm{F}} dk \int_0^R {\varphi_l}^2 dr\end{aligned}$$

$$= \frac{4}{\pi}\sum_l (2l+1)\int_0^{k_F} dk \frac{1}{2k}\left\{\frac{d\varphi_l}{dk}\frac{d\varphi_l}{dr} - \varphi_l \frac{d^2\varphi_l}{drdk}\right\}_{r=R}. \quad (3.9)$$

The last equality of (3.9) can be derived as follows. The Schrödinger equation for $\varphi_l = r\psi_l(r)$ is given by

$$\frac{d^2\varphi_l}{dr^2} + \left[k^2 - \frac{l(l+1)}{\hbar^2 r^2} - \frac{2mV(r)}{\hbar^2}\right]\varphi_l = 0$$

and that for the eigenfunction $\tilde{\varphi}_l$ with eigenvalue $2m\tilde{E}/\hbar^2 = \tilde{k}^2$,

$$\frac{d^2\tilde{\varphi}_l}{dr^2} + \left[\tilde{k}^2 - \frac{l(l+1)}{\hbar^2 r^2} - \frac{2mV(r)}{\hbar^2}\right]\tilde{\varphi}_l = 0.$$

From the above two equations, we obtain

$$(\tilde{k}^2 - k^2)\int_0^R \varphi_l\tilde{\varphi}_l dr = \int_0^R \left(\tilde{\varphi}_l \frac{d^2\varphi_l}{dr^2} - \varphi_l \frac{d^2\tilde{\varphi}_l}{dr^2}\right) dr = \left[\tilde{\varphi}_l \frac{d\varphi_l}{dr} - \varphi_l \frac{d\tilde{\varphi}_l}{dr}\right]_0^R. \quad (3.10)$$

Taking the limit $\tilde{k} \to k$, we obtain the relation

$$\int_0^R \varphi_l{}^2(r)dr = \frac{1}{2k}\left\{\frac{d\varphi_l}{dk}\frac{d\varphi_l}{dr} - \varphi_l \frac{d^2\varphi_l}{drdk}\right\}_{r=R}. \quad (3.11)$$

Here we return to (3.9). Substituting

$$\varphi_l(R) \simeq \sin\left(kR + \delta_l(k) - \frac{l\pi}{2}\right) \quad (3.12)$$

into (3.9), we obtain

$$\int_0^R \rho(r)4\pi r^2 dr = \frac{2}{\pi}\sum_l (2l+1)\int_0^{k_F} dk\left\{\left(R + \frac{d\delta_l}{dk}\right) - \frac{1}{2k}\sin(2kR + 2\delta_l(k) - l\pi)\right\}. \quad (3.13)$$

The change in electron number ΔN is given by

$$\Delta N = \int_0^R [\rho(r) - \rho_0(r)]4\pi r^2 dr$$
$$= \frac{2}{\pi}\sum_l (2l+1)\int_0^{k_F} dk\left\{\frac{d\delta_l}{dk} - \frac{1}{k}\sin\delta_l(k)\cos(2kR - l\pi + \delta_l(k))\right\}. \quad (3.14)$$

Considering the weak k-dependence of $\delta_l(k)$ compared with $2kR$ and partially integrating, we obtain

$$\Delta N = \frac{2}{\pi}\sum_l (2l+1)\left[\delta_l(k_\mathrm{F}) - \frac{1}{2k_\mathrm{F}R}\sin\delta_l(k_\mathrm{F})\sin(2k_\mathrm{F}R - l\pi + \delta_l(k_\mathrm{F}))\right]. \tag{3.15}$$

In the limit $R \to \infty$, ΔN should be equal to the difference of charge ΔZ between the impurity atom and the host atom:

$$\Delta Z = \frac{2}{\pi}\sum_l (2l+1)\delta_l(k_\mathrm{F}). \tag{3.16}$$

The final result is called the Friedel sum rule. Equation (3.15) shows that the effect of a local potential is limited within a finite region and the outside of the region remains unchanged. An impurity cannot affect the total system and the effect remains always a local one. The equality $\Delta N = \Delta Z$ tells us that the electron charge of the impurity atom stays in the neighbourhood of the impurity atom and the impurity atom exists as a neutral atom when we see it from a long distance. This nature arises from the completeness of the screening in metals.

Taking the derivative of (3.15) by R, we obtain the change of electron density $\Delta\rho$ as

$$\begin{aligned}\Delta\rho = \rho(R) - \rho_0(R) &= \frac{1}{4\pi R^2}\frac{d\Delta N(R)}{dR}\\ &= \frac{4k_\mathrm{F}{}^3}{\pi^2}\sum_l (2l+1)\sin\delta_l\left\{-\frac{\cos(2k_\mathrm{F}R + \delta_l - l\pi)}{(2k_\mathrm{F}R)^3} + \frac{\sin(2k_\mathrm{F}R + \delta_l - l\pi)}{(2k_\mathrm{F}R)^4}\right\}.\end{aligned} \tag{3.17}$$

The extra local electron density decays as R^{-3}, oscillating in the period π/k_F. This oscillation is called the Friedel oscillation and is the same as the RKY (Ruderman, Kittel and Yosida) oscillation, which is seen in the spin polarization of conduction electrons due to a magnetic impurity atom, as discussed later. The factor $2k_\mathrm{F}$ appearing in both oscillations can be considered a result of the discontinuity of electron distribution at the Fermi surface. That is, using only the electrons with wavelength longer than $k_\mathrm{F}{}^{-1}$, we cannot screen the space change of charge that is shorter than $(2k_\mathrm{F})^{-1}$.

The Friedel sum rule was derived for the one-body spherical potential but nowadays, it has been shown by Langer and Ambegaokar that the rule holds more generally and can be extended to the case possessing a crystal potential and/or electron interactions [2]. The generalized form of the Friedel sum rule can be written in terms of the scattering matrix (S-matrix) $\hat{S}(\mu)$ at the Fermi surface as

$$\Delta N = \frac{1}{2\pi i} \mathrm{Tr}\, \log \hat{S}(\mu). \tag{3.18}$$

The S-matrix is related to the transition matrix (T-matrix) as

$$\hat{S}_{\alpha\beta}(\mu) = \delta_{\alpha\beta} - 2\pi i \delta(E_\alpha - \mu)\hat{T}_{\alpha\beta}, \tag{3.19}$$

where $\delta_{\alpha\beta}$ and $\delta(E_\alpha - \mu)$ are the Kronecker delta and δ-function, respectively. If the S-matrix has a spherical symmetry, we can expand it in terms of the spherical waves. The l-wave component S_l of $\hat{S}$ can be written using the phase shift δ_l, as

$$S_l = e^{2i\delta_l} \tag{3.20}$$

and (3.18) reproduces (3.16).

The T-matrix is defined for a general scattering potential $\hat{V}$ by

$$\hat{T}(\varepsilon + i\eta) = \hat{V} + \hat{V}\frac{1}{\varepsilon - \mathcal{H}_o + i\eta}\hat{T}(\varepsilon + i\eta) \qquad (\eta \to 0), \tag{3.21}$$

where η is a positive infinitesimal.

3.2 Orthogonality theorem on local perturbation

When a local potential $V(\boldsymbol{r})$ is created in metals owing to an impurity, the distribution of conduction electrons changes to screen it. In this case, the ground state of the N-electron system for $V(\boldsymbol{r}) = 0$ and that for $V(\boldsymbol{r}) \neq 0$ are orthogonal to each other and the overlap integral between the two ground states for different potentials vanishes. This conclusion was obtained by P. W. Anderson in 1967 and is called Anderson's orthogonality theorem [3]. For simplicity, we consider the spherical waves centred at the potential and discuss the change of s-wave ($l = 0$) due to the local potential. For example, when the δ-function type $V(\boldsymbol{r}) = V\delta(\boldsymbol{r})$ is assumed, only the s-wave possessing finite amplitude at the origin is scattered. The s-wave eigenfunctions in $V = 0$ are represented with suffices n as

$$\varphi_n{}^0 = N_{0n} \sin(k_n{}^0 r)/r. \tag{3.22}$$

For $V \neq 0$, a phase shift δ_n is created and φ_n is given by

$$\varphi_n = N_n \sin(k_n r + \delta_n)/r. \tag{3.23}$$

Here N_{0n} and N_n are the normalization constants. For both cases with $V = 0$ and $V \neq 0$, we consider the ground states where N electrons are filled up to the Fermi energy. These many-body states are represented by the Slater determinants:

$$\frac{1}{\sqrt{N!}}\det|\varphi_n(r_n)| = \frac{1}{\sqrt{N!}}\begin{vmatrix} \varphi_1(r_1) & \varphi_1(r_2) & \cdots & \varphi_1(r_N) \\ \varphi_2(r_1) & \varphi_2(r_2) & \cdots & \varphi_2(r_N) \\ \vdots & & & \vdots \\ \varphi_N(r_1) & \varphi_N(r_2) & \cdots & \varphi_N(r_N) \end{vmatrix}. \tag{3.24}$$

The overlap integral S between the gound state $|\mathrm{i}\rangle$ without a potential $V(\boldsymbol{r})$ and the ground state $|\mathrm{f}\rangle$ with $V(\boldsymbol{r})$ is given by

$$\begin{aligned} S &= \frac{1}{N!}\iint \det|\varphi_n{}^0(\boldsymbol{r}_n)|\det|\varphi_n(\boldsymbol{r}_n)|d\boldsymbol{r}_1 d\boldsymbol{r}_2\cdots d\boldsymbol{r}_N \\ &= N^{-(\delta/\pi)^2/2}. \end{aligned} \tag{3.25}$$

This result was derived by P. W. Anderson using the relations among determinants. Here δ is a phase shift of s-wave at the Fermi surface due to a local potential. When we include the spin degeneracy, the overlap integral is given by a product of the overlap integrals with the same phase shift δ for each spin and the exponent is given by twice the exponent of (3.25). The result (3.25) gives $S = 0$, as far as $\delta \neq 0$, since the electron number N is a macroscopic number. This means that the ground state for $V = 0$ and that for $V \neq 0$ are orthogonal to each other. Since δ/π gives a local electron number according to the Friedel sum rule (3.16), when the local electron number changes, the two ground states before and after the change are orthogonal to each other. The phase shift $\delta = \pi$ corresponds to the change of one electron and means the appearance of a bound state due to the potential. In this case the overlap integral between the state for $\delta = \pi$ and that for $V = 0$ is given by the overlap integral between a bound state and a spherical wave extending over the total space. Actually, corresponding to this fact, (3.25) gives $1/\sqrt{N}$ for $\delta = \pi$. According to the theorem of Kohn and Majumdar [4], the energy and electron density of the total system are analytic and change continuously with respect to the strength of V, independently of the existence of bound states.

Now we discuss the reason that the overlap integral S vanishes. Since the orthogonality holds even for a small value of V, let us calculate directly the change in wave-function of the N-electron system by treating the potential as a perturbation. The wave-function Φ_0 denotes that of the Fermi sphere for free electrons when $V = 0$, and a local potential is assumed as

$$\mathcal{H}' = \sum_{kk'} V_{kk'} c_{k'}{}^{\dagger} c_k. \tag{3.26}$$

We neglect the electron spin degeneracy at present and consider it later. Equation (3.26) represents the scattering process where an electron $\boldsymbol{k}$ inside the Fermi

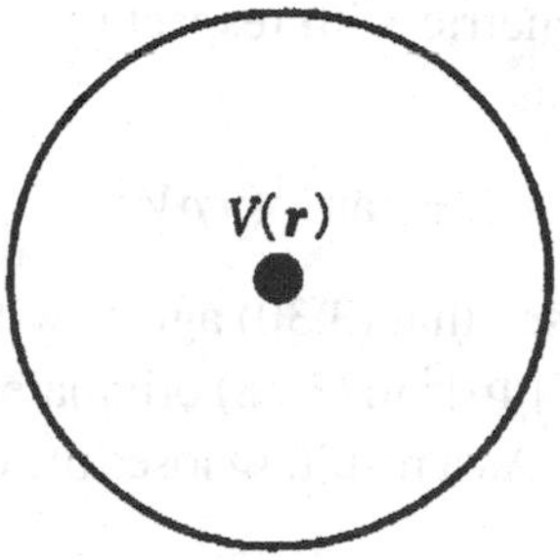

Fig. 3.1 By a local potential $V(\boldsymbol{r})$, the distribution of conduction electrons changes within a finite region.

sphere is scattered into $\boldsymbol{k}'$ outside the sphere. The change of wave-function Φ_1 in the first-order perturbation is given by

$$\Phi_1 = -\sum_{kk'} \frac{V_{kk'}}{\varepsilon_{k'} - \varepsilon_k} c_{k'}{}^\dagger c_k \Phi_0. \tag{3.27}$$

Here we put the Fermi energy $\mu = 0$. In Φ_1, electron–hole pairs $\boldsymbol{k}'(\varepsilon_{k'} > 0)$ and $\boldsymbol{k}(\varepsilon_k < 0)$ are created. To normalize $\Phi = \Phi_0 + \Phi_1 + \cdots$, let us calculate $|\Phi|^2$. Here we assume $|\Phi_0|^2 = 1$ and $V_{kk'} = V$:

$$\begin{aligned} |\Phi|^2 &= |\Phi_0|^2 + |\Phi_1|^2 + \cdots = 1 + \sum_{kk'} \frac{|V_{k'k}|^2}{(\varepsilon_{k'} - \varepsilon_k)^2} + \cdots \\ &= 1 + \rho^2 V^2 \int_{-D}^{0} d\varepsilon \int_0^D d\varepsilon' \frac{1}{(\varepsilon' - \varepsilon)^2} + \cdots \\ &= 1 + \rho^2 V^2 \log(D/0) + \cdots, \end{aligned} \tag{3.28}$$

where we assumed a constant density of states ρ for conduction electrons from $-D$ to D. This assumption and $V_{kk'} = V$ do not lose generality, because the logarithmic divergence arises near the Fermi energy. The logarithmic divergence arises from the lower limit of electron–hole pair excitation energy, $\omega = \varepsilon' - \varepsilon \to 0$. The divergence of (3.28) is called the infrared divergence, since it is the divergence in the low energy limit. If the lower energy limit 0 is replaced by the energy discreteness of conduction band D/N, $|\Phi|^2$ becomes

$$|\Phi|^2 = 1 + \rho^2 V^2 \log N \simeq e^{\rho^2 V^2 \log N} = N^{\rho^2 V^2}. \tag{3.29}$$

Thus, the overlap integral S between Φ_0 and the renormalized $\Phi = (\Phi_0 + \Phi_1) \exp[-(\rho^2 V^2/2) \log N]$ is given by

$$S = N^{-(\rho V)^2/2}. \tag{3.30}$$

For the conduction band symmetric with respect to $\mu = 0$, the phase shift δ at the Fermi energy is written by V as

$$\delta = \tan^{-1}(\pi\rho V). \tag{3.31}$$

Considering this fact, we can see that (3.30) agrees with (3.25) in the small δ limit. The logarithmic divergence of $|\Phi_1|^2$ in (3.28) originates from the infinite number of electron–hole pair excitations. As a result, Φ loses the overlap with the unperturbed Fermi sphere Φ_0.

Here it should be noted that the singularity exists only in the change of wave-function and there is no singularity in the change of energy due to the potential. For example, the second-order perturbation term in energy converges, because the denominator is the first-order term of $(\varepsilon_k - \varepsilon_{k'}) = \omega$, as

$$E^{(2)} = -\sum_{k'k} \frac{|V_{k'k}|^2}{(\varepsilon_{k'} - \varepsilon_k)} = -(\rho V)^2 2D \log 2. \tag{3.32}$$

As seen above, the Fermi surface of metals is infinitely degenerate with respect to the electron–hole pair excitation and possesses the characteristic nature of fragility. As a result the Fermi surface changes drastically by a small perturbation. This nature is related to the occurrence of superconductivity by a weak attractive force.

3.3 Photoemission in metals and the orthogonality theorem

As an application of the orthogonality theorem, let us consider the photoemission of core electrons by the soft X-rays in metals. When we excite a core electron in metallic ions by the X-ray so as to emit it out of metals, in the final state there remains a hole in the core orbital. The hole decays in a finite lifetime. In the case of a long lifetime, the photoemission spectra corresponding to the final state possessing a core hole can be observed. In this case conduction electrons around the core hole are attracted to screen the positive charge of the core hole. Finally, as seen in the previous section, electrons giving unit charge are attracted. Thus the local state of conduction electrons changes before and after the absorption of X-rays, and to this situation Anderson's orthogonality theorem can be applied. The intensity $I(\varepsilon_k)$ of photoelectrons with energy ε_k is given by

$$I(\varepsilon_k) \propto \sum_{\mathrm{f}} |\langle \Psi_{\mathrm{f}} b^{\dagger} c_k{}^{\dagger} | j_k | \Psi_{\mathrm{i}} \rangle|^2 \delta(\omega + \varepsilon_{\mathrm{c}} - \varepsilon_k - E_{\mathrm{f}} + E_{\mathrm{i}} - W), \tag{3.33}$$

where ω is the energy of X-rays and W is the work function. Hereafter we put $\omega - W$ as ω for simplicity. ε_{c}, E_{i} and E_{f} are the energy of core state, energy of initial state of conduction electron Ψ_{i} and that of final state Ψ_{f}, respectively. The matrix element of the transition is obtained from the interaction Hamiltonian $\hat{j}$

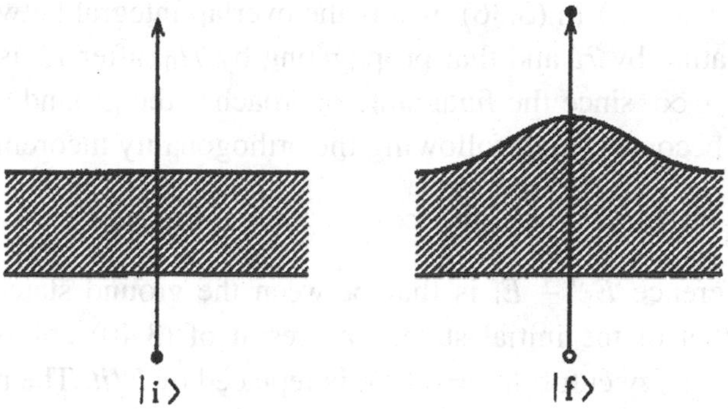

Fig. 3.2 Schematic figure of photoemission in metals. When an electron is emitted from a core state, conduction electrons are attracted by a core hole to screen the core hole charge.

among the core hole $b^\dagger$, the photoelectron $c_k{}^\dagger$ and the X-ray, which is written as

$$\hat{j} = \sum_k j_k(b^\dagger c_k{}^\dagger + c_k b). \tag{3.34}$$

The first term represents the process where the absorbed X-ray emits a photoelectron and creates a core hole and the second term the inverse process. Using the relation

$$\delta(\omega - \varepsilon) = \frac{\mathrm{Re}}{\pi} \int_0^\infty dt e^{i(\omega-\varepsilon)t}, \tag{3.35}$$

we rewrite (3.33) into the following. Since the final states Ψ_f make a complete set, by inserting $\sum_\mathrm{f} |\Psi_\mathrm{f}\rangle\langle\Psi_\mathrm{f}|$ on the right of the operator b in (3.36), we can rewrite (3.33) as

$$\begin{aligned} I(\varepsilon_k) &\propto \frac{\mathrm{Re}}{\pi} \int_0^\infty dt e^{i(\omega+\varepsilon_\mathrm{c}-\varepsilon_k)t} \langle \Psi_\mathrm{i}| e^{iE_\mathrm{i}t} b e^{-i\mathcal{H}t} b^\dagger |\Psi_\mathrm{i}\rangle \\ &= \frac{\mathrm{Re}}{\pi} \int_0^\infty dt e^{i(\omega+\varepsilon_\mathrm{c}-\varepsilon_k)t} \langle \Psi_\mathrm{i}| e^{i\mathcal{H}_0 t} e^{-i\mathcal{H}t} |\Psi_\mathrm{i}\rangle. \end{aligned} \tag{3.36}$$

Here, in the last expression, $|\Psi_\mathrm{i}\rangle$ means an initial state without any core hole and operators for core holes are omitted. The Hamiltonians $\mathcal{H}_0$ and $\mathcal{H}$ are those of conduction electrons for initial and final states, respectively. Here we treat the spin-independent process and ignore spins. The Hamiltonian is written as

$$\mathcal{H} = \mathcal{H}_0 + \mathcal{H}', \tag{3.37}$$

$$\mathcal{H}_0 = \sum_k \varepsilon_k c_k{}^\dagger c_k, \tag{3.38}$$

$$\mathcal{H}' = V \sum_{kk'} c_{k'}{}^\dagger c_k. \tag{3.39}$$

The expectation value $\langle\ \rangle$ in (3.36) means the overlap integral between the wavefunction propagating by $\mathcal{H}$ and that propagating by $\mathcal{H}_0$, after $\mathcal{H}'$ is introduced at $t = 0$. When $t \to \infty$, since the final state approaches the ground state $|\Psi_f{}^0\rangle$, the overlap integral becomes small following the orthogonality theorem, as given by

$$\langle\Psi_i|e^{i\mathcal{H}_0 t}e^{-i\mathcal{H}t}|\Psi_i\rangle \propto e^{-i(E_f{}^0-E_i)t}(iDt)^{-(\delta/\pi)^2}. \qquad (3.40)$$

The energy difference $E_f{}^0 - E_i$ is that between the ground state energy of the final state and that of the initial state. The result of (3.40) corresponds to that where the lower energy cutoff $\omega_s = D/N$ is replaced by $\hbar/it$. The reason why the coefficient of the exponent $(\delta/\pi)^2$ is twice as large as (3.25) is that (3.40) becomes $|\langle\Psi_f{}^0|\Psi_i\rangle|^2$ by inserting $|\Psi_f{}^0\rangle\langle\Psi_f{}^0|$ in the middle left side of (3.40). The result of (3.40) can be obtained directly by the method of Nozières–de Dominicis [5]. By the perturbational calculation we can also obtain it from expanding $(\delta/\pi)^2$ with respect to $(\rho V)^2$. Inserting (3.40) into (3.36), we obtain the intensity of the photoelectron as

$$I(\varepsilon_{\boldsymbol{k}}) \propto \varepsilon^{-1+(\delta/\pi)^2} = \frac{1}{\varepsilon^{1-\alpha}}. \qquad (3.41)$$

Here we have put $\varepsilon = \varepsilon_{\max} - \varepsilon_{\boldsymbol{k}}$, where $\varepsilon_{\max}$ is the maximum value of $\varepsilon_{\boldsymbol{k}}$ corresponding to the ground state of the final states. In this case $\varepsilon = 0$. The intensity $I(\varepsilon)$ in (3.41) is shown by the broken line in Fig. 3.3. If we take the lifetime of the core hole into account and assume a phenomenological damping function $e^{-\gamma t}$, we obtain the asymmetric spectral function shown by the full line in Fig. 3.3. The spectral function for $\varepsilon > 0$ represents the electron–hole pair excitations in the final states. As seen here, the orthogonality theorem not only represents the orthogonality catastrophe between the two ground states, but also gives analytically the power in

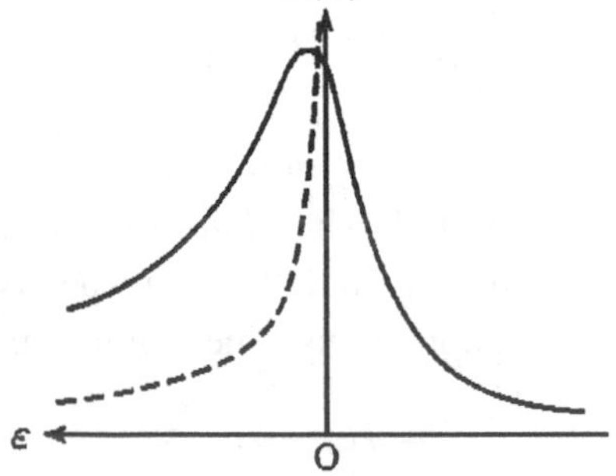

Fig. 3.3 Intensity of photoelectrons excited from core states by a constant energy X-ray. When the damping rate γ of the core hole is taken into account, $I(\varepsilon)$ is changed into an asymmetric form shown by the full line, from the form $I(\varepsilon) \propto \varepsilon^{-1+(\delta/\pi)^2}$.

t or ε as shown in (3.40) and (3.41). It is important in actual cases that the theorem contains the information of excited states.

In (3.41), $\alpha = (\delta/\pi)^2$ is obtained. But in general, considering the spin degeneracy and the screening by general spherical waves l, we obtain

$$\alpha = 2\sum_l (2l+1)(\delta_l/\pi)^2. \tag{3.42}$$

In this case, the phase shift δ_l of the l-wave satisfies the Friedel sum rule:

$$\Delta Z = 1 = 2\sum_l (2l+1)\delta_l/\pi. \tag{3.16$'$}$$

In addition to the final states, when there exist phase shifts $\delta_l{}^{\mathrm{i}}$ also in the initial states, α is given by replacing δ_l in (3.42) by the difference of phase shifts $\delta_l{}^{\mathrm{f}} - \delta_l{}^{\mathrm{i}}$ as the following:

$$\alpha = 2\sum_l (2l+1)(\delta_l{}^{\mathrm{f}} - \delta_l{}^{\mathrm{i}})^2/\pi^2. \tag{3.43}$$

The asymmetric spectra of the photoemission as shown in Fig. 3.3 have been observed in many metals and from the obtained values of α we can estimate the value of the phase shift δ_l with the help of the Friedel sum rule.

Now we will discuss the absorption of soft X-rays by which core electrons are excited to the Fermi surface of the conduction band, and also the emission of soft X-rays. These processes can be understood with the help of the above-mentioned results. Since the absorption and emission are symmetric processes to each other, we confine ourselves to absorption. At $t = 0$, we assume an electron in core states is excited to one of the states with orbital angular momentum l_0. In this case one of $\delta_{l_0}{}^{\mathrm{i}}$ in (3.43) is replaced by $\delta_{l_0}{}^{\mathrm{i}} + \pi$. After $t = 0$, the time development of the system is the same as that of photoemission. For example, when we assume that only the s-wave ($l_0 = l = 0$) is related to the process and ignore the spin degeneracy, we obtain easily the expression for the intensity,

$$I(\varepsilon_k) \propto \varepsilon^{(\delta/\pi-1)^2-1} = \varepsilon^{-2\delta/\pi+(\delta/\pi)^2}. \tag{3.44}$$

In general, including the spin and orbital degeneracy, we obtain the general form

$$I(\varepsilon) \propto \varepsilon^{\beta}, \tag{3.45}$$

$$\beta = -\frac{2\delta_{l_0}}{\pi} + 2\sum_l (2l+1)\left(\frac{\delta_l}{\pi}\right)^2. \tag{3.46}$$

As an example, in the case where a p-electron ($l = 1$) in core states is excited to a state with s-symmetry ($l = 0$), we obtain the exponent β:

$$\beta = -\frac{2\delta_0}{\pi} + 2\sum_l (2l+1)\left(\frac{\delta_l}{\pi}\right)^2. \tag{3.47}$$

In addition to the above assumption, we assume that only the s-wave part contributes to the screening of the core hole. Then,

$$\beta = -\frac{2\delta_0}{\pi} + 2\left(\frac{\delta_0}{\pi}\right)^2. \tag{3.48}$$

On the other hand, from the Friedel sum rule $\delta_0 = \pi/2$ is obtained in this case. Thus, $\beta = -1/2$ and $I(\varepsilon)$ diverges at $\varepsilon = 0$ as seen from (3.45).

As another case, we consider an s-electron of core states excited to a state of p-symmetry ($l = 1$) and the screening for the core hole still done only by the s-wave with $l = 0$. From $\delta_1 = 0$,

$$\beta = 2\left(\frac{\delta_0}{\pi}\right)^2 = \frac{1}{2}. \tag{3.49}$$

As a result, as shown in Fig. 3.4, $I(\varepsilon)$ in (3.45) is suppressed near $\varepsilon = 0$ for $\beta > 0$ and enhanced for $\beta < 0$. When an electron of initial states is excited to a state possessing the same symmetry as that of the final state, the overlap integral is large and the intensity is enhanced. On the other hand, when an electron is excited to a state possessing a symmetry different from that of screening electrons in the final state, the large rearrangement of the electronic state is necessary to screen the core hole and reduces the overlap integral to give a reduced absorption intensity. Anyway,

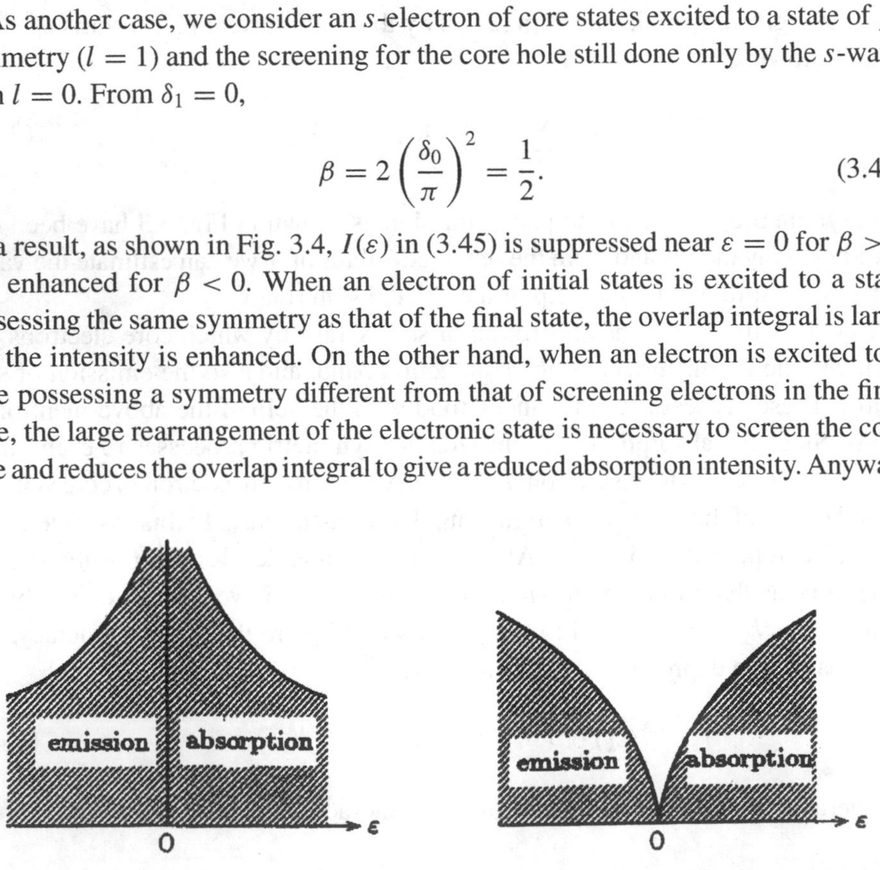

Fig. 3.4 Edge singularities of absorption and emission of soft X-rays in metals. For $\beta < 0$, the edge near $\varepsilon = 0$ is enhanced and for $\beta > 0$, it is weakened. The former is seen in emission and absorption of core p-electrons in Na and Mg. The latter is observed in the absorption and emission from core s-electrons in Li and Be.

in a simple absorption without any screening effect, $\beta = -1$ and the spectrum is given by the δ-function. Compared with this case, we can see that the intensity at $\varepsilon = 0$ is suppressed owing to the overlap integral between the electron clouds. In the emission of soft X-rays, considering the inverse process of the absorption, we can obtain similar results. The above-mentioned conclusion has been confirmed by experiments as an edge singularity in the emission and the absorption of soft X-rays in metals [6]. For example, the singularity corresponding to a negative value of β has been observed for the $2p$ core hole in Na and that corresponding to a positive value of β has been observed for the $1s$ core hole in Li.

3.4 Diffusion of charged particles in metals

When charged particles such as protons and positive muons diffuse in metals, Anderson's orthogonality theorem plays an important role. Let us discuss the diffusion of charged particles as an example of the application of the theorem to a dynamical process. As shown in Fig. 3.5, we assume that a charged particle with mass M stays at an interstitial site. The position of the particle M is denoted as $\boldsymbol{R}$ and the potential as $U(\boldsymbol{R})$, and the Hamiltonian for the particle M is written as

$$H_M = -\frac{\hbar^2 \nabla_R{}^2}{2M} + U(\boldsymbol{R}). \tag{3.50}$$

In addition to this, we should consider the interaction with conduction electrons, because the particle exists in a metal. At the usual temperatures, the charge of the particle should be screened by conduction electrons. At this point, the following problem arises. Even if the charged particle stays at a potential well, the particle is making an oscillating motion within the well. Moreover, the particle makes a tunnelling motion to a neighbouring interstitial site. In this section we consider the role of conduction electrons which screen the moving charged particle.

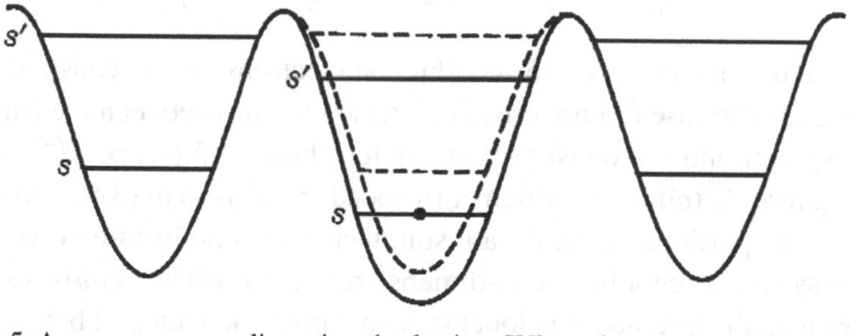

Fig. 3.5 A muon moves, distorting the lattice. When the energy width of μ^+ is sufficiently smaller than the level separation $|\varepsilon_{s'} - \varepsilon_s|$, μ^+ tunnels to the neighbouring same lowest level s from the lowest level s. The fluctuations of lattice distortion adjust two neighbouring levels ε_s to assist the tunnelling.

(a) Interaction with lattice distortion. In discussing the quantum diffusion of charged particles, first of all we need to consider the interaction of the particle with the lattice distortion around it. This problem, which has been studied as a small polaron, is briefly described here. The charged particle (hereafter called the particle) distorts the lattice to make a stable interstitial site. When the particle transfers from an interstitial site to a neighbouring interstitial site, the lattice distortion around the particle also follows it. As a result, at low temperatures we can consider a composite particle composed of the charged particle and the accompanying lattice distortion. In this case we have to estimate the tunnelling matrix element of the composite particle between the neighbouring sites. Here the wave-function of the lattice distortion around site $\boldsymbol{l}$ is denoted as $|\Psi_{\boldsymbol{l}}{}^{i}\rangle$, where i is one of the energy levels of the particle in the potential well. The tunnelling matrix element J^i of the particle from $|\Psi_{\boldsymbol{l}}{}^{i}\rangle$ to $|\Psi_{\boldsymbol{l}+\boldsymbol{g}}{}^{i}\rangle$ at site $\boldsymbol{l}+\boldsymbol{g}$ is given by

$$J^i = J_0{}^i |\langle \Psi_{\boldsymbol{l}}{}^{i} | \Psi_{\boldsymbol{l}+\boldsymbol{g}}{}^{i}\rangle|, \tag{3.51}$$

where $J_0{}^i$ is the tunnelling matrix element without the accompanying lattice distortion. As shown by this result, J is reduced from J_0 by the overlap integral between the neighbouring lattice distortions. Hereafter we consider the tunnelling of the particle between two states with the same energy level i at different sites and ignore that between different levels; we omit the suffix i, hereafter.

Now we consider the overlap integral between lattice distortions with the help of Anderson's orthogonality theorem, by replacing the phonon excitations $\hbar\omega$ with the electron–hole pair excitations, as $\hbar\omega = \varepsilon_{\boldsymbol{k}'} - \varepsilon_{\boldsymbol{k}}$ ($\varepsilon_{\boldsymbol{k}'} > 0$, $\varepsilon_{\boldsymbol{k}} < 0$). The reason the logarithmic divergence appears in the electron–hole pair excitations is that the density of states for the low energy excitations is proportional to ω. This fact can be seen by rewriting (3.28) as

$$V^2 \int_0^{2D} d\omega \int_0^{\omega} d\varepsilon' \frac{\rho^2}{\omega^2} = V^2 \int_0^{2D} d\omega \frac{\rho^2\omega}{\omega^2}. \tag{3.52}$$

Here ρ is the density of states for conduction electrons and is constant near the Fermi energy. In the case of phonons, the interaction $\lambda(\omega)$ between the particle and the lattice system and the density of states for phonons $N(\omega) \propto \omega^{d-1}$ appear in the coupling form $\lambda^2(\omega)N(\omega)$, which corresponds to $V^2\rho^2\omega$ in (3.52). In the usual case $\lambda^2(\omega)$ is proportional to ω. As a result, there exists no infrared divergence in the phonon system, except for a one-dimensional lattice where $N(\omega)\lambda^2(\omega)$ has the ω-linear term. In the three-dimensional system, the overlap integral between lattice distortions is finite and is put as

$$J = J_0 e^{-S}, \tag{3.53}$$

$$e^{-S} = |\langle \Psi_l | \Psi_{l+g} \rangle|. \tag{3.54}$$

Here S is positive and takes a value of the order of unity. At high temperatures, S is not constant but depends on the temperature owing to the excitations of the lattice system. In (3.51), J_0 is the value for the case where the lattice is fixed so that it gives the largest contributions to the tunnelling process of the particle. In this case we assume the lattice distortion cannot follow adiabatically the particle motion and take the lattice distortion into account in the non-adiabatic way. This is because we assume that ions of the lattice are heavier than the tunnelling particle. The non-adiabatic terms play important roles to give the width of the energy level i for the particles, in other words, to give rise to the damping of coherent motion; the coupling with the lattice distortion gives rise to the transition between different energy levels in the potential well.

(b) Interaction with electrons. By the effect of the lattice distortion the tunnelling matrix element J is reduced from J_0 to J in (3.53). Now starting with J, we consider the effect of interaction with conduction electrons. For example, even if we consider a positive muon as a charged particle, the mass is about 200 times as large as the electron mass. One might consider that the electrons can always be treated in an adiabatic way for the heavy charged particle. However, the electron–hole pair excitations in metals possess an infinitely small excitation energy and represent very slow motion. As a result, we have the following conclusion. Among the electron–hole pair excitations, the part possessing excitation energy higher than the kinetic energy of the particle can follow the particle motion and can be treated in an adiabatic way. As the energy scale of the charged particle, we can take the frequency ω_0 of the oscillating motion or the energy level splitting $\hbar\omega_0 \simeq (E_j - E_i)$ in the potential well.

On the other hand, the electron–hole pair excitation possessing energy smaller than $\hbar\omega_0$ cannot follow the motion of the particle and behaves in a non-adiabatic way. The important role of the non-adiabatic effect was stressed by J. Kondo [7]. The interaction between the charged particle $\boldsymbol{R}$, possessing a charge $Z(< 1)$ reduced by an adiabatically accompanying electron cloud, and the conduction electron cloud $\boldsymbol{r}$ following the particle non-adiabatically is written here as $V(\boldsymbol{r}, \boldsymbol{R})$, and the Hamiltonian $\mathcal{H}$ for the total system is written as

$$\mathcal{H} = H_M + H_e + V(\boldsymbol{r}, \boldsymbol{R}). \tag{3.55}$$

Here H_e is the Hamiltonian for the electron system and H_M that for the charged particle given by (3.50).

For the charged particle to keep the transfer motion, in each process of the transfer motion the screening by the conduction electrons should be accomplished. As a result, the charged particle and the conduction electron cloud can be considered to move as a composite particle in a similar way to the case of the lattice distortion. In this case, the overlap integral between neighbouring electron clouds is an important factor. The tunnelling matrix element $\tilde{J}$ of the charged particle accompanying the electron clouds is given by

$$\tilde{J} = J|\langle \varphi_l | \varphi_{l+g} \rangle| = J|\langle \varphi_1 | \varphi_2 \rangle|, \tag{3.56}$$

where $|\varphi_l\rangle$ is the wave-function of the electron cloud at site $\boldsymbol{l}$; since $\boldsymbol{g}$ is a neighbouring site, $\langle \varphi_l | \varphi_{l+g} \rangle$ is given by the overlap integral $\langle \varphi_1 | \varphi_2 \rangle$ between two states separated by distance a.

Here we need to calculate the overlap integral between two electron clouds centred at two sites separated by distance a. The overlap integral arises when an electron cloud around the charged particle staying at site 1 disappears by the transfer of the particle and a new electron cloud is created around the charged particle transferred to site 2. If the creation and annihilation occur independently at the two sites separated by a long distance, the overlap integral is given by the square of the overlap integral at each site described above. However, it is not simple in the case between two sites with a finite distance, because there exists the coherence effect between the two clouds.

At this stage it is necessary to generalize the orthogonality theorem. Since the proof is complicated, only the final result is shown [8]. The most general form of the theorem, which holds without any assumption on the symmetry of potential and on the interaction among electrons, is the following. The overlap integral $\langle \mathrm{f}|\mathrm{i}\rangle$ between two states $|\mathrm{i}\rangle$ and $|\mathrm{f}\rangle$ is written as

$$|\langle \mathrm{f}|\mathrm{i}\rangle| = \left(\frac{\omega_s}{\omega_l}\right)^K, \tag{3.57}$$

where ω_s and ω_l are the cutoff parameters of low and high energy limits, respectively. The important exponent K is given by

$$K = -\frac{1}{8\pi^2}\mathrm{Tr}\{\log[\hat{S}_\mathrm{f}(\mu)\hat{S}_\mathrm{i}(\mu)^{-1}]\}^2, \tag{3.58}$$

where $\hat{S}_\mathrm{i}$ and $\hat{S}_\mathrm{f}$ are the values of S-matrices at the Fermi energy in the initial and final state, respectively. This result corresponds to the generalization of the Friedel sum rule (3.18). When both initial and final states are locally disturbed by a local perturbation, the generalized Friedel sum rule is given by

$$\Delta N = \frac{1}{2\pi i}\mathrm{Tr}\log[\hat{S}_\mathrm{f}(\mu)\hat{S}_\mathrm{i}(\mu)^{-1}]. \tag{3.59}$$

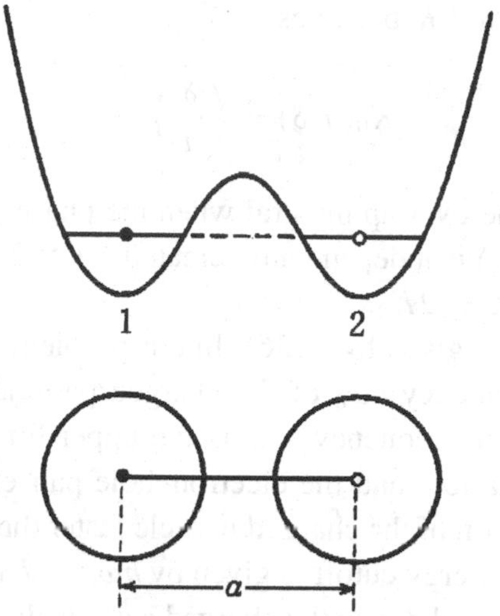

Fig. 3.6 When a charged particle μ^+ tunnels from 1 to 2, conduction electron clouds screening the charge of μ^+ follow μ^+ from 1 to 2. The effective tunnelling matrix element of the composite particle composed of both μ^+ and electron clouds is reduced by the overlap integral between electron clouds.

If $\hat{S}_i(\mu) = \hat{1}$ and $\hat{S}_f(\mu)$ is spherically symmetric, using $S_l = e^{2i\delta_l}$, we obtain

$$K = \frac{1}{\pi^2} \sum_l (2l+1)\delta_l^2, \tag{3.60}$$

which agrees with the result of (3.42). Here we have included the sum of electron spins.

Now, using the generalized orthogonality theorem we calculate the overlap integral between the two N-electron systems possessing potentials centred at two different sites, as shown in Fig. 3.6. For simplicity, we assume that each potential is screened by only the s-wave part of the electrons with respect to the centre of the potential, and their phase shifts are assumed to be δ. The value of K_0 for each spin component in this case, from (3.58) and after a rather complicated calculation, is given by [9]:

$$K_0(x, \delta) = \left[\frac{1}{\pi} \tan^{-1} \frac{\sqrt{1 - x \tan \delta}}{\sqrt{1 + x \tan^2 \delta}} \right]^2. \tag{3.61}$$

Here $x = j_0{}^2(k_F a)$ where the zeroth spherical Bessel function $j_0(z) = \sin z/z$. The reason we have the factor $k_F a$ in (3.61) is that the wave-functions giving the Friedel oscillation in the charge distribution contribute to the overlap integral. If we put

$a = \infty$, we have $x = 0$ and K becomes

$$K_0(0, \delta) = \left(\frac{\delta}{\pi}\right)^2. \tag{3.62}$$

This result represents the overlap integral when the phase shift δ is annihilated at site 1 and the phase shift δ is independently created at site 2. If we consider the spin degeneracy, we obtain $K = 2K_0$.

Now let us calculate $\tilde{J}$ given by (3.56). In our problem, the high energy cutoff $\hbar\omega_l$ is given by the frequency $\hbar\omega_0$ of the charged particle in the potential well. As mentioned before, the frequency $\hbar\omega_0$ is the upper limit of the non-adiabatic electron–hole pair excitation, and the electron–hole pair excitations cannot adiabatically follow the motion of the charged particle faster than the frequency ω_0. On the other hand, the low energy cutoff is given by $\hbar\omega_s = \tilde{J}$ for the low temperature $k_BT < \tilde{J}$, since the potential due to the charged particle disappears within the time $\hbar/\tilde{J}$. In this case $\tilde{J}$ is self-consistently determined by

$$\tilde{J} = J\left(\frac{\tilde{J}}{\hbar\omega_0}\right)^K = J\left(\frac{J}{\hbar\omega_0}\right)^{K/(1-K)}. \tag{3.63}$$

From this equation, we can conclude that when $K > 1$, $\tilde{J} = 0$ and the charged particle is localized. As an example, we consider the case where the electric charge $\Delta Z = 1$ is screened by the s-wave part of the conduction electrons. In this case, the phase shift δ is determined as $\delta = \pi/2$ by the Friedel sum rule, and K takes its maximum value, 1/2. As a result, the particle possessing a unit charge cannot be localized by the above-mentioned mechanism, while the charged particle possessing $\Delta Z = 2$, such as a two-hole bound state, can be localized.

When $k_BT > \tilde{J}$, the low energy cutoff $\hbar\omega_s = k_BT$ and $\tilde{J}$ is given by

$$\tilde{J} = J\left(\frac{k_BT}{\hbar\omega_0}\right)^K. \tag{3.64}$$

This is because the Fermi surface disturbed by the thermal energy k_BT weakens the orthogonality catastrophe.

To see the experimental situation, let us consider the diffusion coefficient D, which is written with the help of the velocity of particle v and the mean free time τ as

$$D = \langle v^2\rangle\tau. \tag{3.65}$$

The charged particle with momentum $\boldsymbol{K}$ and the conduction electron $\boldsymbol{k}$ collide via the interaction $V_{\boldsymbol{kk}'}$ and become $\boldsymbol{K}'$ and $\boldsymbol{k}'$, respectively. The level width $\hbar/\tau$ given by the processes is calculated by the golden rule as

$$\tau^{-1} = \frac{2\pi}{\hbar} \sum_{\substack{kk' \\ \sigma K'}} |V_{k'k}|^2 f_k(1 - f_{k'})\delta(\varepsilon_k - \varepsilon_{k'} + E_K - E_{K'}), \tag{3.66}$$

where ε_k and E_K are the energy of the conduction electron and that of the charged particle, respectively. If the charged particles make a band motion, the momentum conservation $\boldsymbol{k} - \boldsymbol{k}' + \boldsymbol{K} - \boldsymbol{K}' = 0$ holds. When the mass of the particle is large and the band energy of the particle is smaller than $k_B T$, (3.66) becomes

$$\tau^{-1} = \frac{4\pi}{\hbar} \sum_{kk'} |V_{k'k}|^2 f_k(1 - f_{k'})\delta(\varepsilon_k - \varepsilon_{k'}) = \frac{4\pi}{\hbar}(\rho V)^2 k_B T. \tag{3.67}$$

Now we take into account the fact that the forward scattering gives no contribution in the transport phenomena and obtain

$$\tau^{-1} = 2\pi K k_B T/\hbar, \tag{3.68}$$

where $K = 2K_0$ in (3.61) and becomes for a small ρV,

$$K = 2(\rho V)^2(1 - j_0{}^2(k_F a)). \tag{3.69}$$

On the other hand, the velocity of the charged particle is given by $v = \tilde{J}a$. From (3.65), the hop rate ν is given by [7, 9–11]:

$$\nu = \frac{D}{a^2} = \frac{J^2}{2\pi K \omega_0}\left(\frac{k_B T}{\hbar\omega_0}\right)^{2K-1}. \tag{3.70}$$

Since $K < 1/2$ for a single charged particle and $-1 < 2K - 1 < 0$, the hop rate ν increases with decreasing temperature. In Fig. 3.7, we show the hop rate of the positive muon in the Cu metal. This result is obtained by the measurement of the muon spin rotation (μSR). As seen in this figure, the hop rate shows the temperature dependence of $T^{-0.68}$ below 20 K and K is determined as $K = 0.16$. As we have seen above, the screening effect of the conduction electrons gives a unique temperature dependence in the diffusion of charged particles.

Finally, a comment on narrow-band electrons. The d-electrons (holes) in the transition metals and the f-electrons (holes) in the rare earth metals construct narrow bands and behave as charged particles with a large effective mass. Thus, from our preceding argument it is naturally expected that the bandwidth of d- and f-electrons becomes narrower owing to the screening effect of conduction electrons composed of s- and p-electrons. From this point of view, the screening effect and the orthogonality theorem are also important issues in the usual strongly correlated systems such as the transition metals and rare earth metals.

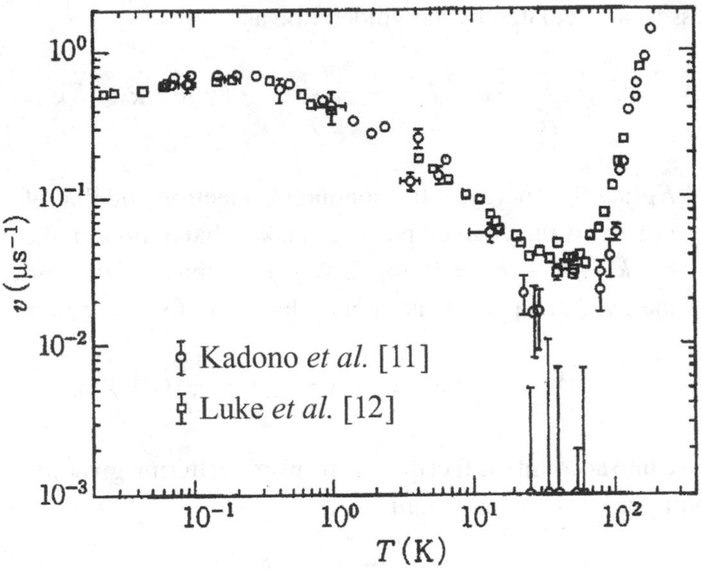

Fig. 3.7 The hop rate of μ^+ in Cu obtained by the spin relaxation method [11].

References

[1] J. Friedel, *Nouovo Cimento Suppl.* **7** (1958) 287.
[2] J. S. Langer and V. Ambegaokar, *Phys. Rev.* **121** (1961) 1090.
[3] P. W. Anderson, *Phys. Rev. Lett.* **18** (1967) 1049; *Phys. Rev.* **164** (1967) 352.
[4] W. Kohn and C. Majumdar, *Phys. Rev.* **138** (1965) A1617.
[5] P. Nozières and C. T. de Dominicis, *Phys. Rev.* **178** (1969) 1097.
[6] G. D. Mahan, *Many-Particle Physics* (Plenum Press, 1981).
[7] J. Kondo, *Physica* **84**B (1976) 40; **124**B (1984) 25; **125**B (1984) 279; **126**B (1984) 377.
[8] K. Yamada and K. Yosida, *Prog. Theor. Phys.* **68** (1982) 1504.
[9] K. Yamada, A. Sakurai and S. Miyazima, *Prog. Theor. Phys.* **73** (1985) 1342.
[10] Yu Kagan and N. V. Prokof'ev, Quantum tunneling diffusion in solids, in *Quantum Tunneling in Condensed Media*, eds. Yu Kagan and A. J. Leggett (North-Holland, 1992) p. 37.
[11] R. Kadono, J. Imazato, T. Matsuzaki *et al.*, *Phys. Rev. B* **39** (1989) 23.
[12] G. M. Luke, J. H. Brewer, S. R. Kreizman *et al.*, *Phys. Rev. B* **43** (1991) 3284.

4

s–d Hamiltonian and Kondo effect

4.1 Spin susceptibility of conduction electrons

Let us consider a free conduction electron system in the presence of a magnetic field $H(\boldsymbol{r}, t)$ oscillating in space and time and calculate the spin susceptibility for this system. We fix the z-axis in the direction of the magnetic field, which is chosen as the spin quantization axis. The Zeeman energy H_Z of the electron possessing spin $\boldsymbol{s} = \sigma/2$, σ being the Pauli matrix, is given by

$$H_Z = -\frac{1}{2}\sigma g \mu_B H(\boldsymbol{r}, t), \tag{4.1}$$

where g and μ_B are the g-value of the electron spin and the Bohr magneton, respectively. By Fourier transformation, $H(\boldsymbol{r}, t)$ is written as

$$H(\boldsymbol{r}, t) = \iint H(\boldsymbol{q}, \omega) e^{i\boldsymbol{q}\cdot\boldsymbol{r} + i\omega t + \alpha t} d\boldsymbol{q} d\omega, \tag{4.2}$$

where α is an infinitesimal positive number and in the limit $t = -\infty$, $H(\boldsymbol{r}, t) = 0$. Now, let us consider the first-order term of the magnetization with respect to the perturbation $H(\boldsymbol{q}, \omega)$. The potential $\delta u_\sigma(\boldsymbol{r}, t)$ applied to the electron with spin σ is given by

$$\begin{aligned} \delta u_\sigma(\boldsymbol{r}, t) &= -\frac{\sigma}{2} g \mu_B H(\boldsymbol{r}, t) \\ &= \iint u_\sigma(\boldsymbol{q}, \omega) e^{i\boldsymbol{q}\cdot\boldsymbol{r} + i\omega t + \alpha t} d\boldsymbol{q} d\omega \\ u_\sigma(\boldsymbol{q}, \omega) &= -\sigma g \mu_B H(\boldsymbol{q}, \omega)/2. \end{aligned} \tag{4.3}$$

This potential depends on the spin. Using the result calculated for the local potential given by (1.68), we can obtain the susceptibility [1, 2]. The change $\delta\rho_\sigma(\boldsymbol{q}, \omega)$ of the electron density with spin σ is given by

$$\delta\rho_\sigma = \frac{1}{\Omega}\sum_{\boldsymbol{k}}\left\{\frac{f(\boldsymbol{k}) - f(\boldsymbol{k}+\boldsymbol{q})}{\varepsilon(\boldsymbol{k}) - \varepsilon(\boldsymbol{k}+\boldsymbol{q}) + \hbar\omega - i\hbar\alpha} e^{i\boldsymbol{q}\cdot\boldsymbol{r}+i\omega t+\alpha t} u_\sigma(\boldsymbol{q}, \omega) + \text{C.C.}\right\}, \quad (4.4)$$

where $f(\boldsymbol{k})$ is the Fermi distribution function. The magnetization $M(\boldsymbol{r}, t)$ is given by

$$\begin{aligned} M(\boldsymbol{r}, t) &= \frac{1}{2} g\mu_{\mathrm{B}}\sigma(\boldsymbol{r}, t) = \frac{1}{2} g\mu_{\mathrm{B}}\sigma\{\delta\rho_\sigma(\boldsymbol{r}, t) - \delta\rho_{-\sigma}(\boldsymbol{r}, t)\} \\ &= -\left(\frac{1}{2} g\mu_{\mathrm{B}}\right)^2 \frac{2}{\Omega}\iint\left\{\sum_{\boldsymbol{k}} \frac{f(\boldsymbol{k}) - f(\boldsymbol{k}+\boldsymbol{q})}{\varepsilon(\boldsymbol{k}) - \varepsilon(\boldsymbol{k}+\boldsymbol{q}) + \hbar\omega - i\hbar\alpha} e^{i\boldsymbol{q}\cdot\boldsymbol{r}+i\omega t+\alpha t}\right. \\ &\qquad \left. \times H(\boldsymbol{q}, \omega) + \text{C.C.}\right\} d\boldsymbol{q}d\omega, \end{aligned} \quad (4.5)$$

where Ω is the volume of the system. Here we introduce the Fourier transform of the susceptibility $\chi(\boldsymbol{r}, t)$ as

$$\chi(\boldsymbol{r}, t) = \iint \chi(\boldsymbol{q}, \omega) e^{i\boldsymbol{q}\cdot\boldsymbol{r}+i\omega t+\alpha t} d\boldsymbol{q}d\omega. \quad (4.6)$$

The susceptibility $\chi(\boldsymbol{q}, \omega)$ is also related to the Fourier transform of $M(\boldsymbol{r}, t)$ as

$$M(\boldsymbol{q}, \omega) = \chi(\boldsymbol{q}, \omega) H(\boldsymbol{q}, \omega). \quad (4.7)$$

From (4.5), $\chi(\boldsymbol{q}, \omega)$ is given by

$$\chi(\boldsymbol{q}, \omega) = -\frac{(g\mu_{\mathrm{B}})^2}{2\Omega}\sum_{\boldsymbol{k}} \frac{f(\boldsymbol{k}) - f(\boldsymbol{k}+\boldsymbol{q})}{\varepsilon(\boldsymbol{k}) - \varepsilon(\boldsymbol{k}+\boldsymbol{q}) + \hbar\omega + i\hbar\alpha}. \quad (4.8)$$

The static susceptibility for a free electron $\chi(\boldsymbol{q}, \omega = 0) = \chi(\boldsymbol{q})$ is given by

$$\begin{aligned} \chi(\boldsymbol{q}) &= \left(\frac{g\mu_{\mathrm{B}}}{2}\right)^2 \frac{2}{\Omega}\sum_{\boldsymbol{k}} \frac{f(\boldsymbol{k}) - f(\boldsymbol{k}+\boldsymbol{q})}{\varepsilon(\boldsymbol{k}+\boldsymbol{q}) - \varepsilon(\boldsymbol{k})} \\ &= \chi_{\text{Pauli}} \frac{1}{2} F_1\left(\frac{q}{2k_{\mathrm{F}}}\right), \end{aligned} \quad (4.9)$$

$$F_1(x) = 1 + \frac{1 - x^2}{2x}\log\left|\frac{1+x}{1-x}\right|. \quad (4.10)$$

Here the Pauli susceptibility χ_{Pauli} for conduction electrons is given by (1.35) and proportional to the density of states $\rho(\varepsilon_{\mathrm{F}})$ at the Fermi surface. The function $F_1(x)$ is shown in Fig. 4.1 and the derivative of $F_1(x)$ diverges at $x = 1$ ($q = 2k_{\mathrm{F}}$). This singularity shows that the change in electron distribution reflects the existence of the Fermi surface and the local change in electron density shorter than the wavelength

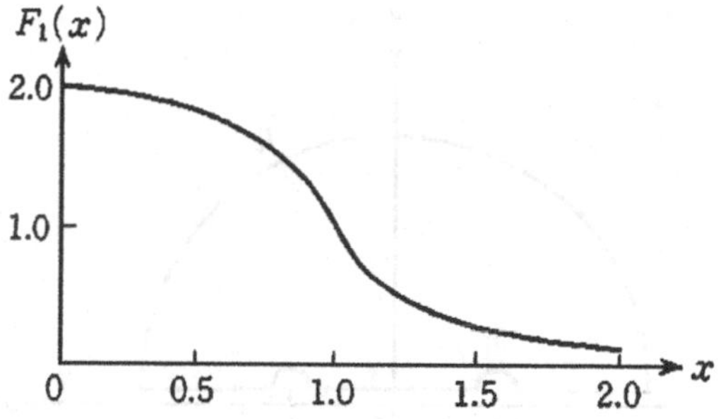

Fig. 4.1 Function $F_1(x)$.

of $1/2k_F$ is difficult, since there exist no electrons possessing wavelength shorter than $1/2k_F$.

4.2 *s–d* Exchange interaction and spin polarization

Now we consider a dilute alloy CuMn, which contains a Mn atom in the Cu metal. The manganese atom Mn possesses spin 5/2 and makes the exchange interaction with conduction electron spins. For simplicity, we assume $S = 1/2$ and the conduction electrons are written as free electrons. This system can be represented by the following Hamiltonian:

$$\mathcal{H} = \mathcal{H}_0 + \mathcal{H}_{s-d} = \sum_{k\sigma} \varepsilon_k c_{k\sigma}{}^{\dagger} c_{k\sigma} - \frac{J}{2N} \sum_{\substack{kk' \\ \sigma\sigma'}} c_{k'\sigma'}{}^{\dagger} \sigma_{\sigma'\sigma} c_{k\sigma} \cdot \mathbf{S}, \tag{4.11}$$

where N is the number of atomic sites. The first term $\mathcal{H}_0$ represents the energy of conduction electrons with N sites and the second term is the Hamiltonian called the *s–d* exchange interaction. The second term can be rewritten as

$$\mathcal{H}_{s-d} = -\frac{J}{2N} \sum_{kk'} \{(c_{k'\uparrow}{}^{\dagger} c_{k\uparrow} - c_{k'\downarrow}{}^{\dagger} c_{k\downarrow}) S_z + c_{k'\uparrow}{}^{\dagger} c_{k\downarrow} S_- + c_{k'\downarrow}{}^{\dagger} c_{k\uparrow} S_+\}. \tag{4.12}$$

If the exchange interaction J is negative, the energy is lowered when the localized spin is antiparallel to the conduction electron spins. Here, we assume the localized spin is fixed by a magnetic field and the average of S_z, $\langle S_z \rangle$, exists. In this case the conduction electrons are polarized by the exchange interaction with the localized spin. Now let us calculate the polarization of the conduction electrons.

The z component $\langle S_z \rangle$ of the localized spin works on the conduction electrons as the following effective field:

$$H_q = \frac{J}{N} \langle S_z \rangle \Big/ \left(\frac{1}{2} g \mu_B \right). \tag{4.13}$$

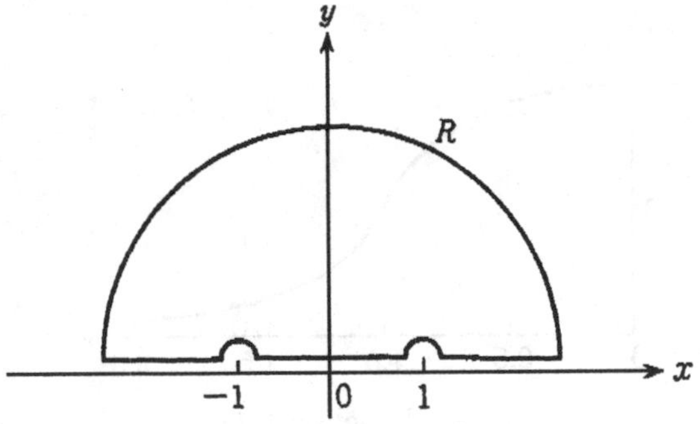

Fig. 4.2 Complex plane shows the integral path. The radius $R \to \infty$.

The polarization of the conduction electron spins due to the magnetic field is calculated as

$$\begin{aligned}\sigma(\boldsymbol{r}) &= \frac{1}{\Omega}\int \chi(\boldsymbol{q})\frac{1}{2}(H_{\boldsymbol{q}}e^{-i\boldsymbol{q}\cdot\boldsymbol{r}} + H_{-\boldsymbol{q}}e^{i\boldsymbol{q}\cdot\boldsymbol{r}})d\boldsymbol{q} \\ &= \frac{1}{2\Omega g\mu_B}\frac{J}{N}\langle S_z\rangle\chi_{\text{Pauli}}\int F_1\left(\frac{q}{2k_F}\right)(e^{-i\boldsymbol{q}\cdot\boldsymbol{r}} + e^{i\boldsymbol{q}\cdot\boldsymbol{r}})d\boldsymbol{q}. \end{aligned} \tag{4.14}$$

Here, we need to calculate the integral in (4.14):

$$\begin{aligned} &\Omega^{-1}\sum_{q} F_1(q/2k_F)(e^{-i\boldsymbol{q}\cdot\boldsymbol{r}} + e^{i\boldsymbol{q}\cdot\boldsymbol{r}}) \\ &= \frac{2\pi}{8\pi^3}\int_0^\infty dq q^2\int_{-1}^1 dz F_1(q/2k_F)(\bar{e}^{iqrz} + q^{iqrz}) \\ &= \frac{(2k_F)^2}{(2\pi^2 r)}(-i)\int_{-\infty}^{\infty} dx x F_1(x)e^{2ik_F xr} \\ &= \frac{(2k_F)^2}{(2\pi^2 r)}\pi\int_{-1}^1 dx\frac{1-x^2}{2}e^{2ik_F xr}, \end{aligned} \tag{4.15}$$

where we have used the following relation for the integration path shown in Fig. 4.2:

$$\log\left(\frac{x+1}{x-1}\right) = \log\left|\frac{1+x}{1-x}\right| - \pi i \qquad (|x| < 1). \tag{4.16}$$

Using the partial integration in (4.15) for a large value of $2k_F r$, we obtain

$$\sigma(r) = \frac{12\pi}{\Omega}\frac{N_e}{N}\frac{J}{g\mu_B}\langle S_z\rangle\chi_{\text{Pauli}}F_2(2k_F r), \tag{4.17}$$

$$F_2(x) = \frac{-x\cos x + \sin x}{x^4}. \tag{4.18}$$

Here N_e is the number of conduction electrons in the volume Ω.

As shown in (4.17), the spin polarization $\sigma(\boldsymbol{r})$ of conduction electrons due to the exchange interaction with the localized spin damps as $(2k_{\mathrm{F}}r)^{-3}$, oscillating in $\cos(2k_{\mathrm{F}}r)$. This behaviour is the same as the Friedel oscillation given by (3.17) and results directly from the singularity of (4.10) at $q = 2k_{\mathrm{F}}$. This oscillation is called the RKY (Ruderman–Kittel–Yosida) oscillation [3].

In the same way as the Friedel oscillation, integrating the spin polarization of conduction electrons $\sigma(\boldsymbol{r})$ over the sphere with radius R, we obtain the result corresponding to (3.15). It is given by

$$\int_0^R \sigma(r) 4\pi r^2 dr = \frac{2J}{Ng\mu_{\mathrm{B}}} \langle S_z \rangle \chi_{\mathrm{Pauli}} \left(1 - \frac{\sin 2k_{\mathrm{F}}R}{2k_{\mathrm{F}}R}\right). \tag{4.19}$$

Equations (4.17) and (4.19) show that the spin polarization induced by the local perturbation due to the localized spin is localized around the localized spin; even in the order of $1/N$, the polarization does not extend over the infinitely long-range region. The $1/r^3$ dependence of the spin polarization in the conduction electrons can be observed by the nuclear magnetic resonance at the Cu nuclei; the width and asymmetry of the resonance peak appear owing to $\sigma(\boldsymbol{r})$, which depends on the distance from the Mn atoms. Since most of the Cu atoms are far from the Mn atom, and the effect of the Mn atom is local, the central line of the resonance does not shift.

The above discussion is not only confined to the conduction electron polarization but is also important as the interaction between localized spins via conduction electrons. Now we assume that a localized spin $\boldsymbol{S}_2$ is situated at a position $\boldsymbol{R}$ from the localized spin $\boldsymbol{S}_1$. The polarization $\sigma(\boldsymbol{r})$ of the conduction electrons at $\boldsymbol{R}$ due to $\boldsymbol{S}_1$ is given by (4.17) and gives the following interaction with the spin $\boldsymbol{S}_2$:

$$\begin{aligned} -\frac{JS_{2z}\Omega}{N} \int d\boldsymbol{r}\, \sigma(\boldsymbol{r} - \boldsymbol{R}_1)\delta(\boldsymbol{r} - \boldsymbol{R}_2) \Big/ \left(\frac{1}{2} g\mu_{\mathrm{B}}\right) \\ = -9\pi \frac{J^2}{\varepsilon_{\mathrm{F}}} \left(\frac{N_{\mathrm{e}}}{N}\right)^2 F_2(2k_{\mathrm{F}}|\boldsymbol{R}_1 - \boldsymbol{R}_2|) S_{1z} S_{2z}. \end{aligned} \tag{4.20}$$

Including the transverse component of spins, we obtain generally the interaction between the localized spins; $\mathcal{H}_{\mathrm{RKKY}}$ is given by

$$\begin{aligned} \mathcal{H}_{\mathrm{RKKY}} &= -9\pi \frac{J^2}{\varepsilon_{\mathrm{F}}} \left(\frac{N_{\mathrm{e}}}{N}\right)^2 F_2(2k_{\mathrm{F}}|\boldsymbol{R}_1 - \boldsymbol{R}_2|) \boldsymbol{S}_1 \cdot \boldsymbol{S}_2 \\ &= -J_{\mathrm{RKKY}} \boldsymbol{S}_1 \cdot \boldsymbol{S}_2. \end{aligned} \tag{4.21}$$

This interaction is called the RKKY interaction, with the name of Kasuya added to RKY for the spin polarization. The RKKY interaction is important in discussing the magnetism of the rare earth metals and the spin glass of the CuMn. Since

$R^{-3} = \Omega^{-1}$, Ω being the volume, the transition temperature T_g of the spin glass is scaled by the concentration of Mn to give a universal temperature independent of the concentration. On the basis of this property, we can have a generalized discussion.

4.3 Kondo effect

In the 1930s the resistance minimum was discovered; the resistance shows a minimum at a temperature as shown in Fig. 4.3, when magnetic impurities such as Mn and Fe are inserted into the normal metals such as Au, Ag and Cu. In ordinary metals the electrical resistance arises from the scattering on lattice vibrations at room temperature. With lowering temperature the resistance due to the lattice vibrations decreases proportionally to T^5. In contrast to this fact, in dilute magnetic alloy systems the resistance increases again with decreasing temperature. In 1964, 30 years after its discovery, J. Kondo presented the theory that explains the resistance minimum. This epoch-making theory not only solved the long-standing problem of the resistance minimum, but also made clear the significance of the many-body problem among electrons in metals. The many-body problem among conduction electrons via a localized spin, elucidated by Kondo, is called the Kondo

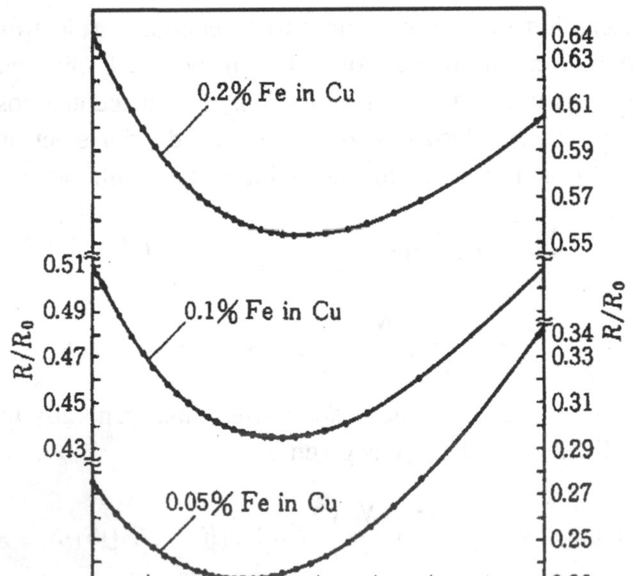

Fig. 4.3 Experimental results show the resistance minimum [5].

problem today. Since his theory, theorists over the world have studied the problem for over 20 years.

First let us introduce the theory of the resistance minimum by Kondo [4]. We use the s–d Hamiltonian (4.11) in which free electrons in metals and a localized spin S interact with each other. For simplicity, the magnitude of the spin S is assumed to be 1/2. The electrical conductivity σ is represented by the mean free time τ_k and the Fermi velocity of electron v_k as

$$\sigma = -\frac{2e^2}{3\Omega}\int d\varepsilon_k \tau_k {v_k}^2 \rho(\varepsilon_k)\frac{\partial f}{\partial \varepsilon_k}, \tag{4.22}$$

where Ω and $\rho(\varepsilon_k)$ are the volume of the system and the density of states for conduction electrons, respectively. In general, the scattering probability $1/\tau(\varepsilon)$ is given by the T-matrix as

$$\frac{1}{\tau(\varepsilon)} = \frac{2\pi}{\hbar}\sum_{\mathrm{f}} |\langle \mathrm{f}|T(\varepsilon)|\mathrm{i}\rangle|^2 \delta(\varepsilon - \varepsilon_{\mathrm{f}} + \varepsilon_{\mathrm{i}}), \tag{4.23}$$

where $|\mathrm{i}\rangle$ and $|\mathrm{f}\rangle$ are the initial and final states, respectively. The T-matrix is defined by (3.21). Here, using $\mathcal{H}_{s-d}$ in (4.11) as the scattering potential $\hat{V}$, we obtain

$$\begin{aligned} T(\varepsilon + i\eta) &= \mathcal{H}_{s-d} + \mathcal{H}_{s-d}\frac{1}{\varepsilon - \mathcal{H}_0 - \mathcal{H}_{s-d} + i\eta}\mathcal{H}_{s-d} \\ &= \mathcal{H}_{s-d} + \mathcal{H}_{s-d}\frac{1}{\varepsilon - \mathcal{H}_0 + i\eta}\mathcal{H}_{s-d} + \cdots, \end{aligned} \tag{4.24}$$

where η is an infinitesimal positive number.

The first-order term of the T-matrix for the scattering of conduction electron $\boldsymbol{k}\uparrow \rightarrow \boldsymbol{k}'\uparrow$ is shown in Fig. 4.4 and given by

$$T^{(1)}(\boldsymbol{k}\uparrow \rightarrow \boldsymbol{k}'\uparrow) = \langle \mathrm{f}|\mathcal{H}_{s-d}|\mathrm{i}\rangle = -\frac{J}{2N}\langle M|S_z|M\rangle, \tag{4.25}$$

where M is the z component of the localized spin S. The term accompanying the spin flip for the scattering $\boldsymbol{k}\uparrow \rightarrow \boldsymbol{k}'\downarrow$ becomes

$$\begin{aligned} T^{(1)}(\boldsymbol{k}\uparrow \rightarrow \boldsymbol{k}'\downarrow) &= -\frac{J}{2N}\langle M+1|S_+|M\rangle \\ &= -\frac{J}{2N}\sqrt{(S-M)(S+M+1)}. \end{aligned} \tag{4.26}$$

Combining (4.25) and (4.26), we obtain

$$T^{(1)} = -\frac{J}{2N}(\boldsymbol{\sigma} \cdot \boldsymbol{S}). \tag{4.27}$$

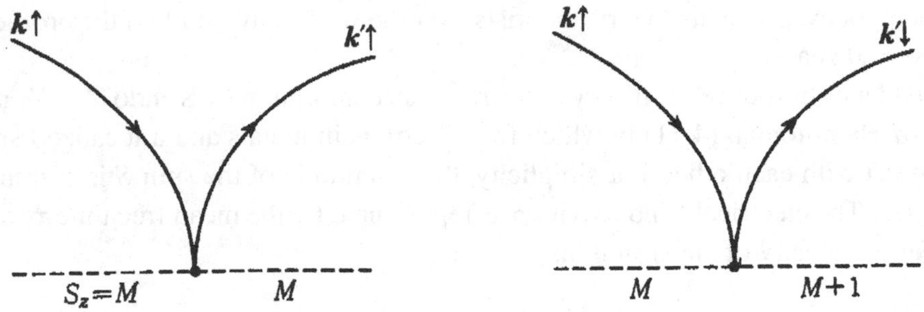

Fig. 4.4 Graphs show the scattering of electrons due to the localized spin. Full lines show the electrons and the dotted line shows the localized spin S. M is the z component of the spin.

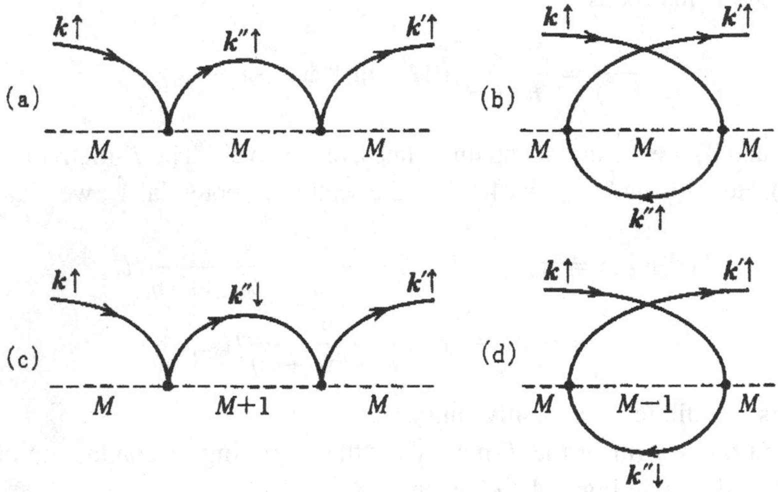

Fig. 4.5 The second-order process of the scattering of electrons due to the localized spin. The arrows directed to the right denote electrons. The arrows directed to the left denote holes.

Using the relation $(\boldsymbol{S}\cdot\boldsymbol{\sigma})^2 = S(S+1) - \boldsymbol{S}\cdot\boldsymbol{\sigma}$, we obtain the resistance given by the Born approximation as

$$R_{\mathrm{B}} = \frac{3}{2}\frac{m\pi}{e^2\hbar}\frac{\Omega}{\varepsilon_{\mathrm{F}}}\left(\frac{J}{2N}\right)^2 S(S+1). \tag{4.28}$$

This result gives the constant resistivity independent of temperature.

As a result, we proceed to the four second-order terms shown by (a)–(d) in Fig. 4.5. For the scattering process (a) $\boldsymbol{k}\uparrow\to \boldsymbol{k}''\uparrow\to \boldsymbol{k}'\uparrow$ without spin flip, the

T-matrix is given by

$$\left(\frac{J}{2N}\right)^2 \sum_{k''} \frac{1 - f_{k''}}{\varepsilon - \varepsilon_{k''} + i\eta} \langle M|S_z{}^2|M\rangle. \tag{4.29}$$

For the process (b), we obtain

$$\left(\frac{J}{2N}\right)^2 \sum_{k''} \frac{-f_{k''}}{\varepsilon_{k''} - \varepsilon - i\eta} \langle M|S_z{}^2|M\rangle, \tag{4.30}$$

where we have fixed the energy of the electrons in the initial and final states at $\varepsilon_k = \varepsilon_{k'} = \varepsilon + i\eta$. For the process (c), in the intermediate state the localized spin is rotated and S_z changes from M to $M + 1$. The matrix element of the T-matrix for the process (c) is given by

$$\left(\frac{J}{2N}\right)^2 \sum_{k''} \frac{1 - f_{k''}}{\varepsilon - \varepsilon_{k''} + i\eta} \langle M|S_- S_+|M\rangle. \tag{4.31}$$

For the process (d), we obtain

$$\left(\frac{J}{2N}\right)^2 \sum_{k''} \frac{-f_{k''}}{\varepsilon_{k''} - \varepsilon - i\eta} \langle M|S_+ S_-|M\rangle. \tag{4.32}$$

Combining the four second-order processes (a)–(d), we obtain

$$\begin{aligned} T^{(2)}(k\uparrow \to k'\uparrow) &= \left(\frac{J}{2N}\right)^2 \left\{ \sum_{k''} \frac{1}{\varepsilon - \varepsilon_{k''} + i\eta} \langle M|S_z{}^2|M\rangle \right. \\ &\quad \left. + \sum_{k''} \frac{1 - f_{k''}}{\varepsilon - \varepsilon_{k''} + i\eta} \langle M|S_- S_+|M\rangle + \sum_{k''} \frac{f_{k''}}{\varepsilon - \varepsilon_{k''} + i\eta} \langle M|S_+ S_-|M\rangle \right\} \\ &= \left(\frac{J}{2N}\right)^2 \left\{ \sum_{k''} \frac{1}{\varepsilon - \varepsilon_{k''} + i\eta} S(S+1) - \sum_{k''} \frac{1 - 2f_{k''}}{\varepsilon - \varepsilon_{k''} + i\eta} \langle M|S_z|M\rangle \right\}, \end{aligned} \tag{4.33}$$

where we have used the relation

$$S_\pm S_\mp = S(S+1) - S_z{}^2 \pm S_z. \tag{4.34}$$

In a similar way to $\boldsymbol{k}\uparrow \to \boldsymbol{k}'\uparrow$, the second-order term of the T-matrix for the scattering from initial state $\boldsymbol{k}\uparrow$ to final state $\boldsymbol{k}'\downarrow$ is obtained as

$$T^{(2)}(k\uparrow \to k'\downarrow) = -\left(\frac{J}{2N}\right)^2 \sum_{k''} \frac{1 - 2f_{k''}}{\varepsilon - \varepsilon_{k''} + i\eta} \langle M+1|S_+|M\rangle. \tag{4.35}$$

Finally, the total T-matrix up to the second-order terms is given by

$$T^{(1)+(2)}(\varepsilon) = \left(\frac{J}{2N}\right)^2 S(S+1) \sum_{k''} \frac{1}{\varepsilon - \varepsilon_{k''} + i\eta} - \frac{J}{2N}(\boldsymbol{S} \cdot \boldsymbol{\sigma}) \left\{ 1 + \frac{J}{2N} \sum_{k''} \frac{1 - 2f(\varepsilon_{k''})}{\varepsilon - \varepsilon_{k''} + i\eta} \right\}. \tag{4.36}$$

The second term in the above bracket is calculated as

$$-\left(\frac{J}{2N}\right)^2 \sum_{k''} \frac{1 - 2f(\varepsilon_{k''})}{\varepsilon - \varepsilon_{k''} + i\delta} = -\left(\frac{J}{2N}\right)^2 \int_{-\infty}^{\infty} \rho(\varepsilon') P \frac{1 - 2f(\varepsilon')}{\varepsilon - \varepsilon'} d\varepsilon' + i\pi \left(\frac{J}{2N}\right)^2 \rho(\varepsilon)[1 - 2f(\varepsilon)]. \tag{4.37}$$

Here, we assume the following density of states for conduction electrons:

$$\begin{aligned} \rho(\varepsilon) &= \rho \qquad (-D \leq \varepsilon \leq D) \\ \rho(\varepsilon) &= 0 \qquad (|\varepsilon| > D). \end{aligned} \tag{4.38}$$

The real part of the first term is obtained as

$$-\left(\frac{J}{2N}\right)^2 \rho \left\{ -\log|D - \varepsilon| - \log|D + \varepsilon| - 2\int_{-D}^{D} \log|\varepsilon - \varepsilon'| \frac{df'}{d\varepsilon'} d\varepsilon' \right\}$$

$$= -\frac{J}{2N}\frac{J\rho}{N} \log\frac{|\varepsilon|}{D} \qquad (|\varepsilon| > kT) \tag{4.39}$$

$$= -\frac{J}{2N}\frac{J\rho}{N} \log\frac{kT}{D} \qquad (|\varepsilon| < kT). \tag{4.40}$$

Thus, the T-matrix is given by

$$T^{(1)+(2)}(\varepsilon) = -\frac{J}{2N}(\boldsymbol{S} \cdot \boldsymbol{\sigma}) \left\{ 1 + \frac{J\rho}{N} \log\frac{\mathrm{Max}(|\varepsilon|, kT)}{D} \right\}. \tag{4.41}$$

Using this result in (4.22) and (4.23), we obtain the resistance as

$$R = R_{\mathrm{B}} \left\{ 1 + \frac{2J\rho}{N} \log\frac{kT}{D} + \cdots \right\}, \tag{4.42}$$

where R_{B} is the result (4.28) obtained by the first Born approximation. In (4.42), the next order term to the first Born term is important. Since we have assumed $J < 0$, for $kT \ll D$ the resistance increases logarithmically with decreasing temperature. This mechanism is the origin of the increase of resistance at low temperatures in the dilute magnetic alloys, and is called the Kondo effect. As the origins giving rise to the logarithmic term $\log(kT/D)$, the following facts are important:

(1) $[S_+, S_-] \neq 0$. That is, the quantum effect that S_+ and S_- do not commute each other.
(2) As seen from the important role of the Fermi distribution function in the intermediate states, one of the origins is the many-body scattering among conduction electrons via the localized spin.
(3) The sharp change of the Fermi distribution function $f(\varepsilon)$, namely, the existence of the Fermi surface.

The combined result of the above three factors gives rise to the Kondo effect, namely the logarithmic terms. The result of (4.42) diverges logarithmically as $T \to 0$. From a high temperature, when the temperature approaches the Kondo temperature defined by

$$kT_{\mathrm{K}} = D \exp\left(\frac{-N}{|J\rho|}\right), \tag{4.43}$$

the contribution from the second Born term becomes equal to the first Born term. As a result, we need the higher order terms with respect to $\rho J/N$ at low temperatures. Abrikosov [6] obtained the result including all the terms given by $[(\rho J/N)\log(kT/D)]^n$ up to $n \to \infty$; the terms become 1 at $T = T_{\mathrm{K}}$. The term $[(\rho J/N)\log kT/D]^n$ gives stronger divergence than the terms $(\rho J/N)^n[\log(kT/D)]^m$ $(n > m)$ and is called the most divergent term. The result for the T-matrix including all the most divergent terms is given by

$$T(\varepsilon) \simeq -\frac{J}{2N}\left[1 - \frac{J\rho}{N}\log\frac{|\varepsilon|}{D}\right]^{-1} \sigma \cdot S. \tag{4.44}$$

Using this result, we obtain the resistivity in the most divergent approximation as

$$R = R_{\mathrm{B}}\left[1 - \frac{J\rho}{N}\log\frac{kT}{D}\right]^{-2}. \tag{4.45}$$

This result also diverges at $T = T_{\mathrm{K}}$. For $J > 0$, the resistance becomes small and converges.

The magnetic susceptibility due to the magnetic impurity was calculated by Yosida and Okiji. The result is

$$\chi_{\mathrm{imp}} = \frac{C}{T}\left\{1 + \frac{J\rho}{N}\left(1 - \frac{J\rho}{N}\log\frac{kT}{D}\right)^{-1}\right\}, \tag{4.46}$$

where the Curie constant $C = (g\mu_{\mathrm{B}})^2 S(S+1)/3$. In (4.46), for $J < 0$, when the temperature is decreased, the term $(J\rho/N)\log(kT/D)$ increases from 0 to 1 and the denominator becomes small. As a result, $\chi_{\rm imp}$ becomes zero at an intermediate temperature and becomes negative below that temperature. Including the factor in the brackets of (4.46) in the Curie constant C, we can consider that the localized spin S tends to zero with decreasing temperature.

4.4 Ground state of dilute magnetic alloy system

What is the meaning of the singularity seen in the temperature dependence of the electrical resistivity and the magnetic susceptibility? What is the low temperature limit? What kind of state is the ground state? From the above results, Yosida presented the following singlet ground state [7]. Writing the localized spin as χ_α or χ_β, and the states of conduction electrons corresponding to each state of the localized spin as $\varphi_{\rm c}{}^\alpha$ or $\varphi_{\rm c}{}^\beta$, we consider the following wave-function:

$$\Psi_{\rm s} = \frac{1}{\sqrt{2}}(\varphi_{\rm c}{}^\alpha \chi_\alpha - \varphi_{\rm c}{}^\beta \chi_\beta). \tag{4.47}$$

The wave-function χ_α denotes the up-spin state, $\varphi_{\rm c}{}^\alpha$ possesses down-spin, χ_β denotes the down-spin state and $\varphi_{\rm c}{}^\beta$ possesses up-spin; in this case $\Psi_{\rm s}$ in (4.47) represents a singlet state. Starting with the singlet state composed of the localized spin and a conduction electron, and including the infinite number of electron–hole pair excitations connected to the initial state, Yosida adopted the following states as $\varphi_{\rm c}{}^\alpha$ and $\varphi_{\rm c}{}^\beta$. For example, for the up-spin component of the localized spin $\varphi_{\rm c}{}^\alpha$ is given by

$$\begin{aligned}\varphi_{\rm c}{}^\alpha = \Bigg\{ &\sum_1 \Gamma_1{}^\alpha c_{1\downarrow}{}^\dagger + \sum_{123} (\Gamma_{12,3}{}^{\alpha\downarrow} c_{1\downarrow}{}^\dagger c_{2\downarrow}{}^\dagger c_{3\downarrow} + \Gamma_{12,3}{}^{\alpha\uparrow} c_{1\downarrow}{}^\dagger c_{2\uparrow}{}^\dagger c_{3\uparrow}) \\ &+ \sum_{12345} (\Gamma_{123,45}{}^{\alpha\downarrow\downarrow} c_{1\downarrow}{}^\dagger c_{2\downarrow}{}^\dagger c_{3\downarrow}{}^\dagger c_{4\downarrow} c_{5\downarrow} + \Gamma_{123,45}{}^{\alpha\downarrow\uparrow} c_{1\downarrow}{}^\dagger c_{2\downarrow}{}^\dagger c_{3\uparrow}{}^\dagger c_{4\downarrow} c_{5\uparrow} \\ &+ \Gamma_{123,45}{}^{\alpha\uparrow\uparrow} c_{1\downarrow}{}^\dagger c_{2\uparrow}{}^\dagger c_{3\uparrow}{}^\dagger c_{4\uparrow} c_{5\uparrow}) + \cdots \Bigg\} |0\rangle,\end{aligned} \tag{4.48}$$

where 1, 2 and 3 mean $\boldsymbol{k}_1, \boldsymbol{k}_2$ and $\boldsymbol{k}_3$ and $\Gamma_1{}^\alpha$ and $\Gamma_{12,3}{}^{\alpha\downarrow}$, etc. are the coefficients to be determined and depend on $\boldsymbol{k}_i$.

Considering also for $\varphi_{\rm c}{}^\beta$ a wave-function similar to (4.48) and inserting it into (4.47), we obtain the Schrödinger equation for (4.11). Finally, the Schrödinger equation can be reduced to the following integral equation for the lowest-order amplitude $\Gamma(\varepsilon_1) = \Gamma_1{}^\alpha$:

$$(\varepsilon_1 - \tilde{E})\Gamma(\varepsilon_1) + \frac{3J}{4N}\sum_2 \Gamma(\varepsilon_2) = \sum_2 K(\varepsilon_1, \varepsilon_2; \tilde{E})\Gamma(\varepsilon_2), \tag{4.49}$$

where the integral kernel $K(\varepsilon_1, \varepsilon_2; \tilde{E})$ is given by the most divergent approximation as

$$K(\varepsilon_1, \varepsilon_2; \tilde{E}) = -\frac{3}{16}\frac{J}{N}\frac{J\rho}{N}\log\left[\frac{\varepsilon_1+\varepsilon_2-\tilde{E}}{D}\right]\bigg/\left(1-\frac{J\rho}{N}\log\frac{\varepsilon_1+\varepsilon_2-\tilde{E}}{D}\right). \tag{4.50}$$

Within the most divergent approximation, (4.49) is solved to give the eigenenergy of the ground state. The result is given by

$$E = \Delta E + \tilde{E}, \tag{4.51}$$

where ΔE is the energy which can be expanded with respect to perturbation J, starting from a doublet state composed of the Fermi sphere and a localized spin. The term $\tilde{E}$ is a binding energy arising only when we start from a singlet state and cannot be expanded with respect to J. The energy $\tilde{E}$ is related to the Kondo temperature as

$$\tilde{E} = -kT_{\mathrm{K}} = -De^{-N/|\rho J|}. \tag{4.52}$$

This relation and (4.51) mean that $|\tilde{E}|$ is the binding energy of the singlet ground state equal to the Kondo temperature. The susceptibility was obtained by H. Ishii and is given at $T = 0$ as

$$\chi_{\mathrm{imp}} = \frac{\frac{1}{2}(g\mu_{\mathrm{B}})^2}{kT_{\mathrm{K}}} = \frac{\frac{1}{2}(g\mu_{\mathrm{B}})^2}{(-\tilde{E})}. \tag{4.53}$$

Now let us discuss the electronic state of the ground state on the basis of Anderson's orthogonality theorem discussed in Section 3.2. First of all it should be noted that the orthogonality theorem also holds for the matrix element of the arbitrary operator $\hat{O}$ with respect to the ground state. For the matrix element $\langle i|\hat{O}|j\rangle$, since the wave-function $\hat{O}|j\rangle$ is a state, we can consider the overlap integral of this state with $|i\rangle$. In the overlap integral, if the local electron numbers for the two states are not equal to each other, the two wave-functions are orthogonal to each other according to the orthogonality theorem. Here we extend the Hamiltonian (4.12) to the case of the anisotropic s–d exchange interaction:

$$\begin{aligned}\mathcal{H} = {} & \sum_{k\sigma}\varepsilon_k c_{k\sigma}{}^{\dagger}c_{k\sigma} + \sum_{kk'}\frac{J_z}{2N}S_z(c_{k'\uparrow}{}^{\dagger}c_{k\uparrow} - c_{k'\downarrow}{}^{\dagger}c_{k\downarrow}) \\ & + \sum_{kk'}\frac{J_\perp}{2N}(S_+c_{k\downarrow}{}^{\dagger}c_{k\uparrow} + S_-c_{k'\uparrow}{}^{\dagger}c_{k\downarrow}).\end{aligned} \tag{4.54}$$

The eigenstates of the first line of (4.54) without the transverse term $J_\perp$ are the two components of (4.47) and degenerate to each other. By connecting the two states by the transverse component $J_\perp$, a singlet state becomes the ground state. As a result, the ground state of (4.54) should possess the finite expectation value of the third term in (4.54). That is, Ψ_s of (4.47) should satisfy

$$\begin{aligned}
&\langle\Psi_s|\sum_{kk'}\frac{J_\perp}{2N}(S_+c_{k'\downarrow}{}^\dagger c_{k\uparrow}+S_-c_{k'\uparrow}{}^\dagger c_{k\downarrow})|\Psi_s\rangle \\
&=\frac{J_\perp}{2N}\sum_{kk'}\frac{1}{2}\{\langle\varphi_c{}^\alpha\chi_\alpha|S_+c_{k'\downarrow}{}^\dagger c_{k\uparrow}|\varphi_c{}^\beta\chi_\beta\rangle+\langle\varphi_c{}^\beta\chi_\beta|S_-c_{k'\uparrow}{}^\dagger c_{k\downarrow}|\varphi_c{}^\alpha\chi_\alpha\rangle\} \\
&=\frac{J_\perp}{4N}\left\{\langle\varphi_c{}^\alpha|\sum_{kk'}c_{k'\downarrow}{}^\dagger c_{k\uparrow}|\varphi_c{}^\beta\rangle+\langle\varphi_c{}^\beta|\sum_{kk'}c_{k'\uparrow}{}^\dagger c_{k\downarrow}|\varphi_c{}^\alpha\rangle\right\}\neq 0. \qquad (4.55)
\end{aligned}$$

It is noted here that $c_{0\sigma}{}^\dagger=\sum_k c_{k\sigma}{}^\dagger/\sqrt{N}$ and $c_{0\sigma}=\sum_k c_{k\sigma}/\sqrt{N}$ are the operator creating one electron at the origin and that annihilating one electron respectively. We write the local electron number with σ spin in the state $\varphi_c{}^\alpha$ as $n_\sigma{}^\alpha$. As the condition for the expectation value of (4.55) not to vanish, we obtain

$$n_\uparrow{}^\alpha=n_\uparrow{}^\beta-1, \qquad (4.56)$$

$$n_\downarrow{}^\alpha=n_\downarrow{}^\beta+1. \qquad (4.57)$$

In addition to this, we can assume that the local electron charge due to the conduction electrons does not change by the exchange interaction between spins, and obtain

$$\sum_\sigma(n_\sigma{}^\alpha+n_\sigma{}^\beta)=0. \qquad (4.58)$$

Since the system is symmetric with exchange of α and β, as a result, the local electron numbers are determined as

$$n_\uparrow{}^\alpha=n_\downarrow{}^\beta=-\frac{1}{2}, \qquad (4.59)$$

$$n_\downarrow{}^\alpha=n_\uparrow{}^\beta=\frac{1}{2}. \qquad (4.60)$$

We show them schematically in Fig. 4.6; $\varphi_c{}^\alpha$ possesses a half electron with down-spin and a half hole with up-spin. As a result, one down-spin exists locally in $\varphi_c{}^\alpha$. In a similar way, $\varphi_c{}^\beta$ possesses one up-spin. Thus, it is confirmed that the state Ψ_s (4.47) is a spin singlet state in which the localized spin and conduction electron spins couple together.

According to the Friedel sum rule, the local electron number $\pm 1/2$ corresponds to the phase shift $\pm\pi/2$. The resistivity due to the impurities, with the impurity concentration n_i and the phase shift of the l-wave at the Fermi surface δ_l, can be written as

$$R=\frac{4\pi\hbar n_i}{N_e e^2 k_F}(2l+1)\sin^2\delta_l, \qquad (4.61)$$

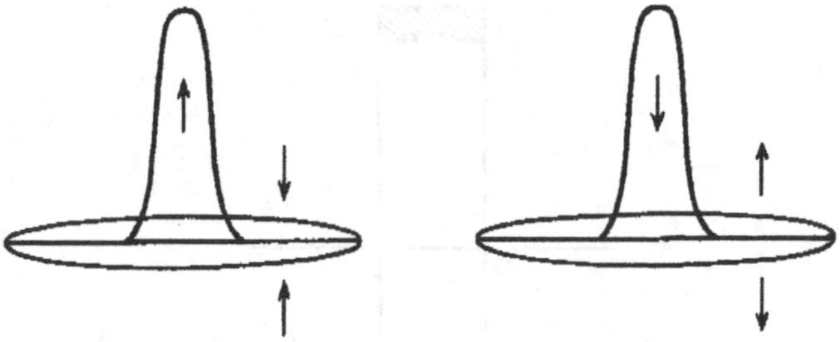

Fig. 4.6 Local electron distribution for each component of a localized spin. For localized up-spin, a half electron with down-spin and a half hole with up-spin are accompanied. For localized down-spin, a half up-spin electron and a half down-spin hole are accompanied.

where N_e is the electron density. The phase shift $\pm\pi/2$ corresponds to the infinite scattering potential and gives the resistance in the unitary limit. As shown by the above discussion, the increase in resistance at low temperatures, shown by the Kondo effect, originates from the transformation to the singlet state composed of the localized spin and the conduction electron spin with decreasing temperature, from the doublet state of an existing localized spin at high temperatures. When the singlet ground state is realized at sufficiently low temperatures, the resistivity takes a universal value corresponding to the unitary limit and does not diverge, in contrast to the divergence obtained by the perturbation calculation starting from a doublet state.

4.5 Scaling law of the *s–d* system

When $\hat{V}$ is an operator of a general perturbation, the T-matrix $T(\omega)$ is defined as

$$T(\omega) = \hat{V} + \hat{V}\frac{1}{\omega - \mathcal{H}_0}T(\omega). \tag{4.62}$$

Here $\mathcal{H}_0$ is an unperturbed Hamiltonian. As shown in Fig. 4.7, we reduce the bandwidth of the conduction electrons by taking out the states within a width ΔD from the top and bottom of the band. If we define $P_{\Delta D}$ as the probability with which excitations by $\hat{V}$ exist within the width ΔD in the intermediate states, for $P_{\Delta D} \ll 1$ we can rewrite $T(\omega)$ as

$$\begin{aligned} T(\omega) &= \hat{V} + \hat{V}\frac{P_{\Delta D}}{\omega - \mathcal{H}_0}T + \hat{V}\frac{(1 - P_{\Delta D})}{\omega - \mathcal{H}_0}T \\ &= \hat{V}' + \hat{V}'\frac{1 - P_{\Delta D}}{\omega - \mathcal{H}_0}T, \end{aligned} \tag{4.63}$$

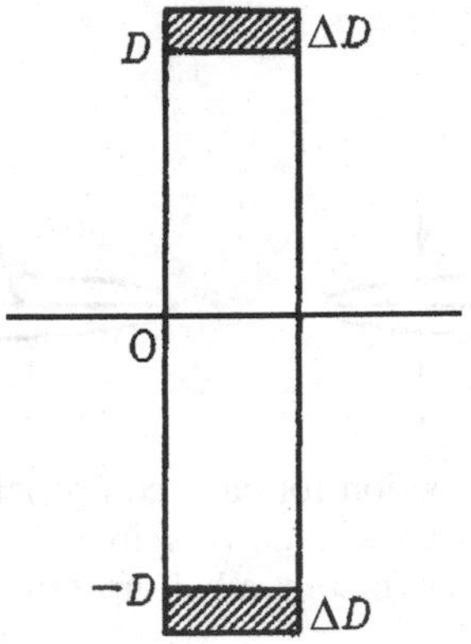

Fig. 4.7 The band is reduced by ΔD from the top and bottom of the conduction band.

$$\hat{V}' = \hat{V} + \hat{V}\frac{P_{\Delta D}}{\omega - \mathcal{H}_0}\hat{V} = \hat{V} + \Delta\hat{V}. \tag{4.64}$$

In the above derivation, we have ignored the second-order term in $P_{\Delta D}$,

$$\hat{V}\frac{P_{\Delta D}}{\omega - \mathcal{H}_0}\hat{V}\frac{P_{\Delta D}}{\omega - \mathcal{H}_0}T.$$

With use of $\hat{V}'$ in place of $\hat{V}$ for the reduced conduction band, $T(\omega)$ is invariant [8].

Here, we assume the density of states ρ is constant for simplicity and put $P_{\Delta D} = \rho\Delta D$. We apply this scaling law to (4.54) and calculate the terms shown in Figs. 4.8 and 4.9. The change of perturbation $\Delta\hat{V}$ can be obtained as

$$\begin{aligned}\Delta\hat{V} = \left(\frac{1}{2N}\right)^2 \sum_{k_1\alpha_1}^{|\varepsilon_{k_1}|<D-\Delta D} \sum_{k_2\alpha_2}^{|\varepsilon_{k_2}|<D-\Delta D} \sum_{k\alpha}^{D>|\varepsilon_k|>D-\Delta D} \frac{1}{\omega - D} \\ \times\Bigg\{ c_{k_2\alpha_2}{}^{\dagger} c_{k\alpha} c_{k\alpha}{}^{\dagger} c_{k_1\alpha_2} \left[J_z S_z \sigma_{\alpha_2\alpha}{}^z + \frac{J_\perp}{2}(S_-\sigma_{\alpha_2\alpha}{}^+ + S_+\sigma_{\alpha_2\alpha}{}^-)\right] \\ \times\left[J_z S_z \sigma_{\alpha\alpha_1}{}^z + \frac{J_\perp}{2}(S_-\sigma_{\alpha\alpha_1}{}^+ + S_+\sigma_{\alpha\alpha_1}{}^-)\right] \\ + c_{k\alpha}{}^{\dagger} c_{k_1\alpha_1} c_{k_2\alpha_2}{}^{\dagger} c_{k\alpha} \left[J_z S_z \sigma_{\alpha\alpha_1}{}^z + \frac{J_\perp}{2}(S_-\sigma_{\alpha\alpha_1}{}^+ + S_+\sigma_{\alpha\alpha_1}{}^-)\right] \\ \times\left[J_z S_z \sigma_{\alpha_2\alpha}{}^z + \frac{J_\perp}{2}(S_-\sigma_{\alpha_2\alpha}{}^+ + S_+\sigma_{\alpha_2\alpha}{}^-)\right]\Bigg\}.\end{aligned} \tag{4.65}$$

Here, we put $|\varepsilon_k| \simeq D$ and ignore the difference between ω and $|\varepsilon_{k_1}|$ or $|\varepsilon_{k_2}|$. Further, we put $c_{k\alpha}c_{k\alpha}{}^{\dagger} = 1$ in the first process of (4.65) and $c_{k\alpha}{}^{\dagger}c_{k\alpha} = 1$ in the

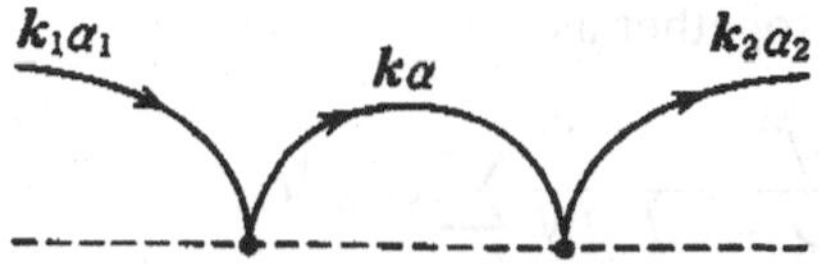

Fig. 4.8 One of two diagrams in the second-order process with respect to the *s–d* exchange interaction.

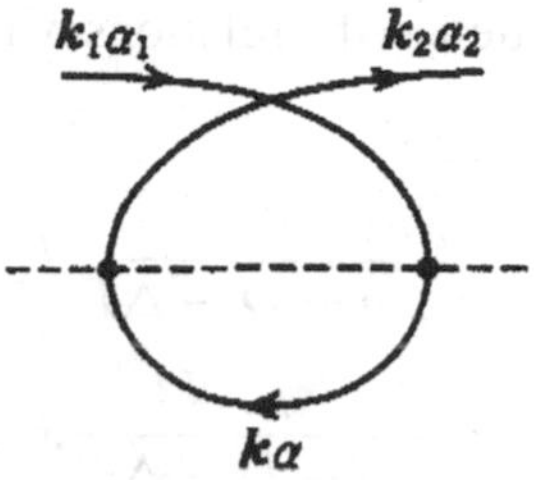

Fig. 4.9 The second of two diagrams in the second-order process with respect to the *s–d* exchange interaction.

second process; $\Delta\hat{V}$ becomes

$$\Delta\hat{V} = \left(\frac{1}{2N}\right)^2 \sum_{k_1\alpha_1}\sum_{k_2\alpha_2} \frac{\rho\Delta D}{\omega - D}\left\{c_{k_2\alpha_2}{}^{\dagger}c_{k_1\alpha_1}\left[\delta_{\alpha_1\alpha_2}\left(\frac{J_z{}^2}{4} + \frac{J_\perp{}^2}{2}\right)\right.\right.$$
$$\left. - J_\perp{}^2 S_z \sigma_{\alpha_2\alpha_1}{}^z - \frac{J_\perp J_z}{2}(S_+\sigma_{\alpha_2\alpha_1}{}^- + S_-\sigma_{\alpha_2\alpha_1}{}^+)\right]$$
$$+ c_{k_1\alpha_1}c_{k_2\alpha_2}{}^{\dagger}\left[\delta_{\alpha_1\alpha_2}\left(\frac{J_z{}^2}{4} + \frac{J_\perp{}^2}{2}\right) + J_\perp{}^2 S_z\sigma_{\alpha_2\alpha_1}{}^z\right.$$
$$\left.\left. + \frac{J_\perp J_z}{2}(S_+\sigma_{\alpha_2\alpha_1}{}^- + S_-\sigma_{\alpha_2\alpha_1}{}^+)\right]\right\}. \tag{4.66}$$

Writing the term possessing the factor $\delta_{\alpha_1\alpha_2}$ as ΔV_0, we obtain

$$\Delta V_0 = \frac{\rho\Delta D}{8N}\frac{\rho D}{N}\frac{1}{\omega - D}(J_z{}^2 + 2J_\perp{}^2).$$

This term gives the energy shift $\Delta E(D)$ brought about by the reduction of the conduction bandwidth from D_0 to D:

$$\Delta E(D) = \frac{1}{8}\int_{D_0}^{D} dD \left\{\left(\frac{J_z\rho}{N}\right)^2 + 2\left(\frac{J_\perp\rho}{N}\right)^2\right\}. \tag{4.67}$$

This shift can be treated by replacing ω with $\omega - \Delta E(D)$.

The spin parts are put together as

$$\frac{\Delta\hat{V}}{\Delta D} = \frac{\rho/N}{\omega - D - \Delta E} \frac{1}{2N} \sum_{\substack{k_2 k_1 \\ \alpha_2\alpha_1}} c_{k_2\alpha_2}{}^{\dagger} c_{k_1\alpha_1}$$

$$\times \left[-J_\perp{}^2 S_z \sigma_{\alpha_2\alpha_1}{}^z - \frac{J_\perp J_z}{2} (S_+ \sigma_{\alpha_2\alpha_1}{}^- + S_- \sigma_{\alpha_2\alpha_1}{}^+) \right]. \tag{4.68}$$

The relation (4.68) can be rewritten as the relations between s–d exchange coupling parameters,

$$\frac{dJ_z}{dD} = -\frac{\rho/N}{\omega - D - \Delta E} J_\perp{}^2, \tag{4.69}$$

$$\frac{dJ_\perp}{dD} = -\frac{\rho/N}{\omega - D - \Delta E} J_\perp J_z. \tag{4.70}$$

From (4.69) and (4.70), we obtain

$$\frac{dJ_z}{dJ_\perp} = \frac{J_\perp}{J_z}. \tag{4.71}$$

By integrating this equation, we obtain the scaling law [9],

$$J_z{}^2 - J_\perp{}^2 = \text{const}. \tag{4.72}$$

This scaling law represents the hyperbolas as shown in Fig. 4.10. In the isotropic case with $J_\perp = J_z$, the scaling equation becomes

$$\frac{dJ}{dD} = -\frac{\rho}{N} \frac{J^2}{\omega - D - \Delta E}. \tag{4.73}$$

Neglecting the D-dependence of $\Delta E(D)$ and using $J = J_0$ at $D = D_0$ as an initial

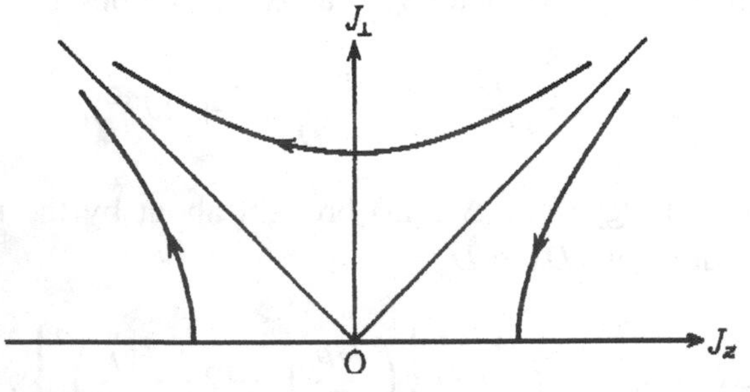

Fig. 4.10 Scaling diagram. With decreasing bandwidth of the conduction band, the coupling constants are scaled along the arrows. Except for the region with $J_z > 0$ and $|J_\perp| < |J_z|$, the system is scaled towards $J \to -\infty$.

condition, we obtain

$$-\frac{N}{\rho J_0}+\frac{N}{\rho J}=\log\left[\frac{D_0}{D-\omega+\Delta E}\right]. \tag{4.74}$$

In this equation, when $D-\omega+\Delta E=-\tilde{E}=kT_{\mathrm{K}}, |J\rho/N|\to\infty$. That is,

$$\frac{J\rho}{N}=\frac{J_0\rho}{N}\left[1+\frac{\rho J_0}{N}\log\frac{D_0}{D-\omega+\Delta E}\right]^{-1}=\frac{J_0\rho}{N}\Big/\left(1+\frac{J_0\rho}{N}\log\frac{D_0}{D}\right). \tag{4.75}$$

Here we have ignored $\omega-\Delta E$. This equation corresponds to the T-matrix (4.44) obtained by Anderson *et al.* [9] by replacing D with ε; the coupling constant $J\rho/N$ tends to $-\infty$ at $D\sim kT_{\mathrm{K}}$ corresponding to $T(\varepsilon)$. This scaling nature means that even if we start from a small coupling constant $J_0\rho/N$, we always arrive at a strongly coupled singlet state as far as low energy excitations are concerned.

4.6 Wilson's theory

By the argument given in the preceding sections, it has been made clear that the ground state of the *s–d* Hamiltonian is the singlet state in which the localized spin of a magnetic atom couples with the spin of conduction electrons. However, it was still difficult to calculate the physical quantities such as the specific heat and the electrical resistivity at low temperatures in a dilute magnetic alloy system. Although the wave-function of the ground state was obtained by Yosida's theory, in order to extend it to a finite temperature, the wave-functions of excited states have to be obtained. Against this difficulty, Wilson [10] transformed the conduction electron system skilfully into the form appropriate to numerical calculation and obtained numerically the coefficient of the specific heat and the magnetic susceptibility. According to his result, the ratio of the two quantities is given by

$$\frac{T\chi_{\mathrm{imp}}}{C_{\mathrm{imp}}}=\frac{g^2{\mu_{\mathrm{B}}}^2}{{k_{\mathrm{B}}}^2}0.1521=\frac{g^2{\mu_{\mathrm{B}}}^2}{{k_{\mathrm{B}}}^2}\frac{3}{2\pi^2}. \tag{4.76}$$

The ratio (4.76), which was obtained by replacing the numerical value, 0.1521, with the guessed value, $3/2\pi^2$, is twice the value in a non-interacting conduction electron system. This fact can be seen by comparing the ratio of (1.35) to (1.34) with (4.76), by putting $g=2$. Today, this ratio $T\chi/C$ is called the Wilson ratio.

We have no space to describe in detail the transformation of the *s–d* Hamiltonian (4.11) done by Wilson, but we will explain it briefly. Expanding the conduction electron states by spherical waves and assuming $\delta(\boldsymbol{r}-\boldsymbol{R})$ for the space dependence of the *s–d* exchange interaction, we can treat the conduction electron as a one-dimensional system composed of only the s-wave, since the δ-function acts only on the s-wave. Here, we assume the linear dispersion, $\varepsilon_k = k$, and that electrons distribute uniformly within the energy width, from -1 to 1. At the next stage, we replace this spectrum by the discrete levels $\varepsilon_n = \Lambda^{-n}$ ($\Lambda \geq 1$), which distribute uniformly in the logarithmic scale. By this procedure, we can expand the low energy levels near the Fermi energy which play an important role in the Kondo effect. Using this transformation, we can make an efficient calculation which is precise for low energy states while it is rough for high energy states. Further, we transform this discrete level system into the tight-binding type of Hamiltonian, where conduction electrons transfer from the origin placed on the localized spin site to neighbouring sites:

$$\mathcal{H}_N = \Lambda^{(N-1)/2}\left\{\sum_{n=0}^{N-1} \Lambda^{-n/2}({f_n}^\dagger f_{n+1} + {f_{n+1}}^\dagger f_n) - \tilde{J}\, {f_0}^\dagger \boldsymbol{\sigma} f_0 \cdot \boldsymbol{S}\right\}. \tag{4.77}$$

In this Hamiltonian the transfer integral for f_n becomes small as $\Lambda^{-n/2}$ for large n, as the site departs far from the origin. In front of this Hamiltonian the factor $\Lambda^{(N-1)/2}$ is multiplied to keep a scale of low energy excitations constant and expand high energy parts. With increasing N, the s–d exchange interaction $\tilde{J}$ multiplied by $\Lambda^{(N-1)/2}$ becomes large and the low energy states approach a spin singlet state. The original Hamiltonian is obtained by the following procedure:

$$\mathcal{H} = \lim_{N\to\infty} \Lambda^{-(N-1)/2}\mathcal{H}_N. \tag{4.78}$$

Wilson showed clearly by numerical calculation that the ground state is a spin singlet state.

4.7 Fermi liquid theory by Nozières

In the ground state of the s–d Hamiltonian, when both conduction electrons and the localized spin are included, there exist locally a half electron with up-spin and a half electron with down-spin. As a result, the phase shifts for this ground state, $\delta(\varepsilon_{\mathrm{F}}) = \pi/2$. Nozières [11] represents the excitations from the ground state as functions of quasi-particle energy ε_σ and density $n_{\sigma'}$ with use of the phase shifts $\delta(\varepsilon_\sigma, n_{\sigma'})$,

$$\delta_\sigma = \delta_\sigma(\varepsilon_\sigma, n_{\sigma'}). \tag{4.79}$$

Here, expanding δ_σ with respect to the deviation of density $n_{\sigma'}$ from that of the ground state, $\delta n_{\sigma'} = n_{\sigma'} - n_{\sigma' 0}$, and taking up to the linear term, we obtain

$$\delta_\sigma(\varepsilon) = \delta_0(\varepsilon) + \sum_{\varepsilon'\sigma'} \phi_{\sigma\sigma'}(\varepsilon, \varepsilon')\delta n_{\sigma'}(\varepsilon'). \tag{4.80}$$

Further, we expand $\delta_0(\varepsilon)$ and the coefficient $\phi_{\sigma\sigma'}$ with respect to ε measured from the Fermi energy $\mu = 0$,

$$\delta_0(\varepsilon) = \delta_0 + \alpha\varepsilon + \beta\varepsilon^2 + \cdots, \tag{4.81}$$

$$\phi_{\sigma\sigma'}(\varepsilon, \varepsilon') = \phi_{\sigma\sigma'} + \varphi_{\sigma\sigma'}(\varepsilon + \varepsilon') + \cdots \tag{4.82}$$

Assuming low temperature and low magnetic field, we discuss physical quantities up to the first-order term in T and H. In this case our system can be described by four parameters δ_0, α and $\phi_{\sigma\pm\sigma} = \phi^s \pm \phi^a$. When we assume the total number of electrons is constant, the parameter $\phi^s = (\phi_{\sigma\sigma} + \phi_{\sigma-\sigma})/2$ does not appear explicitly.

Writing the electron number with up- (down-) spin as $n_\uparrow$ ($n_\downarrow$), we put

$$n_\uparrow - n_\downarrow = m. \tag{4.83}$$

With use of m, δ_σ can be written as

$$\delta_\sigma = \delta_0 + \alpha\varepsilon + \sigma\phi^a m. \tag{4.84}$$

Here σ takes the value 1 for up-spin and -1 for down-spin and $\delta_0 = \pi/2$. If the repulsive force acts among only the electrons with antiparallel spins, we obtain

$$\phi_{\sigma\sigma} = \phi^s + \phi^a = 0. \tag{4.85}$$

Since the electron energy ε shifts by the value of $\delta_\sigma(\varepsilon)/\pi$ divided by ρ, the shifted energy $\tilde{\varepsilon}$ becomes

$$\tilde{\varepsilon} = \varepsilon - \delta_\sigma(\varepsilon)/\pi\rho. \tag{4.86}$$

The change in density of states, since $\alpha/\pi\rho$ is of order $1/N$, becomes

$$\delta\rho = \rho\left[\frac{d\varepsilon}{d\tilde{\varepsilon}} - 1\right] = \frac{\alpha}{\pi}. \tag{4.87}$$

The change in the specific heat becomes

$$\delta C_V/C_V = \alpha/\pi\rho. \tag{4.88}$$

If $H \neq 0$ and $T = 0$, the new energy $\tilde{\varepsilon}_\sigma$ is given by

$$\tilde{\varepsilon}_\uparrow = \varepsilon_\uparrow - \frac{1}{2}g\mu_B H - \frac{1}{\pi\rho}(\delta_0 + \alpha\varepsilon_\uparrow + \phi^a m), \tag{4.89}$$

$$\tilde{\varepsilon}_\downarrow = \varepsilon_\downarrow + \frac{1}{2}g\mu_B H - \frac{1}{\pi\rho}(\delta_0 + \alpha\varepsilon_\downarrow + \phi^a m). \tag{4.90}$$

In the thermal equilibrium, $\tilde{\varepsilon}_\uparrow = \tilde{\varepsilon}_\downarrow = \mu = 0$ and the susceptibility χ is given by

$$\chi = \frac{1}{2} g\mu_B m/H = \frac{1}{2}\rho(g\mu_B)^2\left[1 + \frac{\alpha}{\pi\rho} + \frac{2\phi^a}{\pi}\right]. \tag{4.91}$$

As a result, the Wilson ratio is given by

$$\frac{\delta\chi}{\chi}\bigg/\frac{\delta C_V}{C_V} = 1 + \frac{2\rho\phi^a}{\alpha}. \tag{4.92}$$

This value should be 2 in the limit $\rho J/N \to 0$, as shown by Wilson. Nozières explains this result as follows. When J/N is small compared with the conduction bandwidth D, the spin singlet state is constructed by conduction electrons in the narrow region near the Fermi surface. As a result, by shifting simultaneously ε and μ by the same value in $\delta_\sigma(\varepsilon)$, given by (4.80), $\delta_\sigma(\varepsilon)$ should be invariant, that is

$$\alpha + 2\rho\phi^s = 0 \tag{4.93}$$

should hold. By this result and (4.85), we obtain

$$2\rho\phi^a/\alpha = 1. \tag{4.94}$$

Thus, the Wilson ratio (4.92) becomes 2. As shown above, the excitations of the *s–d* Hamiltonian at low temperatures can be understood on the basis of the Fermi liquid theory. In contrast to the one-body problem, the Wilson ratio becomes 2 owing to the interaction between electrons with antiparallel spins, ϕ^a. In the Anderson Hamiltonian described in the next chapter, the Wilson ratio increases continuously from 1 to 2 with increasing electron interaction between antiparallel spins, U.

The Kondo effect described in this chapter can be concluded as follows [12]. The localized spin that gives rise to the Curie law at high temperatures couples with the conduction electron spins to make a singlet state ($S = 0$) and vanishes. By combining into a spin singlet state the system lowers its energy more than the doublet state possessing an alive spin. This is because the transverse component of the *s–d* interaction works between the degenerate states with respect to the direction of the localized spin and resolves the degeneracy to form the ground state possessing lower energy. This binding energy corresponds to the Kondo temperature. As a result, around the Kondo temperature with decreasing temperature, the transition from the doublet state to the singlet state occurs. In this transition, because the *s–d* interaction conserves the total spin, it is difficult to lead to the singlet state starting with a doublet state using the *s–d* exchange interaction as a perturbation. Since Yosida's theory, Wilson and Nozières have overcome this difficulty by clever ideas to arrive at the ground state. Thus, the Kondo effect can be understood as a general phenomenon in which the ground state is realized by resolving the spin degeneracy. As studied in detail by H. Shiba [13], in the general anisotropic *s–d* exchange interaction with different values of J_x, J_y and J_z, the singlet ground state is always realized. The degeneracy remains in exceptional cases, such as where two of J_x, J_y and J_z are equal. Since the anisotropy in general tends to resolve the degeneracy, the result obtained above is naturally understood.

The *s–d* exchange interaction has been derived on the basis of the existence of localized spin. However, it has been made clear that in the ground state the localized spin vanishes. As a result, it is necessary to investigate further our starting point, namely the localized spin itself. We consider in the next chapter what the magnetic moment is in metals.

References

[1] J. M. Ziman, *Principle of the Theory of Solids* (Cambridge University Press, 1971).
[2] K. Yosida, *Theory of Magnetism* (Springer, 1996).
[3] K. Yosida, *Phys. Rev.* **106** (1957) 893.
[4] J. Kondo, *Prog. Theor. Phys.* **32** (1964) 37; *Solid State Phys.* **23** (1969) 183.
[5] J. P. Franck, F. D. Manchester and D. L. Martin, *Proc. Roy. Soc. A* **263** (1961) 494.
[6] A. A. Abrikosov, *Physics* **2** (1965) 5.
[7] K. Yosida and A. Yoshimori, *Magnetism* V, ed. H. Suhl (Academic Press, 1973).
[8] P. W. Anderson, *J. Phys. C* **3** (1970) 2436.
[9] P. W. Anderson, G. Yuval and D. R. Hamann, *Phys. Rev. B* **1** (1970) 4464.
[10] K. G. Wilson, *Nobel Symposium* **24**, p. 68 (Academic Press, 1974); *Rev. Mod. Phys.* **47** (1975) 793.
[11] P. Nozières, *J. Low Temp. Phys.* **17** (1974) 31.
[12] A. C. Hewson, *The Kondo Problem to Heavy Fermions* (Cambridge University Press, 1993).
[13] H. Shiba, *Prog. Theor. Phys.* **43** (1970) 601.

5

Anderson Hamiltonian

When magnetic impurities such as Mn, Fe and Co are dilutely inserted into non-magnetic metals such as Cu, Ag and Au, susceptibility obeying the Curie–Weiss law is observed. In this case impurity atoms are considered to possess magnetic moments. On the other hand, Mn in Al does not show the Curie–Weiss law and seems not to possess any magnetic moment. In order to describe the magnetic impurities in metals, in 1961 P. W. Anderson presented the following Hamiltonian [1]:

$$\mathcal{H} = \sum_{k\sigma} \varepsilon_k c_{k\sigma}{}^{\dagger} c_{k\sigma} + \sum_{k\sigma} (V_{kd} c_{k\sigma}{}^{\dagger} d_\sigma + V_{kd}{}^{*} d_\sigma{}^{\dagger} c_{k\sigma}) + \sum_{\sigma} \varepsilon_d d_\sigma{}^{\dagger} d_\sigma + U d_\uparrow{}^{\dagger} d_\uparrow d_\downarrow{}^{\dagger} d_\downarrow, \qquad (5.1)$$

where the first term represents the energy of conduction electrons with wave-vector $\boldsymbol{k}$, spin σ and energy ε_k. The third term represents the energy ε_d of the localized d-orbital. Here we have neglected the degeneracy of the d-orbital. When two electrons occupy the d-orbital, the coulomb repulsion U works between them. The second term represents the mixing between the localized d-electron and conduction electrons, with mixing integral V_{kd}. The essential character of the Anderson Hamiltonian is a many-body problem owing to the last electron correlation term. At present, this many-body problem has been solved completely and is instructive as an ideal Fermi liquid system.

5.1 Hartree–Fock approximation

First let us put $U = 0$. In this case the d-level with energy ε_d hybridizes with conduction electrons possessing a continuous energy spectrum and becomes a resonance

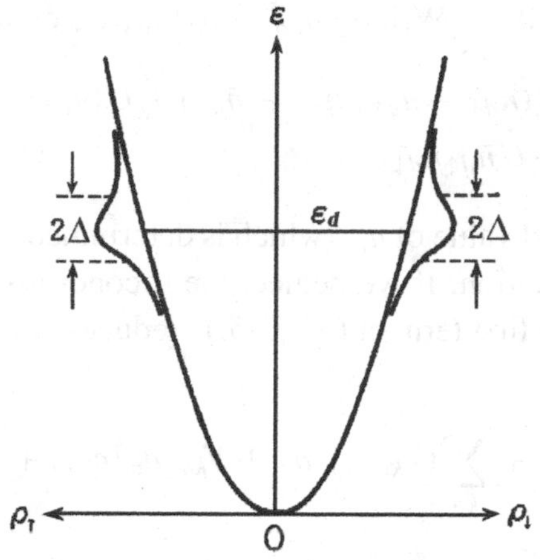

Fig. 5.1 Density of states for a nonmagnetic state is shown schematically. The d-orbital ε_d hybridizes with conduction electrons and is written as a Lorentzian type density of states with width Δ. The quadratic curve shows the density of states for conduction electrons.

level with width 2Δ:

$$2\Delta = 2\pi \sum_k |V_{kd}|^2 \delta(\varepsilon_d - \varepsilon_k) = 2\pi |V|^2 \int d\varepsilon \rho_c(\varepsilon)\delta(\varepsilon_d - \varepsilon) = 2\pi |V|^2 \rho_c(\varepsilon_d), \tag{5.2}$$

where we have assumed V_{kd} is independent of k and $\rho_c(\varepsilon)$ is the density of states per spin for conduction electrons. Equation (5.2) means that the d-electron has lifetime $\tau = \hbar/2\Delta$. As a result, the density of states $\rho_d(\varepsilon)$ for a d-electron is given by the Lorentzian type distribution function with width Δ, as shown in Fig. 5.1;

$$\rho_d(\varepsilon) = \frac{\Delta/\pi}{(\varepsilon - \varepsilon_d)^2 + \Delta^2}. \tag{5.3}$$

Given the Fermi energy μ, we can obtain the number of d-electrons as

$$n_{d\sigma} = \int_{-\infty}^{\mu} d\varepsilon \rho_d(\varepsilon) = \frac{1}{2} + \frac{1}{\pi} \tan^{-1}\left(\frac{\mu - \varepsilon_d}{\Delta}\right). \tag{5.4}$$

When the d-level ε_d is situated at μ, $n_{d\sigma} = 1/2$.

Now, let us assume $U \neq 0$ and consider the case in which the electron interaction exists between d-electrons. Since two electrons staying simultaneously at the same d-orbital possess higher energy by U, the d-electrons tend to avoid each other. First, following Anderson we consider the result given by the Hartree–Fock approximation. In this approximation we decouple the two-body interaction into the following form and neglect the second-order terms of the number fluctuation around their averaged values. Writing $n_{d\sigma} = d_\sigma{}^\dagger d_\sigma$, we obtain

$$U n_{d\uparrow} n_{d\downarrow} = U(n_{d\uparrow} - \bar{n}_{d\uparrow})(n_{d\downarrow} - \bar{n}_{d\downarrow}) + U(\bar{n}_{d\downarrow} n_{d\uparrow} + \bar{n}_{d\uparrow} n_{d\downarrow}) - U\bar{n}_{d\uparrow}\bar{n}_{d\downarrow}. \tag{5.5}$$

Here, $\bar{n}_{d\sigma}$ is the averaged value of $n_{d\sigma}$ which is determined self-consistently by the Hartree–Fock approximation. If we neglect the second-order term in the number fluctuation, which is the first term of (5.5), (5.1) reduces to the following one-body Hamiltonian:

$$\mathcal{H}_{\mathrm{HF}} = \sum_{k\sigma} \varepsilon_k c_{k\sigma}{}^\dagger c_{k\sigma} + \sum_{k\sigma} (V_{kd} c_{k\sigma}{}^\dagger d_\sigma + V_{kd}{}^* d_\sigma{}^\dagger c_{k\sigma}) + \sum_\sigma E_{d\sigma} d_\sigma{}^\dagger d_\sigma, \tag{5.6}$$

$$E_{d\sigma} = \varepsilon_d + U\bar{n}_{d-\sigma}. \tag{5.7}$$

Now let us study the transition from a nonmagnetic solution with $\bar{n}_{d\sigma} = \bar{n}_{d-\sigma} = \bar{n}$ to a magnetic solution with $\bar{n}_{d\sigma} \neq \bar{n}_{d-\sigma}$. Assuming a small deviation of electron number δn, we put

$$n_{d\sigma} = \bar{n} + \sigma\delta n, \tag{5.8}$$

where $\sigma = \pm 1$. Owing to the deviation δn, the magnetic moment $m = g\mu_{\mathrm{B}}\delta n$ arises. Here, $g = 2$. Let us calculate the energy change of the system due to the electron number deviation δn. The change in correlation energy δE_{C} is given by

$$\delta E_{\mathrm{C}} = U(\bar{n} + \delta n)(\bar{n} - \delta n) - U\bar{n}^2 = -U\delta n^2. \tag{5.9}$$

The deviation δn lowers the correlation energy of the system. On the other hand, assuming that the shifts ΔE of d-levels for up- and down-spins are small and using $\delta n = \rho_d(\mu)\Delta E$, we obtain the change in kinetic energy δE_{K} shown in Fig. 5.2 as

$$\delta E_{\mathrm{K}} = \Delta E \delta n = \frac{1}{\rho_d(\mu)} \delta n^2. \tag{5.10}$$

The change in total energy δE is given by

$$\delta E = \delta E_{\mathrm{C}} + \delta E_{\mathrm{K}} = [1 - \rho_d(\mu) U] \delta n^2 / \rho_d(\mu). \tag{5.11}$$

When the coefficient of δn^2 is negative, the energy of the magnetic state is lower than that of the nonmagnetic state. When

$$\rho_d(\mu) U > 1, \tag{5.12}$$

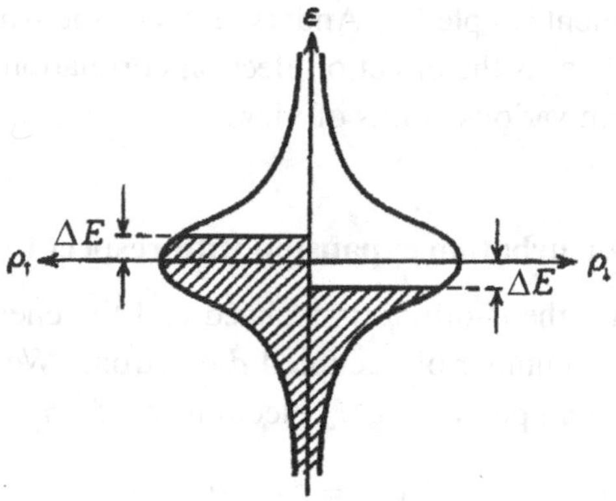

Fig. 5.2 The up-spin electron increases by $\delta n = \rho_d \Delta E$ and the down-spin electron decreases by δn.

the localized magnetic moment of the d-electron appears. From (5.3) $\rho_d(\mu)$ is given by

$$\rho_d(\mu) = \frac{\Delta/\pi}{(\mu - E_{d\sigma})^2 + \Delta^2}, \tag{5.13}$$

$$E_{d\sigma} = \varepsilon_d + U\bar{n}. \tag{5.14}$$

Here if we set $\bar{n} = 1/2$ and $\varepsilon_d - \mu = -U/2$, we obtain $E_{d\sigma} = \mu$. By substituting $E_{d\sigma} = \mu$ in place of ε_d in (5.4), $n = 1/2$. As a result, we can see $\bar{n} = 1/2$ is a nonmagnetic self-consistent solution. When the relation $2\varepsilon_d + U = 2\mu$ holds, the d-orbital possesses electron–hole symmetry and the Hamiltonian is called the symmetric Anderson Hamiltonian. In this case, $\rho_d(\mu)$ in (5.13) becomes

$$\rho_d(\mu) = \frac{1}{\pi\Delta}. \tag{5.15}$$

By (5.12), when $U/\pi\Delta \geq 1$, the nonmagnetic solution with $\bar{n}_{d\uparrow} = \bar{n}_{d\downarrow} = 1/2$ is unstable and the magnetic moment appears.

On the basis of the Hartree–Fock approximation discussed above, the following result is obtained. When U is larger than the width Δ of the d-orbital, a localized magnetic moment arises so as to lower the electron correlation energy. However, when we accept this conclusion, we should be careful of the following point. The states possessing the localized moment are degenerate with respect to the direction of the moment, corresponding to the two solutions given by $\bar{n}_{d\uparrow} - \bar{n}_{d\downarrow} = \pm 2\delta n$. Moreover, this symmetry-broken state arises in a d-orbital, in which an interaction exists among only a few degrees of freedom and large fluctuations spoil the reliability of the mean field approximation. From this point of view, the Anderson Hamiltonian is an important model in the field of the many-body problem, in contrast to the mean field treatment adopted by Anderson. Since the Anderson Hamiltonian is a typical model to discuss the effect of electron correlation U, below we study the model in detail from various points of view.

5.2 Perturbation expansion with respect to V_{kd}

Without a mixing term, the d-orbital is isolated and the energy of the system is determined only by the number of occupied d-electrons. We write the energy of the isolated d-level system possessing n_d electrons as $E^d(n_d)$ and obtain

$$E^d(1) = \varepsilon_d - \mu, \tag{5.16}$$

$$E^d(2) = 2(\varepsilon_d - \mu) + U, \tag{5.17}$$

where μ is the chemical potential. Now we consider the case where the state possessing one electron in the isolated d-orbital is the ground state, i.e., $\varepsilon_d + U/2 > \mu > \varepsilon_d$. We assume that the d-electron has up-spin. Because $V_{kd} = 0$, the conduction electrons make up the Fermi sphere, which we write as $\phi_{\rm v}$. The ground state for $V_{kd} = 0$, $\varphi_\uparrow$ is given by

$$\varphi_\uparrow = d_\uparrow{}^\dagger \phi_{\rm v}. \tag{5.18}$$

This state is degenerate with $\varphi_\downarrow$ possessing a d-electron with down-spin. In the following, we consider the second-order perturbation process of the s–d mixing term by which the state returns to a final state possessing one d-electron:

(i) $\boldsymbol{k}\downarrow$ enters $d\downarrow$ and $d\downarrow$ goes out to $\boldsymbol{k}'\downarrow$

$$\boldsymbol{k}\downarrow \to d\downarrow \to \boldsymbol{k}'\downarrow : \qquad \frac{V_{k'd}V_{dk}}{\varepsilon_k - \varepsilon_d - U} c_{k'\downarrow}{}^\dagger d_\downarrow d_\downarrow{}^\dagger c_{k\downarrow}$$

(ii) $\boldsymbol{k}\downarrow$ enters $d\downarrow$ and $d\uparrow$ goes out to $\boldsymbol{k}'\uparrow$

$$\boldsymbol{k}\downarrow \to d\downarrow$$

$$d\uparrow \to \boldsymbol{k}'\uparrow : \qquad \frac{V_{k'd}V_{dk}}{\varepsilon_k - \varepsilon_d - U} c_{k'\uparrow}{}^\dagger d_\uparrow d_\downarrow{}^\dagger c_{k\downarrow}$$

(iii) $d\uparrow$ goes out to $\boldsymbol{k}'\uparrow$ and $\boldsymbol{k}\uparrow$ enters $d\uparrow$

$$d\uparrow \to \boldsymbol{k}'\uparrow$$

$$\boldsymbol{k}\uparrow \to d\uparrow : \qquad \frac{V_{dk}V_{k'd}}{\varepsilon_d - \varepsilon_{k'}} d_\uparrow{}^\dagger c_{k\uparrow} c_{k'\uparrow}{}^\dagger d_\uparrow$$

(iv) $d\uparrow$ goes out to $\boldsymbol{k}'\uparrow$ and $\boldsymbol{k}\downarrow$ enters $d\downarrow$

$$d\uparrow \to \boldsymbol{k}'\uparrow$$

$$\boldsymbol{k}\downarrow \to d\downarrow : \qquad \frac{V_{dk}V_{k'd}}{\varepsilon_d - \varepsilon_k} d_\downarrow{}^\dagger c_{k\downarrow} c_{k'\uparrow}{}^\dagger d_\uparrow . \tag{5.19}$$

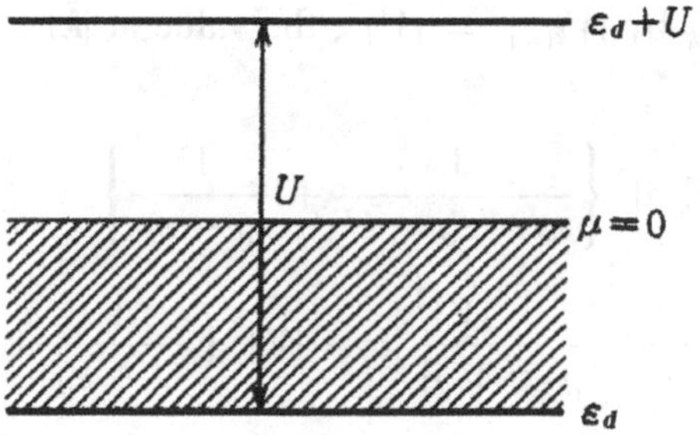

Fig. 5.3 Energy levels of d-electron. The level of the second d-electron becomes $\varepsilon_d + U$.

Adding to the above process the second-order process of V_{kd} starting with $\varphi_\downarrow = d_\downarrow{}^\dagger \varphi_v$, we obtain the effective Hamiltonian within the second-order term with respect to the mixing term, in the space of $n_{d\uparrow} + n_{d\downarrow} = n_d = 1$,

$$\mathcal{H} = \mathcal{H}_{\text{pot}} + \mathcal{H}_{\text{ex}}, \tag{5.20}$$

$$\mathcal{H}_{\text{pot}} = \sum_{kk'} V_{k'd} V_{dk} \left[\frac{1}{\varepsilon_k - U - \varepsilon_d} - \frac{1}{2} n_d \left\{ \frac{1}{\varepsilon_k - \varepsilon_d - U} + \frac{1}{\varepsilon_d - \varepsilon_{k'}} \right\} \right] \times (c_{k'\uparrow}{}^\dagger c_{k\uparrow} + c_{k'\downarrow}{}^\dagger c_{k\downarrow}), \tag{5.21}$$

$$\mathcal{H}_{\text{ex}} = -\sum_{kk'} V_{k'd} V_{dk} \left(\frac{1}{\varepsilon_k - \varepsilon_d - U} + \frac{1}{\varepsilon_d - \varepsilon_{k'}} \right) \{ (c_{k'\uparrow}{}^\dagger c_{k\uparrow} - c_{k'\downarrow}{}^\dagger c_{k\downarrow}) S_z + c_{k'\uparrow}{}^\dagger c_{k\downarrow} S_- + c_{k'\downarrow}{}^\dagger c_{k\uparrow} S_+ \}, \tag{5.22}$$

where

$$S_z = \frac{1}{2}(d_\uparrow{}^\dagger d_\uparrow - d_\downarrow{}^\dagger d_\downarrow), \tag{5.23}$$

$$S_+ = d_\uparrow{}^\dagger d_\downarrow, \tag{5.24}$$

$$S_- = d_\downarrow{}^\dagger d_\uparrow. \tag{5.25}$$

Here, we have used $n_d = 1$. The s–d exchange interaction term (5.22) can be rewritten as

$$\begin{aligned} \mathcal{H}_{\text{ex}} &= -\frac{J}{2N} \sum_{\substack{kk' \\ \sigma\sigma'}} c_{k'\sigma'}{}^\dagger \sigma_{\sigma\sigma'} c_{k\sigma} \cdot \boldsymbol{S} \\ &= -\frac{J}{2N} \sum_{kk'} \{ (c_{k'\uparrow}{}^\dagger c_{k\uparrow} - c_{k'\downarrow}{}^\dagger c_{k\downarrow}) S_z + c_{k'\uparrow}{}^\dagger c_{k\downarrow} S_- + c_{k'\downarrow}{}^\dagger c_{k\uparrow} S_+ \}. \end{aligned} \tag{5.26}$$

Approximating $V_{k'd}V_{dk}$ by $|V_{k_\mathrm{F}d}|^2 = |V|^2$, the value at $|\boldsymbol{k}| = |\boldsymbol{k}'| = k_\mathrm{F}$, we obtain $J/2N$ as

$$\frac{J}{2N} = |V_{k_\mathrm{F}d}|^2 \left\{ \frac{1}{\varepsilon_k - \varepsilon_d - U} + \frac{1}{\varepsilon_d - \varepsilon_{k'}} \right\}$$
$$\simeq -\frac{|V|^2}{U + \varepsilon_d} - \frac{|V|^2}{|\varepsilon_d|} = -|V|^2 \left(\frac{1}{U + \varepsilon_d} + \frac{1}{|\varepsilon_d|} \right), \tag{5.27}$$

where we have neglected ε_k and $\varepsilon_{k'}$, since they are near the Fermi energy $\mu = 0$. The exchange constant J given by (5.27) is always negative. For the symmetric Anderson Hamiltonian with $\varepsilon_d = -U/2$,

$$\frac{J}{N} = -\frac{8|V|^2}{U}. \tag{5.28}$$

The term $\mathcal{H}_\mathrm{pot}$ given by (5.21) is rewritten with $n_d = 1$ as

$$\mathcal{H}_\mathrm{pot} = \frac{1}{2} \sum_{kk'} |V|^2 \left[\frac{1}{\varepsilon_k - \varepsilon_d - U} - \frac{1}{\varepsilon_d - \varepsilon_{k'}} \right]$$
$$\simeq \frac{1}{2} \sum_{kk'} |V|^2 \left(-\frac{1}{\varepsilon_d + U} - \frac{1}{\varepsilon_d} \right). \tag{5.29}$$

$\mathcal{H}_\mathrm{pot}$ vanishes for the symmetric Anderson Hamiltonian with $\varepsilon_d = -U/2$.

The above result can also be derived by a canonical transformation called the Schrieffer–Wolff transformation [2]. By the canonical transformation, $\mathcal{H}$ given by (5.1) is assumed to be transformed into

$$\tilde{\mathcal{H}} = e^S \mathcal{H} e^{-S}. \tag{5.30}$$

We divide the Hamiltonian into $\mathcal{H}_0$ and the mixing term $\mathcal{H}_1$. In the $\tilde{\mathcal{H}}$ given by (5.30), we determine operator S so as to eliminate the first-order term in V_{kd}, that is

$$[\mathcal{H}_0, S] = \mathcal{H}_1. \tag{5.31}$$

The obtained form for S is given by

$$S = \sum_{k\sigma} \left\{ \frac{V_{kd}}{\varepsilon_k - \varepsilon_d - U} n_{d-\sigma} c_{k\sigma}{}^\dagger d_\sigma + \frac{V_{kd}}{\varepsilon_k - \varepsilon_d} (1 - n_{d-\sigma}) c_{k\sigma}{}^\dagger d_\sigma - \mathrm{H.C.} \right\}. \tag{5.32}$$

Here, H.C. means the Hermite conjugate term. By substituting S given by (5.32) into (5.30), we obtain the Hamiltonian up to the order of $|V|^2$ as

$$\mathcal{H}_2 = \frac{1}{2}[S, \mathcal{H}_1] = \mathcal{H}_\mathrm{ex} + \mathcal{H}_\mathrm{pot} + \mathcal{H}_0{}' + \mathcal{H}_\mathrm{ch}. \tag{5.33}$$

The new $\mathcal{H}_0'$ represents the shift of d-level due to the mixing and $\mathcal{H}_{\rm ch}$ is the term changing the d-electron number by 2 and has no effect on the state with $n_d = 1$. As shown above, the effective Hamiltonian obtained up to the order of $|V|^2$ is the s–d Hamiltonian given by (5.26). As a result, when the mixing term V is small and the impurity d-orbital is occupied by one electron, the Anderson Hamiltonian is reduced to the s–d Hamiltonian. Thus, in this limit the Anderson Hamiltonian gives rise to the Kondo effect and arrives at the singlet ground state.

On the other hand, as another standpoint, we first take the mixing term into account and then the electron correlation U by the perturbation expansion. As preparation for this procedure, Green's function is explained below.

5.3 Green's function

The Fermi liquid theory is developed with the help of the correspondence of the quasi-particles near the Fermi surface to the free electron system. Although it is difficult to describe the quasi-particles in terms of the wave-functions in a many-body system, in many cases it is simple to describe the quasi-particles in terms of Green's functions. In developing the Fermi liquid theory on the basis of microscopic grounds, the theory of Green's function is a mathematical instrument of the Fermi liquid theory. Here, Green's function will be explained briefly [3, 4].

Let us consider the creation and annihilation operators of an electron situated at position $\boldsymbol{r}$ and time t in the Heisenberg representation, $\varphi_\sigma{}^\dagger(\boldsymbol{r}, t)$ and $\varphi_\sigma(\boldsymbol{r}, t)$. In this case, the one-particle Green's function $G(\boldsymbol{r}t, \boldsymbol{r}'t')$ at $T = 0$ is defined by taking the average over the ground state $|\Phi_0\rangle$ of an N-electron system as

$$G_{\sigma\sigma'}(\boldsymbol{r}t, \boldsymbol{r}'t') = -i\langle\Phi_0|\mathrm{T}\{\varphi_\sigma(\boldsymbol{r}, t)\varphi_{\sigma'}{}^\dagger(\boldsymbol{r}', t')\}|\Phi_0\rangle. \tag{5.34}$$

Here, T is the time ordering operator that orders the operators from right to left in the order of time. That is,

$$\mathrm{T}\{\varphi_\sigma(\boldsymbol{r}, t)\varphi_{\sigma'}{}^\dagger(\boldsymbol{r}', t')\} = \begin{cases} \varphi_\sigma(\boldsymbol{r}, t)\varphi_{\sigma'}{}^\dagger(\boldsymbol{r}', t') & (t' < t) \\ -\varphi_{\sigma'}{}^\dagger(\boldsymbol{r}', t')\varphi_\sigma(\boldsymbol{r}, t) & (t' > t). \end{cases} \tag{5.35}$$

In the lower part of (5.35), a negative sign is added owing to the exchange between creation and annihilation operators for electrons. In the case of Bose particles, the sign is positive. Except for the factor $-i$, Green's function possesses the following meaning.

When $t > t'$, it means the overlap integral between the two states possessing $N + 1$ electrons, $\varphi_{\sigma'}{}^\dagger(\boldsymbol{r}', t')|\Phi_0\rangle$ and $\varphi_\sigma{}^\dagger(\boldsymbol{r}, t)|\Phi_0\rangle$. Green's function in this case describes the propagation of one electron. On the other hand, when $t < t'$, Green's function is the overlap integral between $\varphi_\sigma(\boldsymbol{r}, t)|\Phi_0\rangle$ and $\varphi_{\sigma'}(\boldsymbol{r}', t')|\Phi_0\rangle$ possessing $N - 1$ electrons and describes the propagation of one hole.

For simplicity, we put $(\boldsymbol{r}', t') = (0, 0)$ and write Green's function as

$$G(\boldsymbol{r}t, 00) = G(\boldsymbol{r}, t). \tag{5.36}$$

Green's function has a discontinuity depending on whether t tends to 0 from the positive or negative side:

$$\begin{aligned} G(\boldsymbol{r}, +0) - G(\boldsymbol{r}, -0) &= -i\langle \Phi_0 | [\varphi_\sigma(\boldsymbol{r}), \varphi_\sigma{}^\dagger(0)]_+ | \Phi_0 \rangle \\ &= -i\delta(\boldsymbol{r}). \end{aligned} \tag{5.37}$$

Now we carry out the Fourier transformation from $\boldsymbol{r}$ into $\boldsymbol{k}$:

$$G(\boldsymbol{r}t, \boldsymbol{r}'t') = \frac{1}{\Omega} \sum_{\boldsymbol{k}\boldsymbol{k}'} G(\boldsymbol{k}t, \boldsymbol{k}'t') e^{i(\boldsymbol{k}\cdot\boldsymbol{r} - \boldsymbol{k}'\cdot\boldsymbol{r}')}, \tag{5.38}$$

$$\begin{aligned} G(\boldsymbol{k}t, \boldsymbol{k}'t') &= \frac{1}{\Omega} \iint d\boldsymbol{r}\, d\boldsymbol{r}'\, G(\boldsymbol{r}t, \boldsymbol{r}'t') e^{-i(\boldsymbol{k}\cdot\boldsymbol{r} - \boldsymbol{k}'\cdot\boldsymbol{r}')} \\ &= -i\langle \Phi_0 | \mathrm{T}\{a_{\boldsymbol{k}}(t) a_{\boldsymbol{k}'}{}^\dagger(t')\} | \Phi_0 \rangle, \end{aligned} \tag{5.39}$$

where $a_{\boldsymbol{k}}$ and $a_{\boldsymbol{k}'}{}^\dagger$ are the Fourier transforms of $\varphi(\boldsymbol{r})$ and $\varphi^\dagger(\boldsymbol{r}')$, respectively,

$$a_{\boldsymbol{k}}(t) = \frac{1}{\sqrt{\Omega}} \int d\boldsymbol{r}\, \varphi(\boldsymbol{r}) e^{-i\boldsymbol{k}\cdot\boldsymbol{r}}, \tag{5.40}$$

$$a_{\boldsymbol{k}'}{}^\dagger(t) = \frac{1}{\sqrt{\Omega}} \int d\boldsymbol{r}\, \varphi^\dagger(\boldsymbol{r}) e^{i\boldsymbol{k}'\cdot\boldsymbol{r}}. \tag{5.41}$$

Since, for the uniform system in space, $G(\boldsymbol{k}t, \boldsymbol{k}'t')$ takes a nonvanishing value only for $\boldsymbol{k}' = \boldsymbol{k}$, Green's function is simplified in the uniform system as

$$G(\boldsymbol{k}, t) = G(\boldsymbol{k}t, \boldsymbol{k}0) = -i\langle \Phi_0 | \mathrm{T}\{a_{\boldsymbol{k}}(t) a_{\boldsymbol{k}}{}^\dagger(0)\} | \Phi_0 \rangle, \tag{5.42}$$

$$G(\boldsymbol{k}, t) = \int d\boldsymbol{r}\, G(\boldsymbol{r}, t) e^{-i\boldsymbol{k}\cdot\boldsymbol{r}}, \tag{5.43}$$

$$G(\boldsymbol{r}, t) = \frac{1}{\Omega} \sum_{\boldsymbol{k}} e^{i\boldsymbol{k}\cdot\boldsymbol{r}} G(\boldsymbol{k}, t). \tag{5.44}$$

$G(\boldsymbol{k}, t)$ is discontinuous as well as $G(\boldsymbol{r}, t)$ given by (5.37). Here, we define $n_{\boldsymbol{k}}$ as the occupation number of electron $\boldsymbol{k}$ in the ground state:

$$n_{\boldsymbol{k}} = \langle \Phi_0 | a_{\boldsymbol{k}}{}^\dagger a_{\boldsymbol{k}} | \Phi_0 \rangle. \tag{5.45}$$

When t tends to zero from a positive value,

$$G(\boldsymbol{k}, +0) = -i(1 - n_{\boldsymbol{k}}). \tag{5.46}$$

When t approaches zero from a negative value,

$$G(\boldsymbol{k}, -0) = i n_{\boldsymbol{k}}. \tag{5.47}$$

From (5.46) and (5.47), the discontinuity of Green's function at $t = 0$ is given by

$$G(\boldsymbol{k}, +0) - G(\boldsymbol{k}, -0) = -i. \tag{5.48}$$

In general, since $|\Phi_0\rangle$ is the ground state of the many-body system, states $a_{\boldsymbol{k}}|\Phi_0\rangle$ and $a_{\boldsymbol{k}}{}^{\dagger}|\Phi_0\rangle$ are not eigenstates of the system. Since

$$a_{\boldsymbol{k}}(t) = e^{i\mathcal{H}t} a_{\boldsymbol{k}} e^{-i\mathcal{H}t}, \tag{5.49}$$

$G(\boldsymbol{k}, t)$ is given by

$$G(\boldsymbol{k}, t) = \begin{cases} -i\langle\Phi_0|e^{i\mathcal{H}t} a_{\boldsymbol{k}} e^{-i\mathcal{H}t} a_{\boldsymbol{k}}{}^{\dagger}|\Phi_0\rangle & (t > 0) \\ i\langle\Phi_0|a_{\boldsymbol{k}}{}^{\dagger} e^{i\mathcal{H}t} a_{\boldsymbol{k}} e^{-i\mathcal{H}t}|\Phi_0\rangle & (t < 0). \end{cases} \tag{5.50}$$

Further, we introduce the Fourier transform of (5.39) with respect to t:

$$G(\boldsymbol{k}\omega, \boldsymbol{k}'\omega') = \frac{1}{2\pi}\iint dt dt' G(\boldsymbol{k}t, \boldsymbol{k}'t') e^{i(\omega t - \omega' t')}, \tag{5.51}$$

$$G(\boldsymbol{k}t, \boldsymbol{k}'t') = \frac{1}{2\pi}\iint d\omega d\omega' G(\boldsymbol{k}\omega, \boldsymbol{k}'\omega') e^{-i(\omega t - \omega' t')}. \tag{5.52}$$

If the system is uniform in space and time, Green's function is written as

$$G(\boldsymbol{k}\omega, \boldsymbol{k}'\omega') = G(\boldsymbol{k}, \omega)\delta_{\boldsymbol{k}\boldsymbol{k}'}\delta(\omega - \omega'). \tag{5.53}$$

In this case Green's function is simplified as

$$G(\boldsymbol{k}, t) = \frac{1}{2\pi}\int d\omega G(\boldsymbol{k}, \omega) e^{-i\omega t}, \tag{5.54}$$

$$G(\boldsymbol{k}, \omega) = \int dt G(\boldsymbol{k}, t) e^{i\omega t}. \tag{5.55}$$

As an example, we consider a free electron system

$$\mathcal{H} = \sum_{k\sigma} \varepsilon_{\boldsymbol{k}}{}^{0} c_{\boldsymbol{k}\sigma}{}^{\dagger} c_{\boldsymbol{k}\sigma}. \tag{5.56}$$

From (5.49),

$$a_{\boldsymbol{k}}(t) = e^{-i\varepsilon_{\boldsymbol{k}}{}^{0}t} a_{\boldsymbol{k}}, \tag{5.57}$$

$$a_{\boldsymbol{k}}{}^{\dagger}(t) = e^{i\varepsilon_{\boldsymbol{k}}{}^{0}t} a_{\boldsymbol{k}}{}^{\dagger}, \tag{5.58}$$

where $\varepsilon_{\boldsymbol{k}}{}^{0} = \hbar^2 k^2/2m$. Depending on $\boldsymbol{k}$ compared with the Fermi wave-number k_{F}, Green's function is written as:

For $|\boldsymbol{k}| > k_F$

$$G_0(\boldsymbol{k}, t) = \begin{cases} ie^{-i\varepsilon_{\boldsymbol{k}}{}^0 t} & (t < 0) \\ 0 & (t > 0). \end{cases} \tag{5.59}$$

For $|\boldsymbol{k}| < k_F$

$$G_0(\boldsymbol{k}, t) = \begin{cases} 0 & (t < 0) \\ -ie^{-i\varepsilon_{\boldsymbol{k}}{}^0 t} & (t > 0). \end{cases} \tag{5.60}$$

The Fourier transform of (5.59) and (5.60) is written simply as

$$G_0(\boldsymbol{k}, \omega) = \frac{1}{\omega - \varepsilon_{\boldsymbol{k}}{}^0 + i\eta}, \tag{5.61}$$

where η is an infinitesimal number and its sign depends on the sign of $(|\boldsymbol{k}| - k_F)$:

$$\eta = \begin{cases} +0 & (|\boldsymbol{k}| > k_F) \\ -0 & (|\boldsymbol{k}| < k_F). \end{cases} \tag{5.62}$$

Equation (5.61) can easily be confirmed by the inverse transformation.

Now let us consider the physical meaning of Green's function in the interacting system. We define $|\Phi_n\rangle$ as an eigenstate for energy $E_n(N+1)$ in the $(N+1)$-particle system; $|\Phi_m\rangle$ is defined as an eigenstate for $E_m(N-1)$ in the $(N-1)$-particle system. $|\Phi_0\rangle$ is the ground state with energy $E_0(N)$ in the N-particle system. We put

$$\omega_{n0} = E_n(N+1) - E_0(N), \tag{5.63}$$

$$\omega_{m0} = E_m(N-1) - E_0(N), \tag{5.64}$$

and Green's function can be written as

$$G(\boldsymbol{k}, t) = \begin{cases} -i\sum_n |\langle\Phi_n|a_{\boldsymbol{k}}{}^\dagger|\Phi_0\rangle|^2 e^{-i\omega_{n0}t} & (t > 0) \\ i\sum_m |\langle\Phi_m|a_{\boldsymbol{k}}|\Phi_0\rangle|^2 e^{i\omega_{m0}t} & (t < 0). \end{cases} \tag{5.65}$$

We represent the energy of the ground state possessing N electrons as $E_0(N)$ and define the chemical potential μ for $N \to \infty$ as

$$E_0(N+1) - E_0(N) = E_0(N) - E_0(N-1) = \mu. \tag{5.66}$$

Putting $\omega_{n0} = \mu + \xi_{n0}$ and $\omega_{m0} = -\mu + \xi_{m0}$, we define the spectral functions A_+ and A_- as the distribution functions of excitation energy for electron and hole, respectively:

$$A_+(\boldsymbol{k}, \omega) = \sum_n |\langle\Phi_n|a_{\boldsymbol{k}}{}^\dagger|\Phi_0\rangle|^2 \delta(\omega - \xi_{n0}), \tag{5.67}$$

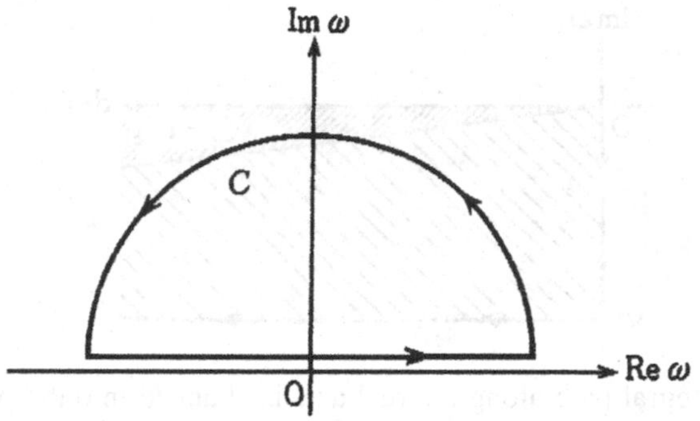

Fig. 5.4 Integral path C. For $t = -0$, to eliminate the contribution of the half-circle with radius $R \to \infty$ we take the upper half-plane.

$$A_-(\boldsymbol{k}, \omega) = \sum_m |\langle \Phi_m | a_{\boldsymbol{k}} | \Phi_0 \rangle|^2 \delta(\omega - \xi_{m0}). \tag{5.68}$$

With the help of these spectral functions, Green's function is written as

$$G(\boldsymbol{k}, t) = \begin{cases} -i \displaystyle\int_0^\infty A_+(\boldsymbol{k}, \omega) e^{-i(\mu+\omega)t} d\omega & (t > 0) \\ i \displaystyle\int_0^\infty A_-(\boldsymbol{k}, \omega) e^{-i(\mu-\omega)t} d\omega & (t < 0) \end{cases}$$

$$G(\boldsymbol{k}, \omega) = \int_0^\infty d\omega' \left\{ \frac{A_+(\boldsymbol{k}, \omega')}{\omega - \omega' - \mu + i\eta} + \frac{A_-(\boldsymbol{k}, \omega')}{\omega + \omega' - \mu - i\eta} \right\}, \tag{5.69}$$

where η is an infinitesimal positive number.

The occupation number of a bare electron with wave-vector $\boldsymbol{k}$ is given by

$$n_{\boldsymbol{k}} = -i[G(\boldsymbol{k}, t)]_{t=-0} = \frac{-i}{2\pi} \int_C d\omega G(\boldsymbol{k}, \omega). \tag{5.70}$$

Here the integral path C is taken as that shown in Fig. 5.4 so as to eliminate the contribution from the upper semicircle for $t < 0$. The total number of particles is given by

$$N = \sum_{\boldsymbol{k}} n_{\boldsymbol{k}} = \frac{-i}{2\pi} \sum_{\boldsymbol{k}} \int_C d\omega G(\boldsymbol{k}, \omega). \tag{5.71}$$

The state $a_{\boldsymbol{k}}^\dagger |\Phi_0\rangle$, in which one particle with wave-vector $\boldsymbol{k}$ is added to the ground state, is not an eigenstate but a linear combination of many eigenstates. Now let us consider the propagation of $a_{\boldsymbol{k}}^\dagger |\Phi_0\rangle$ for $t > 0$:

$$\langle \Phi_0 | a_{\boldsymbol{k}}(t) a_{\boldsymbol{k}}^\dagger | \Phi_0 \rangle = i G(\boldsymbol{k}, t) = e^{-i\mu t} \int_0^\infty A_+(\boldsymbol{k}, \omega) e^{-i\omega t} d\omega. \tag{5.72}$$

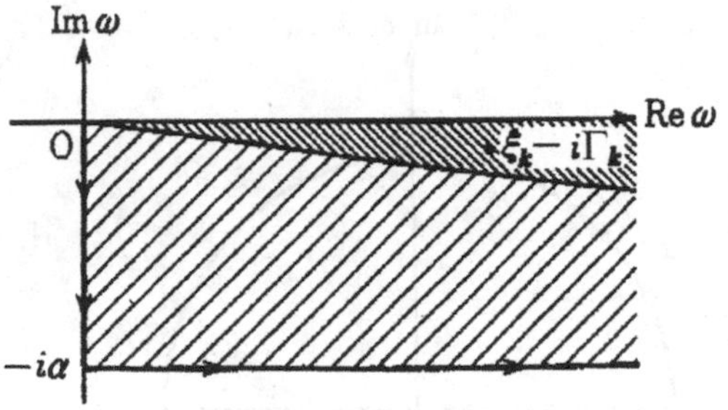

Fig. 5.5 The integral path along the real axis is changed into the path shown by arrows. The value of α is chosen to be large enough to eliminate $e^{-\alpha t}$.

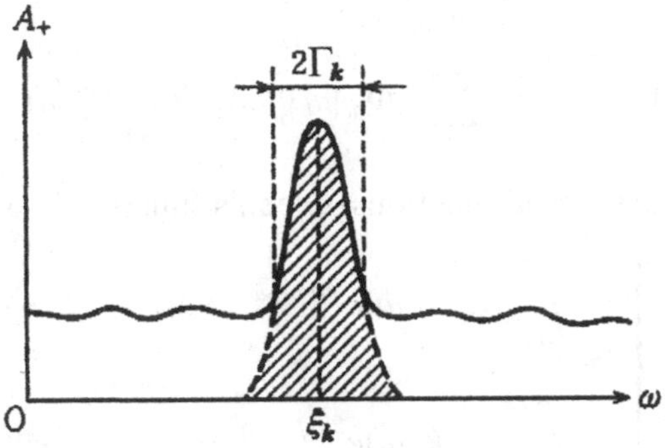

Fig. 5.6 Spectral weight of quasi-particle $A_+(\boldsymbol{k}, \omega)$.

For this expression, we change the integral path into that shown in Fig. 5.5. Here parameter α should be taken so that $e^{-\alpha t}$ is sufficiently small. We obtain

$$\begin{aligned} iG(\boldsymbol{k}, t)e^{i\mu t} &= \int_0^\infty d\omega A_+(\boldsymbol{k}, \omega)e^{-i\omega t} \\ &\simeq \int_0^{-i\alpha} A_+(\boldsymbol{k}, \omega)e^{-i\omega t} d\omega - 2\pi i \sum_j z_j e^{-i\xi_j t}, \end{aligned} \tag{5.73}$$

where the energy distribution of A_+ is shown in Fig. 5.6, and ξ_j and z_j are the pole of $A_+(\boldsymbol{k}, \omega)$ and its residue, respectively. The pole near the real axis is written as $\omega = \xi_{\boldsymbol{k}} - i\Gamma_{\boldsymbol{k}}$. Depending on the time t that has passed since the excitation of electron $\boldsymbol{k}$, the system shows characteristic behaviour in the following three cases:

(a) t is small. In this case we have to take large α so that $\alpha t \gg 1$ and the first term of (5.73) becomes dominant. The time t is too short for quasi-particles to be constructed. That is, $A_+(\boldsymbol{k}, \omega)$ in the whole region of ω contributes to the integration in (5.73).

(b) $t \gg \Gamma_{\boldsymbol{k}}{}^{-1}$. The integral path does not enclose the pole because α is too small. As a result, only the first term in (5.73) contributes. At this time t, quasi-particles have been damped. That is, $A_+(\boldsymbol{k}, \omega)$ within the small ω region contributes dominantly.

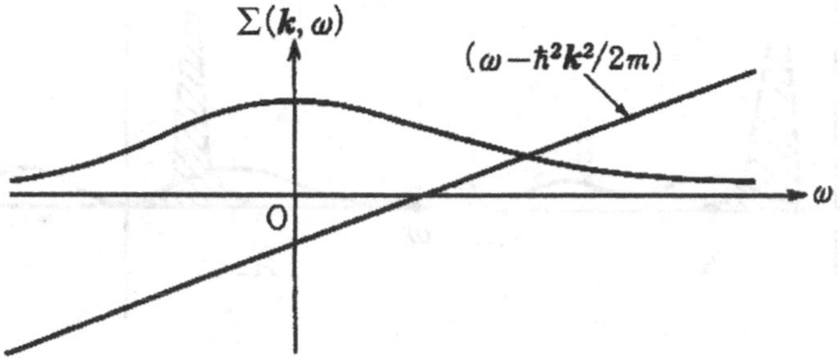

Fig. 5.7 The pole is given by the crossing point between two lines, $\Sigma(\boldsymbol{k}, \omega)$ and $\omega - \hbar^2 \boldsymbol{k}^2/2m$.

(c) Γ_k is sufficiently small. In this case we can choose an intermediate time t between (a) and (b). The time t is large enough for α to be small, but smaller than $\Gamma_k{}^{-1}$. Thus, the main contribution is given by that from the pole since α is small and the integral path encloses the pole:

$$iG(\boldsymbol{k}, t) \simeq 2\pi i z_k \exp[(-i\xi_k - i\mu - \Gamma_k)t]. \tag{5.74}$$

In this case the physical quantities in the system can be described in terms of quasi-particles near the Fermi surface. The damping rate Γ_k is small enough to satisfy the condition (c). This is the essential nature of the Fermi liquid. In general, $a_k{}^\dagger|\Phi_0\rangle$ can be divided into the following two parts:

(1) The coherent part with its norm z_k. This part is the contribution due to the quasi-particles near the Fermi surface.
(2) The incoherent part with norm $1 - n_k - z_k$. This part is the incoherent background.

To make clear the above argument we assume that an electron interaction is introduced into the free electron system and the self-energy of an electron, $\Sigma(\boldsymbol{k}, \omega)$, is given. In this case the one-electron Green's function is given by

$$\left[\omega - \frac{\hbar^2 \boldsymbol{k}^2}{2m} - \Sigma(\boldsymbol{k}, \omega)\right] G(\boldsymbol{k}, \omega) = 1. \tag{5.75}$$

The pole of $G(\boldsymbol{k}, \omega)$ is determined as the crossing point between $\Sigma(\boldsymbol{k}, \omega)$ and $\omega - \hbar^2 \boldsymbol{k}^2/2m$, as shown in Fig. 5.7. In general $\Sigma(\boldsymbol{k}, \omega)$ is a complex number. The residue z_k of $G(\boldsymbol{k}, \omega)$ at the pole $\omega = E_k{}^*$ is given by

$$z_k = \left[1 - \left.\frac{\partial \Sigma(\boldsymbol{k}, \omega)}{\partial \omega}\right|_{\omega = E_k{}^*}\right]^{-1}. \tag{5.76}$$

Further, taking account of the imaginary part, we put

$$\Gamma_k(\omega) = -\mathrm{Im}\,\Sigma(\boldsymbol{k}, \omega), \tag{5.77}$$

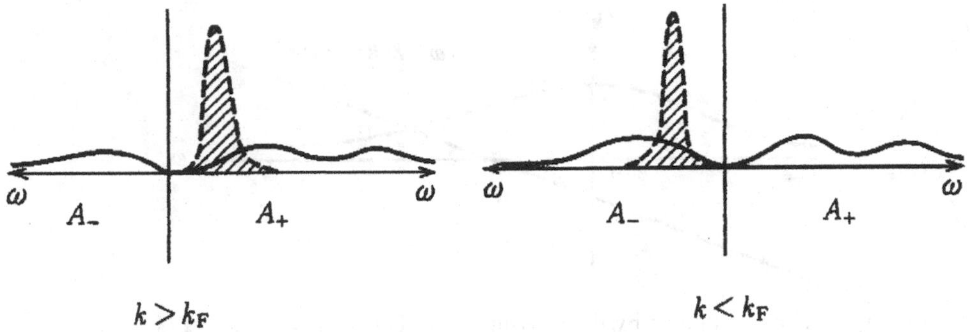

Fig. 5.8 The excitation spectrum that corresponds to the excitation of electrons $k > k_F$ and that of holes $k < k_F$, respectively. The shaded peak is the coherent part and the other parts shown by a full curve are incoherent parts.

and write the pole as $\omega = E_k{}^* - i\Gamma_k{}^*$. From the following equation:

$$(E_k{}^* - i\Gamma_k{}^*)\left(1 - \left.\frac{\partial\Sigma(\boldsymbol{k},\omega)}{\partial\omega}\right|_{\omega=E_k{}^*}\right) - \varepsilon_k - \mathrm{Re}\Sigma(\boldsymbol{k},0) + i\Gamma_k(E_k{}^*) = 0, \tag{5.78}$$

we obtain

$$\Gamma_k{}^*(E_k{}^*) = z_k\Gamma_k(E_k{}^*), \tag{5.79}$$

$$E_k{}^* = z_k(\varepsilon_k + \mathrm{Re}\Sigma(\boldsymbol{k},0)). \tag{5.80}$$

In this case, the excitation spectra of electrons A_+ and of holes A_- are given by Fig. 5.8. Green's function $G(\boldsymbol{k},\omega)$ is given by

$$G(\boldsymbol{k},\omega) = G_{\mathrm{inc}}(\boldsymbol{k},\omega) + \frac{z_k}{(\omega - E_k{}^*) + i\Gamma_k{}^*\mathrm{sgn}(k - k_F)}, \tag{5.81}$$

where G_{inc} is the incoherent part. From (5.81), $A_+(\boldsymbol{k},\omega)$ is given by

$$A_+(\boldsymbol{k},\omega) \simeq \frac{z_k}{\pi}\frac{\Gamma_k{}^*}{(\omega - E_k{}^*)^2 + \Gamma_k{}^{*2}}. \tag{5.82}$$

This is the spectrum of quasi-particles. Here $E_k{}^*$ and $\Gamma_k{}^*$ are the energy and the damping rate of quasi-particles. The factor z_k represents the weight of a bare electron $\boldsymbol{k}$ included in the quasi-particle $\boldsymbol{k}$. As a result, although the occupation of quasi-particles jumps from 1 to 0 at the Fermi energy, that of bare particles jumps with less discontinuity z_k.

We have described above the basic relations among physical quantities related to Green's function, following Nozières [3].

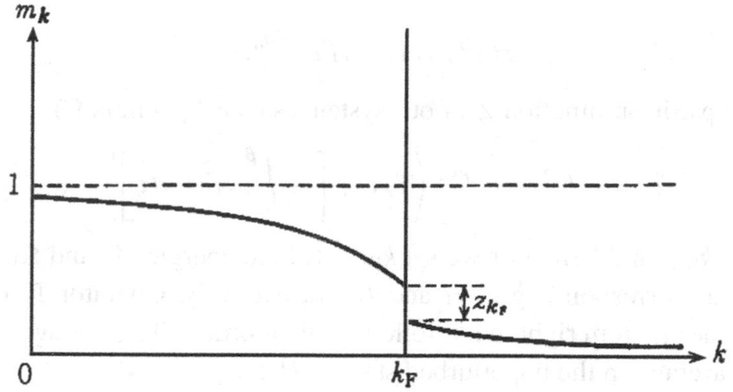

Fig. 5.9 The distribution of bare particles. There is a jump z_{k_F} (< 1) at the Fermi surface.

5.4 Perturbation expansion with respect to U

The ground state of the *s–d* Hamiltonian is the singlet state constructed by the coupling between the localized spin and the conduction electron spin. On the other hand, the state for $J = 0$ is a doublet state constructed by the localized spin and the Fermi sphere. As a result, $J = 0$ is a singular point. For the same reason $V_{kd} = 0$ is a singular point, and the ground state cannot be expanded with respect to J or V_{kd}. On the other hand, the expansion with respect to U is an analytic continuation, keeping the singlet state in the ground state. This is why the expansion with respect to U is simple. The continuation of the ground state with respect to U shows that the Anderson Hamiltonian is an ideal Fermi liquid [5–7].

(a) Electronic specific heat

To treat the correlation energy U between d-electrons as a perturbation, we divide the Hamiltonian (5.1) into the following:

$$\mathcal{H} = \mathcal{H}_0 + \mathcal{H}', \tag{5.83}$$

$$\mathcal{H}_0 = \sum_{k\sigma} \varepsilon_k c_{k\sigma}{}^{\dagger} c_{k\sigma} + \sum_{k\sigma}(V_{kd} c_{k\sigma}{}^{\dagger} d_\sigma + V_{kd} d_\sigma{}^{\dagger} c_{k\sigma}) + \sum_{\sigma} E_{d\sigma}\bar{n}_{d\sigma} - U\bar{n}_{d\uparrow}\bar{n}_{d\downarrow}, \tag{5.84}$$

$$\mathcal{H}' = U(n_{d\uparrow} - \bar{n}_{d\uparrow})(n_{d\downarrow} - \bar{n}_{d\downarrow}). \tag{5.85}$$

Here $\bar{n}_{d\sigma}$ is the average value of the d-electron number over the ground state for $\mathcal{H}_0$. Since $\mathcal{H}_0$ is a one-body Hamiltonian without any electron correlation terms, $\bar{n}_{d\sigma} = \bar{n}_{d-\sigma}$. $E_{d\sigma}$ is given by $\bar{n}_{d\sigma}$ as

$$E_{d\sigma} = \varepsilon_d + U\bar{n}_{d-\sigma}. \tag{5.86}$$

Using

$$\mathcal{H}'(\tau) = e^{\tau\mathcal{H}_0}\mathcal{H}'e^{-\tau\mathcal{H}_0}, \tag{5.87}$$

we write the partition function Z of our system as (see Appendix C)

$$Z = e^{-\beta\Omega} = e^{-\beta\Omega_0}\left\langle \mathrm{T}_\tau \exp\left[-\int_0^\beta \mathcal{H}'(\tau)d\tau\right]\right\rangle. \tag{5.88}$$

Here $\beta = 1/k_{\mathrm{B}}T$ and hereafter we set $k_{\mathrm{B}} = 1$. Free energies Ω and Ω_0 are those of the systems corresponding to $\mathcal{H}$ and $\mathcal{H}_0$, respectively. Operator T_τ orders the imaginary times τ_i from right to left following their orders. The average $\langle A\rangle$ means the thermal average in the unperturbed state of $\mathcal{H}_0$:

$$\langle A\rangle = \mathrm{Tr}/[e^{-\beta\mathcal{H}_0}A]/\mathrm{Tr}\, e^{-\beta\mathcal{H}_0}. \tag{5.89}$$

We define the thermal Green's function of a d-electron as (see Appendix C, (C.21))

$$G_{d\sigma}(\tau_1 - \tau_2) = -\langle\langle \mathrm{T}_\tau d_\sigma(\tau_1)d_\sigma{}^\dagger(\tau_2)\rangle\rangle, \tag{5.90}$$

where $\langle\langle A\rangle\rangle$ is the thermal average for the total Hamiltonian $\mathcal{H}$:

$$\langle\langle A\rangle\rangle = \mathrm{Tr}[e^{-\beta\mathcal{H}}A]/\mathrm{Tr}\, e^{-\beta\mathcal{H}}, \tag{5.91}$$

$$A(\tau) = e^{\tau\mathcal{H}}Ae^{-\tau\mathcal{H}}. \tag{5.92}$$

Green's function (5.90) can be rewritten in the interaction representation, from (C.10) and (C.25), as

$$G_{d\sigma}(\tau_1 - \tau_2) = \\ -\left\langle \mathrm{T}_\tau d_\sigma(\tau_1)d_\sigma{}^\dagger(\tau_2)\exp\left[-\int_0^\beta \mathcal{H}'(\tau)dt\right]\right\rangle \Big/ \left\langle \mathrm{T}_\tau \exp\left[-\int_0^\beta \mathcal{H}'(\tau)d\tau\right]\right\rangle. \tag{5.93}$$

In this expression, since we use the interaction representation, $A(\tau)$ is given by replacing $\mathcal{H}$ with $\mathcal{H}_0$ in (5.92). Let us introduce the Fourier transform of the thermal Green's function (5.90):

$$G_{d\sigma}(\omega_l) = -\int_0^\beta d\tau \langle\langle \mathrm{T}_\tau d_\sigma(\tau)d_\sigma{}^\dagger\rangle\rangle e^{i\omega_l\tau}. \tag{5.94}$$

Here l is an integer and $\omega_l = (2l+1)\pi T$. The d-electron Green's function in the unperturbed state without $\mathcal{H}'$, $G_{d\sigma}{}^0(\omega_l)$, is given by

$$G_{d\sigma}{}^0(\omega_l) = \left[i\omega_l + \mu - E_{d\sigma} - \sum_k \frac{|V_k|^2}{i\omega_l + \mu - \varepsilon_k}\right]^{-1} \\ = [i\omega_l + \mu - E_{d\sigma} + i\Delta\, \mathrm{sgn}\,\omega_l]^{-1}, \tag{5.95}$$

$$\Delta = \pi\rho\langle|V_k|^2\rangle, \qquad \text{sgn}\ \omega_l = \omega_l/|\omega_l|.$$

The last expression of (5.95) is the result obtained by assuming a wide bandwidth and constant density of states ρ for conduction electrons; $\langle|V_k|^2\rangle$ is the average over the Fermi surface.

Introducing the self-energy $\Sigma_\sigma(\omega_l)$ for a d-electron due to $\mathcal{H}'$, we write the thermal Green's function of a d-electron as

$$G_{d\sigma}(\omega_l) = [i\omega_l + \mu - E_{d\sigma} - \Sigma_\sigma(\omega_l) + i\Delta\ \text{sgn}\ \omega_l]^{-1}. \tag{5.96}$$

The self-energy means the 'proper' self-energy that cannot be separated by cutting one electron line. Now let us introduce the nth-order 'improper' self-energy that permits repetition of proper self-energy, and write it as $\Sigma'_{n\sigma}(\omega_l)$. The thermodynamic potential Ω is given by (C.37) as

$$\Omega - \Omega_0 = 2\sum_n \frac{T}{2n} \sum_{\omega_l} G_{d\sigma}{}^0(\omega_l)\Sigma'_{n\sigma}(\omega_l). \tag{5.97}$$

Here Ω_0 is the thermodynamic potential of the unperturbed state. The factor 2 is due to spin and $1/2n$ is multiplied to avoid over-counting, since $2n$ equivalent diagrams appear corresponding to $2n$ Green's functions, when Ω is separated into $G_d{}^0$ and Σ'_n. With the help of the method derived by Luttinger to obtain the lowest-order correction of the free energy (see Appendix C), within accuracy up to the T^2 term, Ω is given by

$$\Omega = 2T\sum_{\omega_l} e^{i\omega_l 0_+} \log G_{d\sigma}(\omega_l). \tag{5.98}$$

Now, using the expansion of a singular function at $\omega = 0$,

$$2\pi T\sum_l F(\omega_l) = \int_{-\infty}^{\infty} d\omega F(\omega) - \left(\frac{\pi^2 T^2}{6}\right)\delta F'(0) + \cdots, \tag{5.99}$$

$$\delta F'(0) = (\partial F/\partial\omega)_{\omega=0_-} - (\partial F/\partial\omega)_{\omega=0_+}, \tag{5.100}$$

we obtain the electronic specific heat C from the T^2 term of Ω given by (5.98) as

$$C = -T(\partial^2\Omega/\partial T^2) = \gamma T. \tag{5.101}$$

Here, by restoring the Boltzmann constant,

$$\gamma = \frac{2\pi^2 k_B{}^2}{3}\rho_d(0)\tilde{\gamma}, \tag{5.102}$$

$$\tilde{\gamma} = 1 - \left.\frac{\partial\Sigma(\omega)}{\partial i\omega}\right|_{\omega=0}, \tag{5.103}$$

$$\rho_d(0) = \frac{1}{\pi}\text{Im}\, G_d(0_-) = \frac{1}{\pi}\text{Im}[\mu - E_d - \Sigma(0) - i\Delta]^{-1}. \tag{5.104}$$

As seen above, the specific heat due to the d-electron is enhanced $\tilde{\gamma}$ times owing to the electron correlation U.

(b) Friedel sum rule in the interacting system

For generality, we apply a magnetic field to our system and represent the energy of a d-electron, $E_{d\sigma}$, as

$$E_{d\sigma} = E_d - \frac{1}{2} g\mu_B \sigma H = E_d - h_\sigma, \tag{5.105}$$

where h_σ is the Zeeman term

$$h_\sigma = \frac{\sigma}{2} g\mu_B H = \sigma \mu_B H \qquad (g = 2). \tag{5.106}$$

Let us introduce the thermal Green's function of conduction electrons $G_{kk\sigma}(\tau)$ defined as

$$G_{kk\sigma}(\tau) = -\langle\langle T_\tau c_{k\sigma}(\tau) c_{k\sigma}{}^\dagger \rangle\rangle. \tag{5.107}$$

The Fourier transform of Green's function is given by

$$G_{kk\sigma}(\omega_l) = G_{kk\sigma}{}^0(\omega_l) + G_{kk\sigma}{}^0(\omega_l) V_{kd} G_{d\sigma}(\omega_l) V_{dk} G_{kk\sigma}{}^0(\omega_l). \tag{5.108}$$

Here the thermal Green's function of a d-electron, $G_{d\sigma}(\omega_l)$, is given by the original form of (5.96) as

$$G_{d\sigma}(\omega_l) = \left[i\omega_l + \mu - E_{d\sigma} - \Sigma_\sigma(\omega) - \sum_k \frac{|V_k|^2}{i\omega_l + \mu - \varepsilon_k} \right]^{-1}. \tag{5.109}$$

Green's function for a free electron, $G_{kk\sigma}{}^0(\omega_l)$, is given by

$$G_{kk\sigma}{}^0(\omega_l) = [i\omega_l + \mu - \varepsilon_k]^{-1} = G_{k\sigma}{}^0(\omega_l). \tag{5.110}$$

The local change of total electron number $\Delta n_{d\sigma}$ due to the impurity is obtained by adding to (5.109) the second term of (5.108) arising from the change of conduction electrons ($\delta = 0_+$):

$$\begin{aligned}
\Delta n_{d\sigma} &= -\frac{1}{\pi} \mathrm{Im} \int_{-\infty}^{\infty} d\omega f(\omega) G_{d\sigma}(\omega + i\delta) \\
&\quad - \sum_k \frac{1}{\pi} \mathrm{Im} \int_{-\infty}^{\infty} d\omega f(\omega) G_{k\sigma}{}^0(\omega_+) V_{kd} G_{d\sigma}(\omega_+) V_{dk} G_{k\sigma}{}^0(\omega_+) \\
&= \int_{-\infty}^{\infty} d\omega f(\omega) \left(-\frac{1}{\pi}\right) \mathrm{Im} \left\{ \frac{\partial}{\partial\omega} \log \left[E_{d\sigma} + \Sigma_\sigma(\omega_+) - \mu - \omega_+ \right.\right. \\
&\quad \left.\left. + \sum_k |V_k|^2 G_{k\sigma}{}^0(\omega_+) \right] + G_{d\sigma}(\omega_+) \frac{\partial}{\partial\omega} \Sigma_\sigma(\omega_+) \right\}.
\end{aligned} \tag{5.111}$$

Using the fact that the energy is conserved at $T = 0$ when electron interactions do not depend on their spins, we can show (see Appendix C)

$$\int_{-\infty}^{\infty} d\omega f(\omega)\left(-\frac{1}{\pi}\right)\mathrm{Im}\left\{G_{d\sigma}(\omega+i\delta)\frac{\partial}{\partial\omega}\Sigma_\sigma(\omega+i\delta)\right\} = 0. \tag{5.112}$$

As a result, the change of local electron number is represented as

$$\Delta n_{d\sigma} = -\frac{1}{\pi}\mathrm{Im}\log\left[E_{d\sigma} + \Sigma_\sigma(\omega_+) - \mu + \sum_k |V_k|^2 G_{k\sigma}{}^0(\omega_+)\right]_{\omega=0}. \tag{5.113}$$

Here, using the approximation

$$\sum_k |V_k|^2 G_{k\sigma}{}^0(i\delta) = -i\pi\rho\langle|V_k|^2\rangle = -i\Delta, \tag{5.114}$$

we obtain

$$\Delta n_{d\sigma} = -\frac{1}{\pi}\mathrm{Im}\log[E_{d\sigma} + \Sigma_\sigma(i\delta) - \mu - i\Delta]. \tag{5.115}$$

Defining the phase shift δ_σ as

$$\delta_\sigma = \tan^{-1}\frac{\Delta}{E_{d\sigma} + \Sigma_\sigma(i\delta) - \mu} = \frac{\pi}{2} - \tan^{-1}\frac{E_{d\sigma} + \Sigma_\sigma(0) - \mu}{\Delta}, \tag{5.116}$$

we can simply write $\Delta n_{d\sigma}$ as

$$\Delta n_{d\sigma} = \delta_\sigma/\pi. \tag{5.117}$$

This is the Friedel sum rule extended to the interacting system [6, 8].

(c) Magnetic susceptibility

By shifting the d-level E_d, the d-electron number is changed. We consider the charge susceptibility χ_c:

$$\chi_c = -\sum_\sigma \frac{\partial n_{d\sigma}}{\partial E_d} = \sum_\sigma \int_0^\beta d\tau\, \langle\langle (n_{d\sigma}(\tau) - \bar{n})(n_{d\sigma}(0) - \bar{n})\rangle\rangle. \tag{5.118}$$

Using (5.115), we obtain

$$\chi_c = \frac{1}{\pi}\sum_\sigma \frac{\Delta(1 + \partial\Sigma_\sigma(0)/\partial E_d)}{(\mu - E_d - \Sigma_\sigma(0))^2 + \Delta^2} = \sum_\sigma \rho_{d\sigma}(0)(1 + \partial\Sigma_\sigma(0)/\partial E_d). \tag{5.119}$$

In a similar way, the spin susceptibility χ_s is obtained as

$$\begin{aligned}\chi_s &= \partial M/\partial H|_{H=0} = \mu_B(\Delta n_{d\uparrow} - \Delta n_{d\downarrow})/H|_{H=0} \\ &= 2\mu_B{}^2\rho_d(0)[1 - \partial\Sigma_\sigma(0)/\partial h_\sigma + \partial\Sigma_\sigma(0)/\partial h_{-\sigma}]|_{H=0} \\ &= 2\mu_B{}^2\rho_d(0)\tilde{\chi}_s,\end{aligned} \tag{5.120}$$

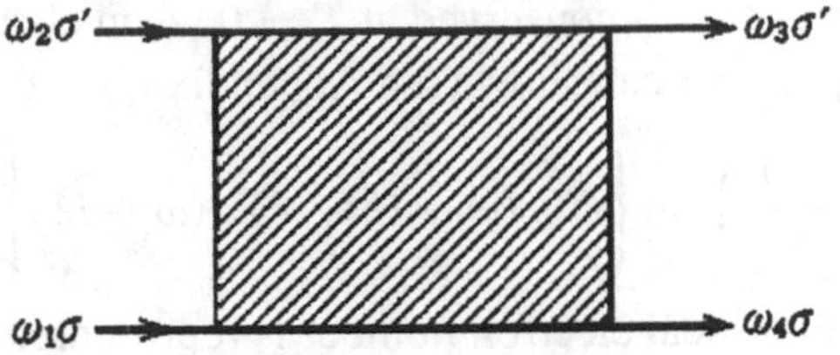

Fig. 5.10 Four-point vertex $\Gamma_{\sigma\sigma';\sigma'\sigma}(\omega_1, \omega_2; \omega_3, \omega_4)$.

$$\tilde{\chi}_s = \tilde{\chi}_{\uparrow\uparrow} + \tilde{\chi}_{\uparrow\downarrow}, \tag{5.121}$$

$$\tilde{\chi}_{\uparrow\uparrow} = 1 - \partial\Sigma_\sigma(0)/\partial h_\sigma|_{h_\sigma=0}, \tag{5.122}$$

$$\tilde{\chi}_{\uparrow\downarrow} = \partial\Sigma_\sigma(0)/\partial h_{-\sigma}|_{h_\sigma=0}. \tag{5.123}$$

The spin susceptibility is enhanced $\tilde{\chi}_s$ times owing to the electron correlation as well as the specific heat. The derivatives of self-energy by h_σ and $h_{-\sigma}$ introduced in (5.120) are related to the derivative of self-energy by E_d, as

$$-\frac{\partial\Sigma_\sigma(0)}{\partial E_d} = \frac{\partial\Sigma_\sigma(0)}{\partial h_\sigma} + \frac{\partial\Sigma_\sigma(0)}{\partial h_{-\sigma}}. \tag{5.124}$$

(d) Ward's identity

Let us write the interaction vertex between electrons with spin σ and σ' as $\Gamma_{\sigma\sigma';\sigma'\sigma}(\omega_1, \omega_2; \omega_3, \omega_4)$. This four-point vertex and various derivatives of self-energy mentioned above are related to each other by Ward's identity. We assume for simplicity $T = 0$ and consider ω-integration along the imaginary axis. Taking the derivative by E_d, we obtain

$$-\frac{\partial\Sigma_\sigma(\omega)}{\partial E_d} = -\frac{1}{2\pi}\int_{-\infty}^{\infty} d\omega' \sum_{\sigma'} \Gamma_{\sigma\sigma';\sigma'\sigma}(\omega, \omega'; \omega', \omega) {G_{d\sigma'}}^2(\omega'). \tag{5.125}$$

Shifting the frequency of every closed loop in the self-energy $\Sigma_\sigma(\omega)$ by ω and taking the derivative by $i\omega$, we obtain

$$\frac{\partial\Sigma_\sigma(\omega)}{\partial i\omega} = -\frac{\delta G_d}{2\pi i}\sum_{\sigma'} \Gamma_{\sigma\sigma';\sigma'\sigma}(\omega, 0; 0, \omega)$$
$$-\frac{1}{2\pi}\int d\omega' \sum_{\sigma'} \Gamma_{\sigma\sigma';\sigma'\sigma}(\omega, \omega'; \omega', \omega) {G_{d\sigma'}}^2(\omega'). \tag{5.126}$$

Here we have used the following result for the derivative of Green's function:

$$\frac{\partial G_d(\omega)}{\partial i\omega} = -{G_d}^2(\omega) + \frac{\delta G_d}{i}\delta(\omega), \tag{5.127}$$

$$\delta G = G(i\delta) - G(-i\delta) = 2i \,\mathrm{Im}\, G^R(0). \tag{5.128}$$

The second term of (5.127) arises from the discontinuity of $G(\omega)$ at $i\omega = 0$.

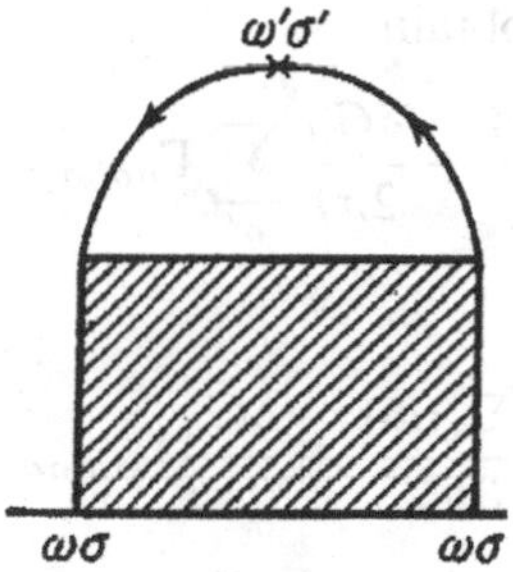

Fig. 5.11 The derivative of self-energy $\Sigma_\sigma(\omega)$ by $h_{\sigma'}$. $G_{d\sigma(\omega')}$ becomes $-[G_{d\sigma'}(\omega')]^2$ by the derivative. The derivative of self-energy by E_d is equal to the sum of the derivatives of self-energy by h_σ and $h_{\sigma'}$.

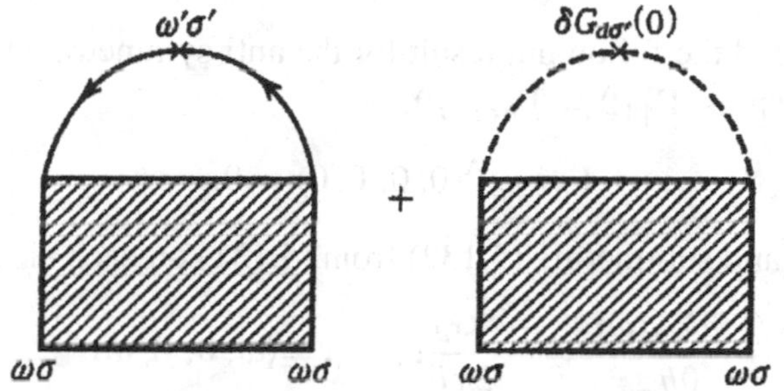

Fig. 5.12 The ω-derivative of self-energy. When frequencies of all closed loops are shifted, we take the sum over σ'. When frequencies of closed loops with only the σ spin are shifted, only the terms with $\sigma' = \sigma$ are obtained. The terms given by the right figure are added owing to the discontinuity of $G_{d\sigma}(\omega)$ at $\omega = 0$.

The derivative of $\Sigma_\sigma(0)$ by $h_{\sigma'}$ is given by

$$\frac{\partial \Sigma_\sigma(0)}{\partial h_{\sigma'}} = -\frac{1}{2\pi}\int_{-\infty}^{\infty} d\omega' \Gamma_{\sigma\sigma';\sigma'\sigma}(\omega, \omega'; \omega', \omega) G_{d\sigma'}{}^2(\omega'). \tag{5.129}$$

In (5.126) we have shifted every frequency of every Green's function by ω, in all the closed loops possessing up- or down-spin. Now we shift the frequency of closed loops possessing only σ spin by ω and don't shift that possessing $-\sigma$ spin. Taking the derivative by ω, we obtain

$$\frac{\partial \Sigma_\sigma(\omega)}{\partial i\omega} = \frac{\delta G_d}{2\pi i}\Gamma_{\sigma\sigma;\sigma\sigma}(\omega, 0;\ 0, \omega) - \frac{1}{2\pi}\int_{-\infty}^{\infty} d\omega' \Gamma_{\sigma\sigma;\sigma\sigma}(\omega, \omega'; \omega', \omega) G_{d\sigma}{}^2(\omega'). \tag{5.130}$$

From (5.125) and (5.126), we obtain

$$\frac{\partial \Sigma_\sigma(\omega)}{\partial i\omega} = -\frac{\partial \Sigma_\sigma(\omega)}{\partial E_d} + \frac{\delta G_d}{2\pi i}\sum_{\sigma'} \Gamma_{\sigma\sigma';\sigma'\sigma}(\omega, 0; 0, \omega). \tag{5.131}$$

From (5.129) and (5.130),

$$\frac{\partial \Sigma_\sigma(\omega)}{\partial i\omega} = \frac{\partial \Sigma_\sigma(\omega)}{\partial h_\sigma} + \frac{\delta G_d}{2\pi i}\Gamma_{\sigma\sigma;\sigma\sigma}(\omega, 0; 0, \omega). \tag{5.132}$$

These relations among the derivatives of self-energy and vertex functions are called Ward's indentities.

Here, putting the external frequency $\omega = 0$, we obtain

$$1 - \left.\frac{\partial \Sigma_\sigma(\omega)}{\partial i\omega}\right|_{\omega=0} = 1 - \left.\frac{\partial \Sigma_\sigma(0)}{\partial h_\sigma}\right|_{h_\sigma=0} = \tilde{\chi}_{\uparrow\uparrow}, \tag{5.133}$$

where we have used the following result for the anti-symmetrized vertex function between parallel spins, $\Gamma_{\uparrow\uparrow}{}^{A} = \Gamma_{\sigma\sigma;\sigma\sigma}{}^{A}$:

$$\Gamma_{\sigma\sigma;\sigma\sigma}{}^{A}(0, 0; 0, 0) = 0. \tag{5.134}$$

Using (5.124) and subtracting (5.132) from (5.131) on each side, we obtain

$$\frac{\partial \Sigma_\sigma(\omega)}{\partial h_{-\sigma}} = -\frac{\delta G_d}{2\pi i}\Gamma_{\sigma-\sigma;-\sigma\sigma}(\omega, 0; 0, \omega). \tag{5.135}$$

Setting $\omega = 0$, we obtain

$$\tilde{\chi}_{\uparrow\downarrow} = -\frac{\delta G_d}{2\pi i}\Gamma_{\sigma-\sigma;-\sigma\sigma}(0, 0; 0, 0) = \rho_d(0)\Gamma_{\uparrow\downarrow}(0), \tag{5.136}$$

where we have used the relation

$$-\delta G_d/2\pi i = -(1/2\pi i)[G_d(i\delta) - G_d(-i\delta)] = \rho_d(0). \tag{5.137}$$

From (5.103) and (5.133), we obtain the following identity:

$$\tilde{\gamma} = \tilde{\chi}_{\uparrow\uparrow}. \tag{5.138}$$

(e) Symmetric case ($\varepsilon_d = -U/2$)

The symmetric case with $\varepsilon_d = -U/2$ is simple and instructive. In this case $\bar{n}_{d\sigma} = 1/2$ and $E_{d\sigma} = 0$ in (5.86). Green's function of the d-electron in the unperturbed state, $G_{d\sigma}{}^{0}(\omega)$, is given by

$$G_{d\sigma}{}^{0}(\omega) = [i\omega + i\Delta\,\mathrm{sgn}\,\omega]^{-1}. \tag{5.139}$$

Green's function $G_{d\sigma}{}^{0}(\omega)$ in the symmetric case is an odd function of ω. The self-energy $\Sigma_\sigma(\omega)$, which is given by the integration of odd number products of odd functions, vanishes at $\omega = 0$:

$$\Sigma_\sigma(0) = 0. \tag{5.140}$$

In this case, the density of state $\rho_d(0)$ in (5.137) becomes

$$\rho_d(0) = \frac{1}{\pi\Delta}. \tag{5.141}$$

The odd-order terms of self-energy with respect to U, $\Sigma_\sigma(\omega)$, possess odd number products of Green's functions $G_{d-\sigma}{}^0(\omega_i)$ in closed loops with $-\sigma$ spin. If we change the sign of the frequency in every closed loop with $-\sigma$ spin, the total functions given by odd number products of Green's functions have a minus sign because of the odd function $G_{d-\sigma}{}^0(\omega)$. In the total diagrams there should exist closed loops rotating in the opposite direction to each other. The diagrams with opposite closed loops cancel each other. As a result, the odd-order terms in U of the self-energy $\Sigma_\sigma(\omega)$ vanish. $\partial\Sigma_\sigma(0)/\partial h_\sigma$ also has no odd-order terms, since the property of closed loops with $-\sigma$ spin is not changed by the derivative of h_σ. Thus, putting $u = U/\pi\Delta$, we obtain

$$1 - \frac{\partial\Sigma_\sigma(0)}{\partial h_\sigma} = \tilde{\chi}_{\uparrow\uparrow} = \tilde{\chi}_{\text{even}} = \sum_{m=0}^{\infty} a_{2m}u^{2m}. \tag{5.142}$$

On the other hand, in $\partial\Sigma_\sigma(0)/\partial h_{-\sigma}$, only the odd-order terms with respect to U exist since $G_{d-\sigma}{}^0(\omega_i)$ becomes $G^0_{d-\sigma}{}^2(\omega_i)$ by the derivative of $h_{-\sigma}$:

$$\frac{\partial\Sigma_\sigma(0)}{\partial h_{-\sigma}} = \tilde{\chi}_{\uparrow\downarrow} = \tilde{\chi}_{\text{odd}} = \sum_{m=0}^{\infty} a_{2m+1}u^{2m+1}. \tag{5.143}$$

Thus, we have

$$\tilde{\chi}_{\text{s}} = \sum_{n=0}^{\infty} a_n u^n = \tilde{\chi}_{\uparrow\uparrow} + \tilde{\chi}_{\uparrow\downarrow} = \tilde{\chi}_{\text{even}} + \tilde{\chi}_{\text{odd}}. \tag{5.144}$$

From (5.124) the charge susceptibility χ_{c} is obtained as

$$\chi_{\text{c}} = \sum_\sigma \rho_{d\sigma}(0)\tilde{\chi}_{\text{c}}, \tag{5.145}$$

$$\tilde{\chi}_{\text{c}} = 1 + \frac{\partial\Sigma_\sigma(0)}{\partial E_d} = \tilde{\chi}_{\uparrow\uparrow} - \tilde{\chi}_{\uparrow\downarrow}, \tag{5.146}$$

$$= \tilde{\chi}_{\text{even}} - \tilde{\chi}_{\text{odd}} = \sum_{n=0}^{\infty} a_n(-u)^n = \tilde{\chi}_{\text{s}}(-u). \tag{5.147}$$

With increasing $u = U/\pi\Delta$, the charge susceptibility is suppressed:

$$\lim_{u\to\infty} \tilde{\chi}_c(u) = 0. \tag{5.148}$$

In this limit, independent parameters coincide with each other, $\tilde{\chi}_{\uparrow\uparrow} = \tilde{\chi}_{\uparrow\downarrow} = \tilde{\gamma}$, and only one independent parameter exists:

$$\tilde{\chi}_{\uparrow\uparrow} = \tilde{\chi}_{\rm even} = \tilde{\chi}_{\rm odd} = \tilde{\chi}_{\uparrow\downarrow}. \tag{5.149}$$

The ratio of χ_s to γ called the Wilson ratio (or Sommerfeld ratio) is given by

$$R_W = \left(\frac{\chi_s}{2{\mu_B}^2}\right) \Big/ \left(\gamma \Big/ \frac{2\pi^2}{3}{k_B}^2\right) = \tilde{\chi}_s/\tilde{\gamma} = \frac{\tilde{\chi}_{\uparrow\uparrow} + \tilde{\chi}_{\uparrow\downarrow}}{\tilde{\chi}_{\uparrow\uparrow}} = 1 + \tilde{\chi}_{\uparrow\downarrow}/\tilde{\chi}_{\uparrow\uparrow}. \tag{5.150}$$

When u is large enough for $\tilde{\chi}_c$ to be zero, R_W becomes 2. This value $R_W = 2$ agrees with the Wilson ratio in the singlet ground state of the *s–d* Hamiltonian. When $u = 0$, the system is written as a one-body Hamiltonian and $R_W = 1$. With increasing u, R_W increases from 1 and approaches 2.

By the perturbation calculation, the following results are obtained:

$$\chi_s = \frac{(g\mu_B)^2}{2}\frac{1}{\pi\Delta}\left\{1 + u + \left(3 - \frac{\pi^2}{4}\right)u^2 + \left(15 - \frac{3\pi^2}{2}\right)u^3 + 0.055u^4 + \cdots\right\}, \tag{5.151}$$

$$\gamma = \frac{2\pi^2}{3}{k_B}^2\frac{1}{\pi\Delta}\left\{1 + \left(3 - \frac{\pi^2}{4}\right)u^2 + 0.055u^4 + \cdots\right\}, \tag{5.152}$$

where the coefficient of the u^4 term was obtained by a numerical calculation. These results are shown in Fig. 5.13.

The ground state energy E_g can also be calculated by the perturbation expansion with respect to U. The result is

$$E_g = E(u = 0) + \pi\Delta\left\{-\frac{1}{4}u - 0.0369u^2 + 0.0008u^4 + \cdots\right\}, \tag{5.153}$$

where except for the first-order term only even-order terms exist because of the electron–hole symmetry. This ground state energy is lower than that obtained by the Hartree–Fock approximation even beyond $U/\pi\Delta = 1$, as shown in Fig. 5.14. As a result, the magnetic phase transition at $u = 1$ on the basis of the Hartree–Fock approximation loses its justification.

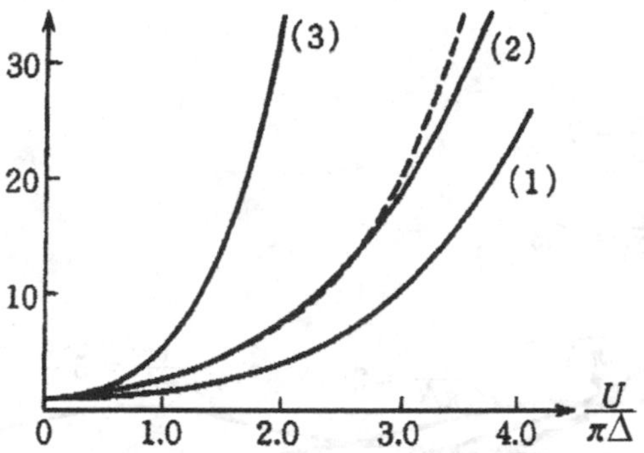

Fig. 5.13 This figure shows (1) $\tilde{\gamma}$, (2) $\tilde{\chi}_s$ and (3) $\tilde{R} = \tilde{\chi}_{\uparrow\uparrow}{}^2 + \tilde{\chi}_{\uparrow\downarrow}{}^2$, obtained up to the fourth-order terms, as functions of $u = U/\pi\Delta$. The dotted curve near (2) shows the function $\exp[u]$.

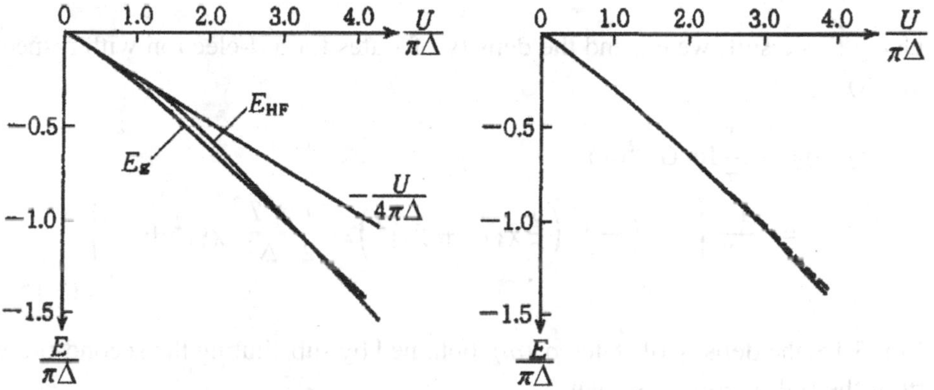

Fig. 5.14 The ground state energy obtained by the perturbation expansion up to the fourth-order term is shown as a function of $u = U/\pi\Delta$, in comparison with the Hartree–Fock approximation. In the right figure the full curve shows the ground state energy obtained by the exact solution, and the broken curve shows the result of perturbation calculation up to u^4. The Hartree–Fock calculation cannot be reliable around a magnetic transition point, $u \simeq 1$.

In the symmetric case, the self-energy $\Sigma_\sigma(\omega)$ of the thermal Green's function is expanded at low temperature and low energy as follows:

$$\Sigma_\sigma(\omega) = -(\tilde{\chi}_{\uparrow\uparrow} - 1)i\omega - \frac{i\Delta}{2}\tilde{\chi}_{\uparrow\downarrow}{}^2 \left\{ -\left(\frac{\omega}{\Delta}\right)^2 + \left(\frac{\pi k_B T}{\Delta}\right)^2 \right\} \operatorname{sgn}\omega + \cdots \tag{5.154}$$

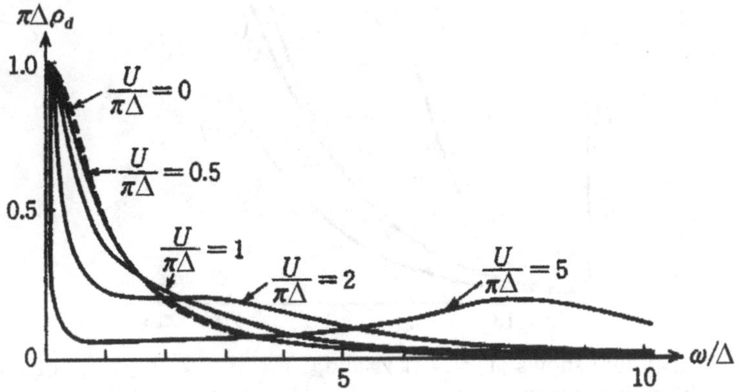

Fig. 5.15 Density of states for d-electrons. The values of u are 0, 1/2, 1, 2 and 5. With increasing u, the peak at the Fermi energy becomes sharp, keeping its height constant $(1/\pi\Delta)$, and peaks at $\omega = \pm U/2$ develop.

Using this result, we expand the density of states for a d-electron with respect to (ω/Δ) as

$$\begin{aligned}\rho_d(\omega) &= -\frac{1}{\pi}\mathrm{Im}\, G_d{}^R(\omega)\\ &= \frac{1}{\pi\Delta}\left\{1-\left(\frac{\omega}{\Delta}\right)^2\left(\frac{1}{2}\tilde{\chi}_{\uparrow\downarrow}{}^2+\tilde{\chi}_{\uparrow\uparrow}{}^2\right)-\frac{1}{2}\frac{\pi^2T^2}{\Delta^2}\tilde{\chi}_{\uparrow\downarrow}{}^2+\cdots\right\}.\end{aligned} \tag{5.155}$$

In Fig. 5.15, the density of states $\rho_d(\omega)$, obtained by substituting the second-order term of the self-energy, is shown.

The relaxation time $\tau_{\boldsymbol{k}}(\omega)$ of the conduction electron $\boldsymbol{k}$ is given by

$$\frac{1}{\tau_{\boldsymbol{k}}(\omega)} = -2n_{\mathrm{i}}\ \mathrm{Im}\ t_{\boldsymbol{k}}(\omega), \tag{5.156a}$$

$$t_{\boldsymbol{k}}(\omega) = V_{\boldsymbol{k}d}G_d{}^R(\omega)V_{d\boldsymbol{k}}, \tag{5.156b}$$

where n_{i} is the concentration of impurity atoms. Substituting $\Sigma_\sigma(\omega)$ given by (5.154) into $G_d{}^R(\omega)$ in the T-matrix (5.156b), we can obtain the electrical resistivity at low temperatures. Writing the resistivity corresponding to the unitary limit as R_0, we can express the resistivity as

$$R = R_0\left\{1-\frac{\pi^2}{3}\left(\frac{k_{\mathrm{B}}T}{\Delta}\right)^2(2\tilde{\chi}_{\uparrow\uparrow}{}^2+\tilde{\chi}_{\uparrow\downarrow}{}^2)+\cdots\right\}. \tag{5.157}$$

In the limit $u \to \infty$, where the Anderson Hamiltonian is reduced to the s–d

Hamiltonian, we can put $\tilde{\chi}_{\uparrow\uparrow} = \tilde{\chi}_{\uparrow\downarrow} = \tilde{\chi}_s/2$ and R becomes

$$R = R_0\left\{1 - \frac{\tilde{\chi}_s{}^2}{4}\left(\frac{\pi k_B T}{\Delta}\right)^2 + \cdots\right\}. \tag{5.158}$$

As shown in this expression, the resistivity decreases from the unitary limit value R_0 as $-T^2$ with the coefficient $(\pi k_B \tilde{\chi}_s/2\Delta)^2$ with increasing temperature. The scaling temperature of the T^2-dependence in the resistivity is $T_K \simeq \Delta/\tilde{\chi}_s$.

The extension of the above theory to the degenerate d-orbital system has been developed by H. Shiba [6] and A. Yoshimori [7]. Shiba proved that the coefficient of the T-linear term of the spin-lattice relaxation time $T_1{}^{-1}$ is proportional to $\chi_s{}^2$. Since the Knight shift is proportional to χ_s, the so-called Korringa relation holds in the locally interacting systems as well as the Anderson Hamiltonian. Yoshimori derived the Wilson ratio for the generalized impurity system, taking account of the orbital degeneracy and the crystal field splittings.

5.5 Exact solution of the Anderson Hamiltonian

In the 1980s, the exact solution of the s–d Hamiltonian was obtained on the basis of the Bethe ansatz by Andrei [9] and Wiegmann [10]. We introduce it briefly following the review article [11]. Succeeding to this, the exact solution of the Anderson Hamiltonian was obtained by Wiegmann [10] and Kawakami and Okiji [12].

The exact solution for the model with $\varepsilon_d = -U/2$ and conduction bandwidth $D \to \infty$ has been obtained analytically. This model agrees completely with that used for the perturbation calculation mentioned above.

Here we put

$$\chi_s = \frac{(g\mu_B)^2}{2}\frac{1}{\pi\Delta}\tilde{\chi}_s, \tag{5.159}$$

$$\chi_c = \frac{2}{\pi\Delta}\tilde{\chi}_c, \tag{5.160}$$

$$\gamma = \frac{2\pi^2 k_B{}^2}{3}\frac{1}{\pi\Delta}\tilde{\gamma}. \tag{5.161}$$

The results obtained by the exact solution are the following:

$$\tilde{\chi}_s = \sqrt{\frac{\pi}{2u}}\exp\left[\frac{\pi^2 u}{8} - \frac{1}{2u}\right] + \frac{1}{\sqrt{2\pi u}}\int_{-\infty}^{\infty}\frac{e^{-x^2/2u}}{1 + \left(\frac{\pi u}{2} + ix\right)^2}dx, \tag{5.162}$$

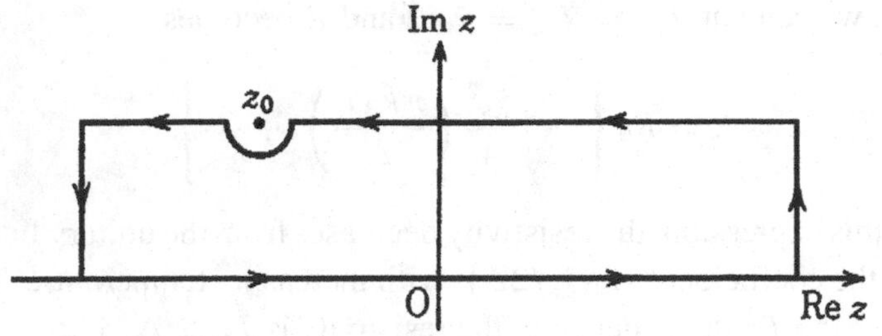

Fig. 5.16 The result of $\tilde{\chi}_s$ obtained by the exact solution is transformed along this integral path. As a result, singular terms cancel out and the analytic expression of $\tilde{\chi}_s$ is obtained.

$$\tilde{\chi}_c = \frac{1}{\sqrt{2\pi u}} \int_{-\infty}^{\infty} \frac{e^{-x^2/2u}}{1 + \left(\frac{\pi u}{2} + x\right)^2} dx, \tag{5.163}$$

$$\tilde{\gamma} = \frac{1}{2}(\tilde{\chi}_s + \tilde{\chi}_c). \tag{5.164}$$

The first term of (5.162) has the factor $\exp(-1/u)$ and seems to be singular at $u = 0$. However, Zlatić and Horvatić [13] showed that this is not the case. They analysed carefully the second terms given by the integral I_s:

$$I_s = \frac{1}{\sqrt{2\pi u}} \int_{-\infty}^{\infty} \frac{e^{-x^2/2u}}{1 + \left(\frac{\pi u}{2} + ix\right)^2} dx = \frac{1}{\sqrt{2\pi u}} \mathrm{Re} \int_{-\infty}^{\infty} \frac{e^{-x^2/2u}}{x - z_0} dx, \tag{5.165}$$

$$z_0 = -1 + \frac{\pi u}{2} i.$$

Then, we shift the integral path upwards by $\pi u i/2$ as shown in Fig. 5.16. By this procedure, a new term possessing sign opposite to the first term of (5.162) arises from the pole at $z = z_0$ and cancels the first term of (5.162). After the cancellation of singular terms, the remaining term gives the following expression for $\tilde{\chi}_s$:

$$\tilde{\chi}_s = e^{\pi^2 u/8} \sqrt{\frac{2}{\pi u}} \int_0^{\infty} e^{-x^2/2u} \frac{\cos(\pi x/2)}{1 - x^2} dx. \tag{5.166}$$

Concerning $\tilde{\chi}_c$, changing the integral variable x in (5.163) into $x - \pi u/2$, we obtain

$$\tilde{\chi}_c = e^{-\pi^2 u/8} \sqrt{\frac{2}{\pi u}} \int_0^{\infty} e^{-x^2/2u} \frac{\cosh(\pi x/2)}{1 + x^2} dx. \tag{5.167}$$

The final expressions, (5.166) and (5.167), are analytic and can be expanded with respect to u as

$$\tilde{\chi}_s = \sum_{n=0}^{\infty} a_n u^n, \tag{5.168}$$

$$\tilde{\chi}_c = \sum_{n=0}^{\infty} (-1)^n a_n u^n, \tag{5.169}$$

$$\tilde{\gamma} = \frac{1}{2}(\tilde{\chi}_s + \tilde{\chi}_c) = \sum_{m=0}^{\infty} a_{2m} u^{2m}. \tag{5.170}$$

The coefficient a_n can be obtained by setting $x = \sqrt{2u}y$ in (5.166) and (5.167). The result is given by the following equations:

$$a_n = (2n-1)a_{n-1} - \left(\frac{\pi}{2}\right)^2 a_{n-2} \qquad (n \geq 2), \tag{5.171}$$

$$a_0 = a_1 = 1. \tag{5.172}$$

The coefficients a_n thus obtained agree with those obtained by the perturbation calculation up to fourth-order terms. Generally, the solution of (5.171) is given by

$$a_n = \left[\left(\frac{\pi}{2}\right)^{2n+1} \Big/ (2n+1)!!\right] P_n, \tag{5.173}$$

$$P_n = \sum_{k=0}^{\infty} \frac{(-1)^k}{k!} \frac{(2n+1)!!}{[2(n+k)+1]!!} \left(\frac{\pi^2}{8}\right)^k. \tag{5.174}$$

Here P_n satisfies

$$\frac{2}{\pi} = P_0 \leq P_n \leq P_\infty = 1. \tag{5.175}$$

When n is large, from (5.173) we obtain

$$a_n \simeq \left(\frac{\pi}{2}\right)^{2n+1} \Big/ (2n+1)!! \tag{5.176}$$

From the above result, the series expansion converges smoothly for $|u| \leq \infty$; the convergence radius is infinite. For example, the Wilson ratio $R_W(u) = \tilde{\chi}_s/\tilde{\gamma}$ tends to 2 for $u \to \infty$, as mentioned above. At $u = 2$, $R_W = 1.962$ and this value means that $u = 2$ is in a sufficiently strong coupling region. In this case the series expansions, including up to the nth-order term, $R_W{}^{(n)}(u)$, give the following values at $u = 2$:

$$R_W{}^{(4)}(2) = 1.889, \qquad R_W{}^{(6)}(2) = 1.952, \qquad R_W{}^{(8)}(2) = 1.961.$$

As shown above, $R_W^{(n)}(u)$ approaches the exact value smoothly. In a similar way, we can confirm that the perturbation expansions for the other physical quantities also give reasonable results owing to the rapid convergence.

The above result shows that the analyticity of physical quantities with respect to the electron interaction is confirmed with use of the exact solution. This analyticity is the basis of the Fermi liquid theory. Although one might simply consider that the Fermi liquid is an assumption without any proof, the Fermi liquid theory describes exactly the physical phenomena, as far as the condition of continuity is satisfied.

If readers wish to know more about the Kondo problem, I recommend the book by A. C. Hewson [14].

References

[1] P. W. Anderson, *Phys. Rev.* **124** (1961) 41.
[2] J. R. Schrieffer and P. A. Wolff, *Phys. Rev.* **149** (1966) 491.
[3] P. Nozières, *Theory of Interacting Fermi Systems* (Benjamin, 1964).
[4] A. A. Abrikosov, L. P. Gor'kov and E. Dzyaloshinskii, *Method of Quantum Field Theory in Statistical Physics* (Prentice-Hall, 1963).
[5] K. Yosida and K. Yamada, *Prog. Theor. Phys. Suppl.* **46** (1970) 254; K. Yamada, *Prog. Theor. Phys.* **53** (1975) 970; K. Yosida and K. Yamada, *Prog. Theor. Phys.* **53** (1975) 1286; K. Yamada, *Prog. Theor. Phys.* **54** (1975) 316.
[6] H. Shiba, *Prog. Theor. Phys.* **54** (1975) 967.
[7] A. Yoshimori, *Prog. Theor. Phys.* **55** (1976) 67.
[8] J. S. Langer and V. Ambegaokar, *Phys. Rev.* **121** (1961) 1090.
[9] N. Andrei, *Phys. Rev. Lett.* **45** (1980) 379.
[10] P. B. Wiegmann, *Sov. Phys. JETP Lett.* **3** (1980) 392.
[11] A. M. Tsvelik and P. B. Wiegmann, *Adv. Phys.* **32** (1983) 453.
[12] N. Kawakami and A. Okiji, *Phys. Lett.* **86**A (1981) 483; A. Okiji and N. Kawakami, *J. Appl. Phys.* **55** (1984) 1931.
[13] V. Zlatić and B. Horvatić, *Phys. Rev. B* **28** (1983) 6904.
[14] A. C. Hewson, *The Kondo Problem to Heavy Fermions* (Cambridge University Press, 1993).

6

Hubbard Hamiltonian

Let us consider a tight binding model, in which electrons occupying atomic orbitals transfer to neighbouring atomic orbitals. We introduce the creation (annihilation) operator $a_{i\sigma}{}^{\dagger}$ ($a_{i\sigma}$) of the electron bound to lattice point i. E_0 and t_{ij} are the energy of the atomic orbital and the transfer matrix between sites i and j, respectively. Further, when two electrons stay at the same atomic orbital, the coulomb repulsion U works between the two electrons. This Hamiltonian, introduced by Hubbard, is written as

$$\mathcal{H} = \sum_{i\sigma} E_0 a_{i\sigma}{}^{\dagger} a_{i\sigma} + \sum_{i \neq j, \sigma} t_{ij} a_{i\sigma}{}^{\dagger} a_{j\sigma} + \frac{1}{2} \sum_{i\sigma} U a_{i\sigma}{}^{\dagger} a_{i\sigma} a_{i-\sigma}{}^{\dagger} a_{i-\sigma}. \tag{6.1}$$

The Hubbard Hamiltonian is simple but important in leading to various phenomena related to electron correlation [1]. At the present time this Hamiltonian still remains an important subject to be studied.

6.1 Basic properties

In (6.1) we have neglected the degeneracy of atomic orbitals. For simplicity, we assume t_{ij} is finite between only the nearest neighbouring lattice points, and is equal to t. Hereafter, we set the number of lattice points as N and the total electron number as N_e. Introducing wave-vector $\boldsymbol{k}$ and using

$$a_{\boldsymbol{k}} = \frac{1}{\sqrt{N}} \sum_i a_i e^{-i\boldsymbol{k}\cdot\boldsymbol{R}_i}, \tag{6.2}$$

we rewrite (6.1) as

$$\mathcal{H} = \sum_{\boldsymbol{k}\sigma} \varepsilon_{\boldsymbol{k}} a_{\boldsymbol{k}\sigma}{}^{\dagger} a_{\boldsymbol{k}\sigma} + \frac{U}{2N} \sum_{\substack{\boldsymbol{k}\boldsymbol{k}'\boldsymbol{q} \neq 0 \\ \sigma}} a_{\boldsymbol{k}-\boldsymbol{q}\sigma}{}^{\dagger} a_{\boldsymbol{k}'+\boldsymbol{q}-\sigma}{}^{\dagger} a_{\boldsymbol{k}'-\sigma} a_{\boldsymbol{k}\sigma}, \tag{6.3}$$

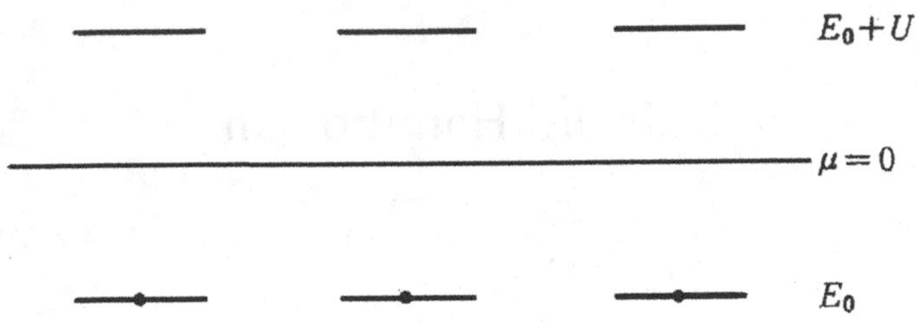

Fig. 6.1 Energy levels in Hubbard Hamiltonian for $t = 0$. The level of the second electron on the same site becomes $E_0 + U$ owing to the coulomb repulsion U.

$$\varepsilon_k = t \sum_{\delta} e^{ik\cdot\delta} + E_0. \tag{6.4}$$

Here, the summation δ is taken over the nearest neighbouring lattice points. Concerning the electron number per site, $N_e/N = 1$ is the half-filling. When $N_e/N > 1$, by considering holes in place of electrons we can transform to the case $N_e/N < 1$. Now we consider the physical meaning of this model by taking simple limiting cases.

(a) $N_e/N = 1$ (half-filling case)

(i) $t = 0$. At chemical potential μ and temperature T, which satisfy $\mu - E_0 \gg k_B T$ and $E_0 + U - \mu \gg k_B T$, the system is an insulator possessing one electron with up- or down-spin on each lattice site.

(ii) Small transfer matrix t ($|t| \ll \mu - E_0$ and $E_0 + U - \mu \gg |t|$). Starting with $t = 0$, we consider the perturbation expansion up to t^2. When the spins of electrons on the neighbouring sites are antiparallel to each other, the electrons can transfer to the neighbouring sites, but electrons with parallel spins cannot transfer owing to the Pauli principle. As a result, it is the electrons with antiparallel spins that lower the energy by the second-order perturbation. The energy gain ΔE due to the second-order perturbation with respect to t between two sites i and j is given by

$$\Delta E = -2t^2/U. \tag{6.5}$$

The factor 2 corresponds to the two processes in which electrons at i and j transfer to j and i, respectively and return to the original sites. Thus, writing an electron spin at each site as $\boldsymbol{s}_i$, we can write the Hamiltonian (6.1) as

$$\mathcal{H} = -\sum_{\langle i,j \rangle} J \left(\boldsymbol{s}_i \cdot \boldsymbol{s}_j - \frac{1}{4} \right), \tag{6.6}$$

where $\langle i, j \rangle$ means the sum over the i, j pair. From (6.5) and (6.6), the exchange interaction J is given by

$$J = -2t^2/U. \tag{6.7}$$

Since $s_i \cdot s_j$ is $-3/4$ for the singlet and $1/4$ for the triplet, the value of (6.6) becomes J and 0, respectively. In the system described by the spin Hamiltonian (6.6), the ground state is generally an antiferromagnetic state in the three-dimensional case. When the long-range exchange interactions are added to the nearest neighbour ones, we can determine the ground state, considering the Fourier transform of J_{ij}, $J(\boldsymbol{q})$. In general, screw (helical) spin structures with $\boldsymbol{q} = \boldsymbol{Q}$ giving the maximum of $|J(\boldsymbol{q})|$ are realized, as far as the spins can be treated as classical spins.

(iii) $U = 0$ or small U. The ground state is given by the state where two electrons possessing energy $\varepsilon_{\boldsymbol{k}}$, wave-vector $\boldsymbol{k}$ and antiparallel spins occupy up to the Fermi energy μ. If we exclude a special case where the unperturbed charge susceptibility and/or magnetic susceptibility $\chi_0(\boldsymbol{q})$ diverge owing to the nesting effect, the Fermi liquid state remains even in the presence of weak coulomb interaction. In this case the electron interaction U enhances the electronic specific heat and the magnetic susceptibility.

(iv) $U \neq 0, t \neq 0$. As noted above, since case (ii) is the insulating state and case (iii) is the metallic state, in case (iv) there exists a metal–insulator transition at a value of U/t. That is, the metal–insulator transition is determined by the competition between the gain of the kinetic energy arising from t and the increase of the coulomb interaction energy due to the double occupancy of two electrons on the same site. When U becomes larger than zt, z being the number of the nearest neighbouring sites, a transition called the Mott transition to the insulator occurs owing to the electron correlation. Here we have assumed $N_e/N = 1$. In the insulating state, one electron occupies each lattice point. When one more electron occupies the lattice point, the energy becomes high by U and there exists a gap called the Hubbard gap. This transition is an important subject to be studied, because the electron correlation plays an important role in the transition and, actually, most of the transition metal oxides are Mott insulators. This point will be discussed in Section 6.4.

(v) Two-electron problem. Let us consider a simple example of the Hubbard model, where two electrons occupy the two sites, 1 and 2. We assume $E_0 = 0$:

$$\mathcal{H} = \sum_{\sigma} t(a_{1\sigma}{}^{\dagger} a_{2\sigma} + a_{2\sigma}{}^{\dagger} a_{1\sigma}) + U(n_{1\uparrow}n_{1\downarrow} + n_{2\uparrow}n_{2\downarrow}). \tag{6.8}$$

Putting $t = 0$, we start with the following two singlet states:

$$\varphi_1 = \frac{1}{\sqrt{2}}(a_{1\uparrow}{}^{\dagger} a_{2\downarrow}{}^{\dagger} + a_{2\uparrow}{}^{\dagger} a_{1\downarrow}{}^{\dagger}), \tag{6.9a}$$

$$\varphi_2 = \frac{1}{\sqrt{2}}(a_{1\uparrow}{}^{\dagger} a_{1\downarrow}{}^{\dagger} + a_{2\uparrow}{}^{\dagger} a_{2\downarrow}{}^{\dagger}). \tag{6.9b}$$

For $t = 0$, the energy of φ_1 is 0 and that of φ_2 is U. For $t > 0$, we assume the eigenfunction and eigen-energy of $\mathcal{H}$ as $\Phi = (\varphi_1 + g\varphi_2)/\sqrt{1+g^2}$ and E, respectively:

$$(\mathcal{H} - E)\Phi = 0. \tag{6.10}$$

From the Schrödinger equation, parameter g and eigenvalue E are determined by

$$-E + 2tg = 0, \tag{6.11a}$$

$$2t + (U - E)g = 0, \tag{6.11b}$$

$$\begin{vmatrix} -E & 2t \\ 2t & U - E \end{vmatrix} = 0. \tag{6.12}$$

The ground state energy is determined as

$$E = \frac{-1}{2}(\sqrt{U^2 + 16t^2} - U), \tag{6.13}$$

$$g = E/2t.$$

When $U \ll t,\ E \simeq -2t + U/2 - U^2/16t$. When $U \gg t,\ E \simeq -4t^2/U$. This result means that with increasing U the transfer motion of an electron is suppressed and the weight of φ_2, $g^2/(1+g^2)$, is reduced from 1/2 to $4t^2/U^2$.

(b) Case of $N_e/N \neq 1$

Even in the presence of U, electrons or holes can move so as to avoid each other and the system remains in a metallic state, although the bandwidth is narrowed. Because $N_e/N \neq 1$, electrons (holes) can transfer through the unoccupied sites. However, in the vicinity of the half-filled lattice ($N_e/N = 1$), electrons and holes are localized accompanying the lattice distortion, and the system does not stay in the metallic state. We don't discuss this problem here. As is known from Nagaoka's theorem [2], the tendency for ferromagnetism appears (see Section 6.6). In the following, we introduce the Kanamori theory on electron correlation, which is appropriate to a system with low density of electrons or holes.

6.2 Theory of electron correlation

(a) Kanamori theory on electron correlation

The Kanamori theory [3], which applies the multiple scattering theory of Brueckner to electron correlation, contains an important concept on electron correlation. Let us consider a system possessing a small number of holes, such as Ni and Pd. Generally, by the screening effect due to the s-electrons, the coulomb interaction between d-electrons situated at different atoms becomes weak. As a result, it is the intra-atomic coulomb interaction U which is important as the electron correlation. When the intra-atomic coulomb interaction U is larger than the bandwidth W, the effective coulomb interaction, U_{eff}, is reduced to W. This is because the electrons transfer avoiding each other, not to stay at the same atom. In this motion, the kinetic energy increases by a magnitude of U_{eff} that corresponds to the bandwidth W. As a result, for any large value of U, the increase of energy is limited to a finite value as far as the system remains in the metallic state.

Now let us consider the multiple scattering of two electrons in the Hamiltonian (6.3).

We represent the two electrons as $\boldsymbol{k}_1\sigma_1$ and $\boldsymbol{k}_2\sigma_2$ and put the anti-symmetrized two-particle wave-function as $|\boldsymbol{k}_1\sigma_1, \boldsymbol{k}_2\sigma_2\rangle$. Using the Hartree–Fock approximation, we obtain the interaction energy ΔE_{HF} between the two electrons as

$$\Delta E_{\text{HF}}(\boldsymbol{k}_1\sigma_1, \boldsymbol{k}_2\sigma_2) = \frac{U}{N}(1 - \delta_{\sigma_1\sigma_2}). \tag{6.14}$$

As shown by (6.14), the interaction energy between two electrons with parallel spins is 0 and that between two electrons with antiparallel spins is U/N. Since the two electrons possessing parallel spins cannot enter the same atomic orbital because of the Pauli principle, U does not work. As a result, for the case $\sigma_1 = \sigma_2 = \sigma$, $|\boldsymbol{k}_1\sigma, \boldsymbol{k}_2\sigma\rangle$ is an eigenstate of $\mathcal{H}$. For two electrons with opposite spins, we consider the following wave-function:

$$\Psi(1,2) = \sum_{\boldsymbol{k}_1\boldsymbol{k}_2} \Gamma(\boldsymbol{k}_1, \boldsymbol{k}_2)\varphi(1, \boldsymbol{k}_1)\varphi(2, \boldsymbol{k}_2). \tag{6.15}$$

This orbital wave-function should satisfy, for the spin singlet state,

$$\Gamma(\boldsymbol{k}_1, \boldsymbol{k}_2) = \Gamma(\boldsymbol{k}_2, \boldsymbol{k}_1). \tag{6.16}$$

Substituting (6.15) into the Schrödinger equation

$$\mathcal{H}\Psi(1,2) = E\Psi(1,2), \tag{6.17}$$

we obtain

$$[\varepsilon(\boldsymbol{k}_1) + \varepsilon(\boldsymbol{k}_2) - E]\Gamma(\boldsymbol{k}_1, \boldsymbol{k}_2) + \frac{U}{N}\sum_{\boldsymbol{k}'} \Gamma\left(\frac{\boldsymbol{Q}}{2} + \boldsymbol{k}', \frac{\boldsymbol{Q}}{2} - \boldsymbol{k}'\right) = 0, \tag{6.18}$$

where $\boldsymbol{Q} = \boldsymbol{k}_1 + \boldsymbol{k}_2$. When $\boldsymbol{Q}/2 \pm \boldsymbol{k}$ exists outside the first Brillouin zone, by adding a proper reciprocal lattice vector it is assumed to be reduced into the first Brillouin zone. Assuming the wave-vector $\boldsymbol{Q}$ for the centre of mass and putting

$\Gamma(\boldsymbol{k}_1, \boldsymbol{k}_2) = \Gamma(\boldsymbol{Q}/2 + \boldsymbol{k}, \boldsymbol{Q}/2 - \boldsymbol{k})$, we can determine E from (6.18) as

$$-\frac{1}{U} = \frac{1}{N} \sum_{\boldsymbol{k}} \frac{1}{\varepsilon(\boldsymbol{Q}/2 + \boldsymbol{k}) + \varepsilon(\boldsymbol{Q}/2 - \boldsymbol{k}) - E}. \tag{6.19}$$

As shown by (6.19), the eigen-energy for the singlet state is shifted by the electron correlation U. By representing the shift as $\Delta E(\boldsymbol{k}_1, \boldsymbol{k}_2)$, we put

$$E = \varepsilon(\boldsymbol{k}_1) + \varepsilon(\boldsymbol{k}_2) + \Delta E(\boldsymbol{k}_1, \boldsymbol{k}_2). \tag{6.20}$$

As the unperturbed state for $U = 0$, we assume $\Psi(\boldsymbol{k}_1, \boldsymbol{k}_2) = \varphi(1, \boldsymbol{k}_1)\varphi(2, \boldsymbol{k}_2)$. Starting with this state, we put $\Gamma(\boldsymbol{k}_1, \boldsymbol{k}_2) = 1$ to obtain the solution of (6.18). Putting $\boldsymbol{q} = \boldsymbol{Q}/2 = (\boldsymbol{k}_1 + \boldsymbol{k}_2)/2$, we obtain

$$\varepsilon(\boldsymbol{k}_1) + \varepsilon(\boldsymbol{k}_2) - E + \frac{U}{N}\left(1 + \sum_{\boldsymbol{k}'}{}' \Gamma(\boldsymbol{q} + \boldsymbol{k}', \boldsymbol{q} - \boldsymbol{k}')\right) = 0, \tag{6.21}$$

where in the sum over $\boldsymbol{k}'$, $\boldsymbol{k}' = \boldsymbol{k}$ is excluded, because $\boldsymbol{q} + \boldsymbol{k}' = \boldsymbol{k}_1 = \boldsymbol{q} + \boldsymbol{k}$ and $\boldsymbol{q} - \boldsymbol{k}' = \boldsymbol{k}_2 = \boldsymbol{q} - \boldsymbol{k}$ for $\boldsymbol{k}' = \boldsymbol{k}$. The coefficient $\Gamma(\boldsymbol{q} + \boldsymbol{k}', \boldsymbol{q} - \boldsymbol{k}') \equiv \Gamma(\boldsymbol{k}')$ for $\boldsymbol{k}' \neq \boldsymbol{k}$ is determined by (6.18):

$$[\varepsilon(\boldsymbol{q} - \boldsymbol{k}') + \varepsilon(\boldsymbol{q} + \boldsymbol{k}') - E]\Gamma(\boldsymbol{k}') + \frac{U}{N}\left(1 + \sum_{\boldsymbol{k}''}{}' \Gamma(\boldsymbol{k}'')\right) = 0. \tag{6.22}$$

From (6.21), $E = \varepsilon(\boldsymbol{k}_1) + \varepsilon(\boldsymbol{k}_2)$ holds in the zeroth order of $1/N$. Substituting this result into (6.22), for $\boldsymbol{k}' \neq \boldsymbol{k}$ we obtain

$$\Gamma(\boldsymbol{k}') = -\frac{U}{N}\left(1 + \sum_{\boldsymbol{k}''}{}' \Gamma(\boldsymbol{k}'')\right) \Big/ [\varepsilon(\boldsymbol{q} - \boldsymbol{k}') + \varepsilon(\boldsymbol{q} + \boldsymbol{k}') - \varepsilon(\boldsymbol{k}_1) - \varepsilon(\boldsymbol{k}_2)]. \tag{6.23}$$

From this equation,

$$\sum_{\boldsymbol{k}'}{}' \Gamma(\boldsymbol{k}') = -UG(\boldsymbol{k}_1, \boldsymbol{k}_2)/[1 + UG(\boldsymbol{k}_1, \boldsymbol{k}_2)], \tag{6.24}$$

$$G(\boldsymbol{k}_1, \boldsymbol{k}_2) = \frac{1}{N} \sum_{\boldsymbol{k}'}{}' \frac{1}{\varepsilon(\boldsymbol{q} - \boldsymbol{k}') + \varepsilon(\boldsymbol{q} + \boldsymbol{k}') - \varepsilon(\boldsymbol{k}_1) - \varepsilon(\boldsymbol{k}_2)}. \tag{6.25}$$

Substituting these results into (6.21), we obtain $\Delta E(\boldsymbol{k}_1, \boldsymbol{k}_2)$:

$$\Delta E(\boldsymbol{k}_1, \boldsymbol{k}_2) = \frac{U}{N}\left[1 - \frac{UG(\boldsymbol{k}_1, \boldsymbol{k}_2)}{1 + UG(\boldsymbol{k}_1, \boldsymbol{k}_2)}\right] = \frac{U}{N} \frac{1}{1 + UG(\boldsymbol{k}_1, \boldsymbol{k}_2)}. \tag{6.26}$$

Comparing this result with (6.14) given by the Hartree–Fock approximation, we obtain the effective value of the correlation energy:

$$U_{\text{eff}} = U/[1 + UG(\boldsymbol{k}_1, \boldsymbol{k}_2)]. \tag{6.27}$$

The value of $G(\boldsymbol{k}_1, \boldsymbol{k}_2)$ is generally of the magnitude of the inverse of the bandwidth W, as seen from (6.25). As a result, when U is larger than the bandwidth W, U_{eff} becomes of order W.

The calculation described above includes the higher-order terms due to the repeated scattering of the two electrons $\boldsymbol{k}_1$ and $\boldsymbol{k}_2$. The denominator of (6.27) arises from the scattering process. For the system with a small number of particles, it is valid to neglect the other particles and take only the two-particle scattering into account. As a result, this approximation is valid for the low-density system and is called the low-density approximation or the ladder approximation. This reduction of electron correlation due to the avoiding motion is generally seen for any density $N_e/N \neq 1$.

Using the above result, Kanamori discussed the ferromagnetism of the transition metals such as Ni. According to the Hartree–Fock approximation, the condition for the appearance of ferromagnetism is given by $U_{\text{eff}}\rho(0) > 1$, $\rho(0)$ being the density of states for electrons at the Fermi energy. As a result of the Kanamori approximation, U is reduced to U_{eff} in (6.27) and the ferromagnetism hardly appears. Kanamori showed that the ferromagnetism is realized for the case where the density of states at the Fermi energy $\rho(0)$ is relatively large compared with the inverse of the total bandwidth. Actually, the density of states for Ni metal possesses a peak structure near the band top.

(b) Variational theory by Gutzwiller

In the Hubbard Hamiltonian, when the coulomb repulsion U on the same site is large, the wave-function for the many-body system reduces the possibility of double occupancy to avoid energy loss due to the coulomb repulsion U. The variational wave-function taking this point into account was proposed by Gutzwiller [4]. Let us consider the basic principle of the variational theory. We assume $T = 0$ and the number of lattice points is L. The number of electrons with spin σ and the number of sites occupied by two electrons are denoted as N_σ and D, respectively. We introduce $n_\sigma = N_\sigma/L$ and $d = D/L$. The ground state for $U = 0$ is denoted as $|\Psi_0\rangle$. For this case the number of doubly occupied lattice points D_0 is given by $D_0 = n_\uparrow n_\downarrow L$. When the coulomb interaction U is introduced, D decreases and $D < D_0$, since the interaction energy is given by UD. Writing the wave-function for $U \neq 0$ as $|\Psi\rangle$ and putting

$$|\Psi\rangle = \prod_{i=1}^{L} [1 - (1 - g)n_{i\uparrow}n_{i\downarrow}]|\Psi_0\rangle = g^D|\Psi_0\rangle, \tag{6.28}$$

we determine g so as to minimize the energy E, which is given by

$$E = \frac{\langle\Psi|\mathcal{H}|\Psi\rangle}{\langle\Psi|\Psi\rangle} = \left[\left\langle\Psi\middle|\sum_{ij}\sum_{\sigma} t_{ij}a_{i\sigma}{}^{\dagger}a_{j\sigma}\middle|\Psi\right\rangle + \left\langle\Psi\middle|U\sum_{i} n_{i\uparrow}n_{i\downarrow}\middle|\Psi\right\rangle\right] \Big/ \langle\Psi|\Psi\rangle. \tag{6.29}$$

Here $|\Psi\rangle$ is the eigenfunction of $\sum_i n_{i\uparrow}n_{i\downarrow}$ and the second term of (6.29) is UD. For the calculation of the kinetic energy given by the first term of (6.29), Gutzwiller used the following approximation. The electrons with opposite spins are assumed to be independent of each other. The decrease in kinetic energy for each spin is calculated statistically for the randomly distributed electrons. By the variational calculation for the parameter g, the minimized energy E_g is given by

$$E_g/L = q_{\uparrow}(d, n_{\uparrow}, n_{\downarrow})\bar{\varepsilon}_{\uparrow} + q_{\downarrow}(d, n_{\uparrow}, n_{\downarrow})\bar{\varepsilon}_{\downarrow} + Ud. \tag{6.30}$$

Here, q_σ representing the jump of the occupation number $\langle a_{k\sigma}{}^{\dagger}a_{k\sigma}\rangle$ at the Fermi surface is given by

$$q_\sigma = \frac{\{[(n_\sigma - d)(1 - n_\sigma - n_{-\sigma} + d)]^{1/2} + [(n_{-\sigma} - d)d]^{1/2}\}^2}{n_\sigma(1 - n_\sigma)}, \tag{6.31}$$

and the band energy $\bar{\varepsilon}_\sigma$ is given by

$$\bar{\varepsilon}_\sigma = L^{-1}\left\langle\Psi_0\middle|\sum_{ij} t_{ij}a_{i\sigma}{}^{\dagger}a_{j\sigma}\middle|\Psi_0\right\rangle = \sum_{|k|<k_{\mathrm{F}}} \varepsilon_k < 0. \tag{6.32}$$

Since q_σ is smaller than unity for $U \neq 0$, (6.30) means a reduction of the kinetic energy gain owing to the narrowing of the bandwidth as a result of reducing the double occupancy.

Gutzwiller reached the conclusion that the appearance of ferromagnetism is more difficult than suggested by the result obtained from the Hartree–Fock approximation, in accordance with the Kanamori theory. Moreover, Gutzwiller's theory is important in the metal–insulator transition for the half-filled case with $n = 1$. As shown by Brinkman and Rice [5], for the paramagnetic state (6.31) becomes

$$q = 8d(1 - 2d). \tag{6.33}$$

Substituting this result into (6.30) and taking the derivative of E_g by d, we obtain the following solutions:

$$d = \frac{1}{4}\left(1 - \frac{U}{U_{\mathrm{c}}}\right), \tag{6.34}$$

$$q = 1 - \left(\frac{U}{U_c}\right)^2, \tag{6.35}$$

$$\frac{E_g}{L} = -|\bar{\varepsilon}_0| \left[1 - \frac{U}{U_c}\right]^2. \tag{6.36}$$

Here $\bar{\varepsilon}_0 = 2\bar{\varepsilon}_\uparrow = 2\bar{\varepsilon}_\downarrow$ and $U_c = 8|\bar{\varepsilon}_0|$. As U approaches U_c, $d = q = E_g = 0$ and the system becomes insulating. Strictly speaking, when the system becomes the insulator, the exchange interaction (6.6) exists and the ground state becomes an antiferromagnetic insulating state. It should be noted that the metal–insulator transition is obtained by the Gutzwiller theory.

In this case the electron mass m^* is given by

$$\frac{m^*}{m} = q^{-1} = \left[1 - \left(\frac{U}{U_c}\right)^2\right]^{-1}. \tag{6.37}$$

Using the Bohr magneton μ_B and the density of state $\rho(0)$ per spin on the Fermi surface, the spin susceptibility χ_s is given by

$$\chi_s = 2\mu_B{}^2\rho(0)\left\{\left[1 - \left(\frac{U}{U_c}\right)^2\right] \times \left[1 - \rho(0)U\frac{1 + U/(2U_c)}{(1 + U/U_c)^2}\right]\right\}^{-1}, \tag{6.38}$$

where $\chi_s{}^0 - 2\mu_B{}^2\rho(0)$ is the Pauli susceptibility in the non-interacting system. Here the important point is that, as discussed in the Fermi liquid theory, when U approaches a critical value U_c, m^* and χ_s diverge. However, the ratio of χ_s to m^* remains finite and this fact means that their divergences are not due to the instability of the ferromagnetic state. The Gutzwiller theory assumes a variational wave-function and uses an approximation in calculating the first term of (6.29). Without using this approximation Yokoyama and Shiba [6] calculated correctly the expectation value of the first term using a numerical computation and showed the absence of the metal–insulator transition, in contrast to the result by Brinkman and Rice. This fact means that the variational wave-function is not a good approximation for the metal–insulator transition. However, for the infinite-dimensional ($d = \infty$) Hubbard Hamiltonian, the Gutzwiller approximation becomes correct and the Mott transition derived by Brinkman and Rice becomes valid. This result agrees with the Mott transition of the infinite-dimensional Hubbard Hamiltonian discussed in the next section. In a general dimension, the variational theory by Gutzwiller is considered to be appropriate to the metallic state.

6.3 Infinite-dimensional Hubbard Hamiltonian

Compared with a general dimension model, the infinite-dimensional Hubbard Hamiltonian [7] is easily treated. As an extension of the square lattice and the cubic lattice, we consider the hypercubic lattice in the infinite dimension. Using a transfer matrix t between the nearest neighbour lattice points and lattice spacing a as unit of length, the energy dispersion of the electron with the wave-vector $\boldsymbol{k} = (k_1, k_2, \dots, k_n, \dots, k_d)$ is given by

$$\varepsilon_{\boldsymbol{k}} = -2t \sum_{n=1}^{d} \cos k_n \qquad (-\pi \le k_n \le \pi). \tag{6.39}$$

The density of states $\rho_d(\varepsilon)$ for electrons with this dispersion is given by the Gauss distribution function for $d \to \infty$:

$$\rho_d(\varepsilon) = \sum_{\boldsymbol{k}} \delta(\varepsilon - \varepsilon_{\boldsymbol{k}}) = \frac{1}{2t(\pi d)^{1/2}} \exp\left[-\left(\frac{\varepsilon}{2t\sqrt{d}}\right)^2\right]. \tag{6.40}$$

This result is derived by expanding the Fourier transform $\Phi_d(s)$ of $\rho_d(\varepsilon)$ with respect to s and neglecting higher-order terms than $1/d$. The density of states $\rho_d(\varepsilon)$ given by (6.40) expands in its width with increasing dimension d. To avoid this and keep the density of states invariant we reduce t to $t^* = t/\sqrt{2d}$. That is, instead of (6.39), we assume the model

$$\varepsilon_{\boldsymbol{k}} = -\frac{2t}{\sqrt{2d}} \sum_{n=1}^{d} \cos k_n. \tag{6.41}$$

In addition, the form of $\varepsilon_{\boldsymbol{k}}$ in (6.41) should be modified to avoid complete nesting. Since $\cos(k_n + \pi) = -\cos k_n$,

$$\varepsilon_{\boldsymbol{k}+\boldsymbol{Q}} = -\varepsilon_{\boldsymbol{k}}, \qquad \boldsymbol{Q} = \pi(1, 1, \dots, 1). \tag{6.42}$$

In the case of half-filling, $N_e/N = 1$, so-called complete nesting is realized and the unperturbed susceptibility $\chi^0(\boldsymbol{Q})$ diverges. In this case an infinitesimal coulomb repulsion U gives rise to an antiferromagnetic ground state, in which the antiferromagnetic long-range order exists and the energy gap opens on all the Fermi surface ($\varepsilon_{\boldsymbol{k}} = \mu = 0$). Although this feature is also one of the characteristics seen in the Hubbard Hamiltonian, to avoid the special case and generalize the argument, Müller-Hartmann [8] extended the model to include the transfer matrix t_m to the mth nearest neighbouring sites:

$$\varepsilon_{\boldsymbol{k}} = \sum_{m=1}^{\infty} t_m \varepsilon_m(\boldsymbol{k}), \tag{6.43}$$

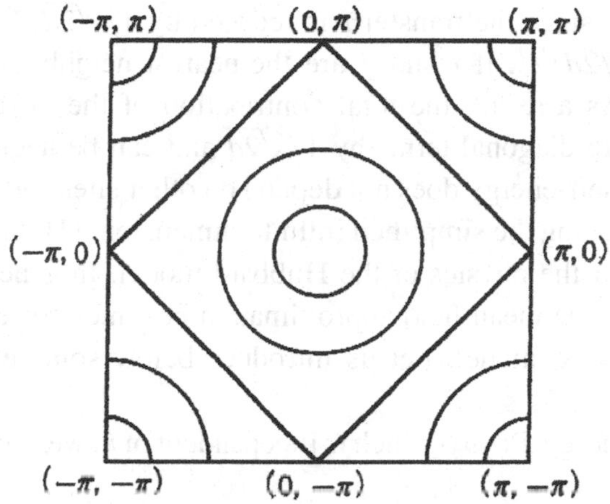

Fig. 6.2 Schematic figure of the equi-energy lines for the square lattice, $\varepsilon_k = \cos k_x + \cos k_y$. The solid lines given by $\varepsilon_k = 0$ are the Fermi lines. The Fermi lines coincide with each other by the translation of $(\pm\pi, \pm\pi)$.

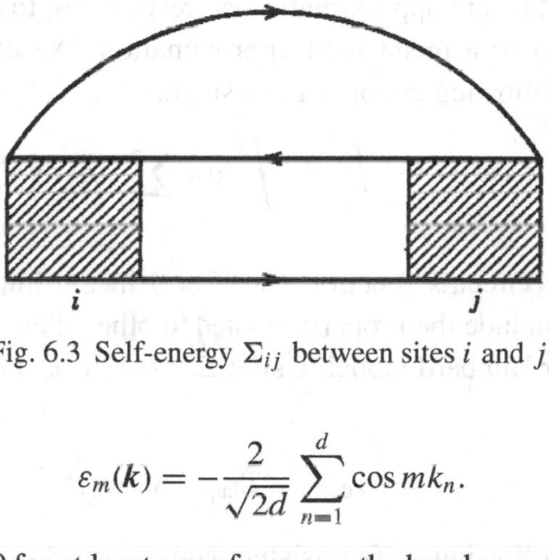

Fig. 6.3 Self-energy Σ_{ij} between sites i and j.

$$\varepsilon_m(\boldsymbol{k}) = -\frac{2}{\sqrt{2d}} \sum_{n=1}^{d} \cos mk_n. \tag{6.44}$$

In this case, if $t_m \neq 0$ for at least one of even m, the band ε_k excludes the properties of complete nesting. Fortunately, in this case also the density of states is simply given by

$$\rho_d(\varepsilon) = \exp[-(\varepsilon/t)^2/2 + O(d^{-1/2})]/(2\pi t^2)^{1/2}. \tag{6.45}$$

As long as the terms higher than $1/\sqrt{d}$ are ignored, $\rho_d(\varepsilon)$ can be given by the Gaussian distribution function. Hereafter, we consider the $d = \infty$ system without complete nesting properties.

Another characteristic of the $d = \infty$ model is that the self-energy $\Sigma(\omega)$ is independent of the momentum. The self-energy $\Sigma_{ij}(\omega)$ connecting different sites i and j contains at least three transfer matrices t connecting i and j as shown in Fig. 6.3.

Here in our model, since the transfer t is reduced by $1/\sqrt{2d}$, Σ_{ij} is smaller than Σ_{ii} by a factor $(1/2d)^{3/2}$. If i and j are the nearest neighbours, the number of terms Σ_{ij} is $2d$. As a result, the total contribution of the off-diagonal terms is smaller than the site-diagonal terms by $1/\sqrt{2d}$ and can be neglected in the limit $d = \infty$. Thus, the self-energy does not depend on other sites and is independent of the momentum $\boldsymbol{k}$. Using the simplified infinite-dimensional Hubbard Hamiltonian, we have made clear the physics of the Hubbard model. In general, with increasing dimensionality, the mean field approximation becomes correct. This is an advantage of the $d = \infty$ model. Let us introduce below some examples of these studies.

Using the self-energy $\Sigma(i\omega_n)$ which is independent of $\boldsymbol{k}$, we can write the thermal Green's function as

$$G(\boldsymbol{k}, i\omega_n) = [i\omega_n + \mu - \varepsilon_{\boldsymbol{k}} - \Sigma(i\omega_n)]^{-1}. \tag{6.46}$$

Here, although we can calculate $\Sigma(i\omega_n)$ using the perturbation theory with respect to U, we adopt a single site approximation corresponding to a coherent potential approximation (CPA) or a mean field approximation. We consider an impurity model given by the following action S for a single site:

$$S[G_0] = U \int_0^\beta d\tau n_\uparrow(\tau) n_\downarrow(\tau) - \int_0^\beta d\tau \int_0^\beta d\tau' \sum_\sigma c_\sigma{}^\dagger(\tau)[G_0(\tau - \tau')]^{-1} c_\sigma(\tau'). \tag{6.47}$$

Here G_0 is the bare Green's function for $U = 0$ in the impurity model and is assumed already to include the property related to other sites, within the one-body approximation in our Hubbard model. Using $\Sigma_{\text{imp}}(G_0, i\omega_n)$ determined by (6.47), $G(i\omega_n)$ is given by

$$G(i\omega_n) = [G_0{}^{-1} - \Sigma_{\text{imp}}(G_0, i\omega_n)]^{-1}. \tag{6.48}$$

Here we adopt the following self-consistent equation: $G(i\omega_n)$ given by (6.48) is assumed to be equal to the site-diagonal term of Green's function for the Hubbard Hamiltonian, which is given by taking the sum over $\boldsymbol{k}$ of (6.46). To satisfy this, we put $\Sigma(i\omega_n) = \Sigma_{\text{imp}}(i\omega_n)$ and obtain the following equation:

$$G(i\omega_n) = \int_{-\infty}^{\infty} d\varepsilon \frac{\rho(\varepsilon)}{i\omega_n + \mu - \Sigma_{\text{imp}}(i\omega_n) - \varepsilon}. \tag{6.49}$$

The density of states $\rho(\varepsilon)$ is the only quantity that reflects the property of the lattice and is treated as an external parameter. As a result, our task is to determine the unperturbed impurity Green's function G_0 that approximates the $d = \infty$ Hubbard Hamiltonian, using (6.48) and (6.49).

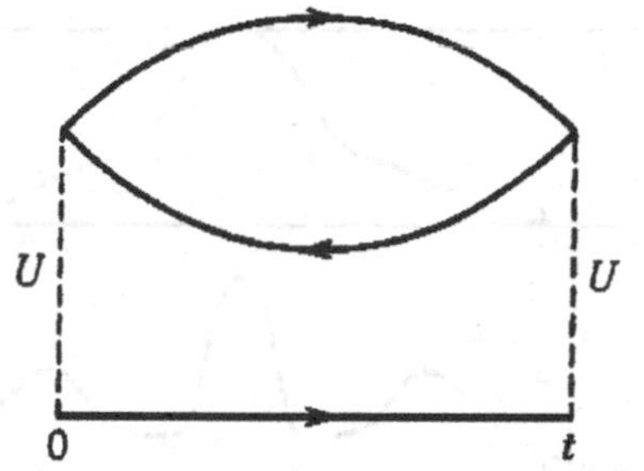

Fig. 6.4 The second-order term of the self-energy.

At first, assuming translational symmetry and the paramagnetic state, we solve the equation. Here we consider the following Anderson Hamiltonian as a general model for the single-site impurity models:

$$\mathcal{H}_{\mathrm{AM}} = \sum_{k\sigma} \varepsilon_k c_{k\sigma}{}^{\dagger} c_{k\sigma} + \varepsilon_d \sum_{\sigma} d_\sigma{}^{\dagger} d_\sigma + U n_{d\uparrow} n_{d\downarrow} + \sum_{k\sigma} [V_k c_{k\sigma}{}^{\dagger} d_\sigma + \mathrm{H.C.}]. \tag{6.50}$$

For this Hamiltonian, we can eliminate the operators for conduction electrons and obtain

$$[G_0{}^{\mathrm{AM}}(i\omega_n)]^{-1} = i\omega_n - \varepsilon_d + \int_{-\infty}^{\infty} \frac{d\varepsilon}{\pi} \frac{\Delta(\varepsilon)}{i\omega_n - \varepsilon}, \tag{6.51}$$

$$\Delta(\varepsilon) = \pi \sum_k |V_k|^2 \delta(\varepsilon - \varepsilon_k). \tag{6.52}$$

The solution for the Fermi liquid state is obtained by assuming that $\mathrm{Im}\, G_0(i\omega_n = \omega + i\delta)$ does not vanish at $\omega \to 0$. As the best case for the Fermi liquid state, let us consider the Anderson Hamiltonian possessing electron–hole symmetry, $\varepsilon_d - \mu = -U/2$. In this case the electron interaction term can be written as

$$\mathcal{H}' = U \left(n_{d\uparrow} - \frac{1}{2} \right) \left(n_{d\downarrow} - \frac{1}{2} \right). \tag{6.53}$$

Using this interaction Hamiltonian, we calculate the second-order term of the self-energy for the unperturbed Green's function $\tilde{G}_0{}^{-1} = G_0{}^{-1} - Un/2 (n = N_e/N)$. The result for Σ shown in Fig. 6.4 is given by

$$\Sigma(t) = U^2 \tilde{G}_0{}^2(t) \tilde{G}_0(-t). \tag{6.54}$$

Substituting this result into Σ_{imp} in (6.49), we obtain Green's function G and determine $\tilde{G}_0$ self-consistently from $\tilde{G}_0{}^{-1} = G^{-1} + \Sigma$. The density of states thus obtained is shown in Fig. 6.5. It is important that the height of the central peak at the Fermi energy is constant independently of U, as far as the system remains in

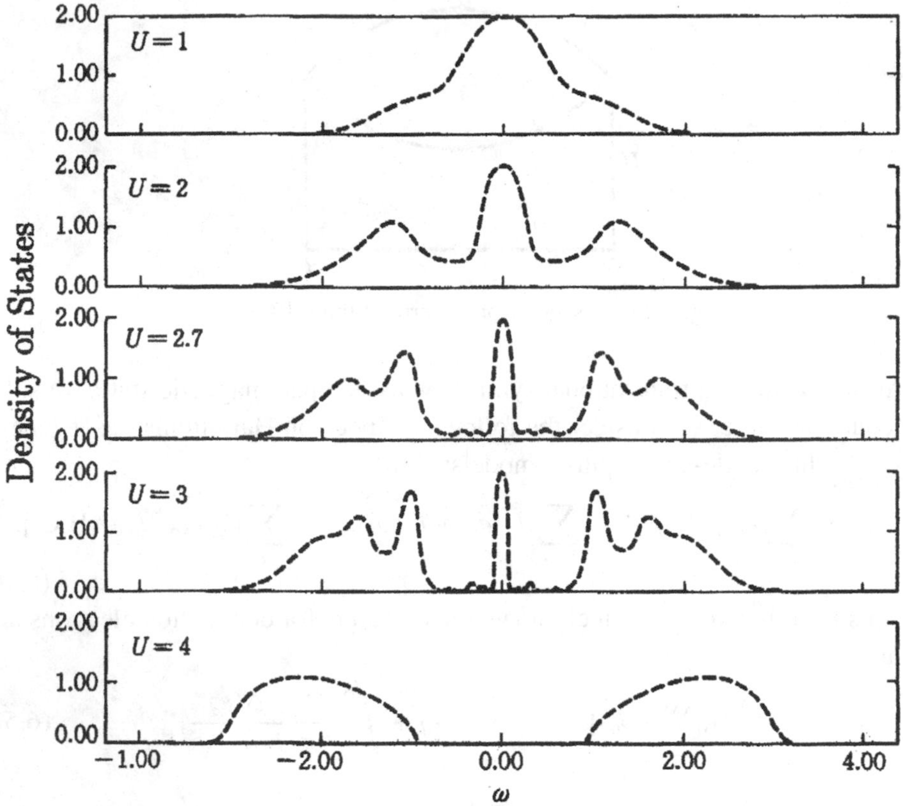

Fig. 6.5 Energy spectrum of $d = \infty$ Hubbard Hamiltonian for typical values of U. At $U = U_c$, the metal–insulator transition occurs and an energy gap opens. The values of U are given by the unit of $D = 1$ [9].

the Fermi liquid state. Finally, $\Sigma(0) = 0$ and for a small ω the real part of $\Sigma(\omega)$ is given by the linear term of ω. The imaginary part of $\Sigma(\omega)$ starts with the ω^2 term. As a result, $\mathrm{Im}\, G(i\delta)$ at $\omega = 0$ is independent of U and is determined by $\Delta(0)$. This conclusion can be understood by the properties of the symmetric Anderson Hamiltonian shown in Fig. 5.15. For $\Delta(0)$ not to vanish is the condition necessary for a Fermi liquid. The existence of the mixing term in the Fermi liquid ground state is explained on the basis of Anderson's orthogonality theorem, as discussed in Chapters 4 and 7.

Figure 6.5 shows the spectral density for typical values of U. For large values of U two peaks below and above the Fermi energy appear corresponding to the Hubbard gap. The two peaks represent the spectrum for electrons with spin σ, when the electron number with opposite spin is fixed to $n_{-\sigma} = 0$ and $n_{-\sigma} = 1$, respectively. The central peak at the Fermi energy is the excitation spectrum when the number of opposite spin electrons is fixed to the long-time average value corresponding to $\omega \simeq 0$. This level is nothing but the level $\varepsilon_d + U(n/2) = 0$. As shown by the calculation for the Anderson Hamiltonian, the coefficients of the ω-linear term of the self-energy $\mathrm{Re}\,\Sigma$ and the ω^2 term of $\mathrm{Im}\,\Sigma$ increase with increasing U. From these effects, the peak at the Fermi energy becomes narrow. This is because the charge fluctuation is suppressed with increasing U and it takes a long time to observe the spectrum corresponding to the average value, $n_{-\sigma} = 1/2$. This property is universal in the Fermi liquid. The narrow peak with constant height at the Fermi energy originates from the enhancement of effective mass and the increase of damping rate for quasi-particles. It is naturally expected that in a limit of increasing U the Mott transition will be realized. Now let us discuss how the Mott transition can be described in the $d = \infty$ model.

As noted above, the necessary condition to be a Fermi liquid is $\Delta(0) \neq 0$. Taking the contra position of the above statement, we obtain the conclusion that if $\Delta(0) = 0$, the system is not a Fermi liquid. If we put $\Delta(0) = 0$ in the Anderson Hamiltonian, it represents the high temperature state above T_{K} in which the coherent part of the hybridization vanishes and the spin of the d-electron fluctuates freely. This state corresponds to a paramagnetic state in the Mott insulator. Using the second-order term of self-energy given by (6.54), we can obtain a self-consistent solution. As far as we assume a finite bandwidth, we can derive the transition to the Mott insulator from the Fermi liquid. This transition is general in the $d = \infty$ Hubbard model independently of the approximation, and is confirmed by the numerical calculation based on the quantum Monte Carlo method.

6.4 Mott transition

In the Hubbard Hamiltonian, for the half-filling, $N_{\mathrm{e}}/N = 1$, there exists one electron at each lattice point. When the coulomb repulsion between two electrons on the same atomic site becomes larger than the band energy proportional to the transfer matrix t, a single electron is localized at each lattice point and the system becomes an insulator. This transition is called the Mott transition. The insulator due to the electron correlation U is called the Mott insulator. The Mott transition is the important issue among the physical phenomena related to the electron correlation. Recently, the process of the transition from the Fermi liquid to the Mott insulator has been made clear through the study of the infinite-dimensional Hubbard Hamiltonian.

Let us consider the $d = \infty$ Hubbard model [10]. For the density of states possessing the Gaussian or Lorentzian energy distribution, we need the infinite coulomb repulsion U in order to realize the Mott transition, since the bandwidth extends over the infinite energy region. In real systems, the density of states is confined to a finite energy range. For this reason, we consider the $d = \infty$ model possessing finite bandwidth. Here we substitute the following form into $\rho(\varepsilon)$ in (6.49):

$$\rho(\varepsilon) = \frac{2}{\pi D^2}\sqrt{D^2 - \varepsilon^2}. \tag{6.55}$$

This is a semicircle with radius D. Using the relation

$$\int_{-D}^{D} \frac{\rho(\varepsilon)d\varepsilon}{z - \varepsilon} = \frac{2}{z + \sqrt{z^2 - D^2}}, \tag{6.56}$$

we can rewrite (6.49) as

$$[G_0{}^{-1} - \Sigma]^{-1} = \frac{2}{i\omega_n - \Sigma + i\ \mathrm{sgn}(\omega_n)\sqrt{D^2 + (\omega_n + i\Sigma)^2}}, \tag{6.57}$$

where we have put $\mu = 0$. Here we can use the second-order self-energy $\Sigma^{(2)}(i\omega_n)$ for the symmetric Anderson Hamiltonian. The $\Sigma^{(2)}(i\omega_n)$ is given for $|\omega_n| \gg \Delta$ by

$$\Sigma^{(2)}(i\omega_n) = \frac{U^2}{4} G_0(i\omega_n), \tag{6.58}$$

which agrees with the expression of the self-energy without hybridization term in the symmetric Anderson Hamiltonian. Hence, the solution corresponding to the Mott insulator is obtained by extending (6.58) to $\omega_n \to 0$. The U^2 term of the self-energy in the Fermi liquid approaches zero as ω_n tends to zero, in contrast with (6.58). On the other hand, in the Mott insulator $\Delta(0) \to 0$ as $\omega_n \to 0$, and $G_0(i\omega_n)$ is given by

$$G_0(i\omega_n) \simeq 1/i\omega_n. \tag{6.59}$$

Actually, by substituting (6.57) into (6.58) we obtain, for $|\omega_n| \ll U$,

$$G_0(i\omega_n)^{-1} = i\omega_n \frac{U^2}{U^2 - D^2}. \tag{6.60}$$

For $U \gg D$ and $\omega_n \sim \pm U$, $G_0(i\omega_n)$ is given by

$$G_0(i\omega_n)^{-1} = i\omega_n - i\omega_n \frac{4(i\omega_n)^2 - U^2 - \sqrt{[4(i\omega_n)^2 - U^2]^2 - 4[4(i\omega_n)^2 + U^2]D^2}}{2(4(i\omega_n)^2 + U^2)}. \tag{6.61}$$

Combining this result with (6.58), we obtain

$$G = [G_0{}^{-1} - \Sigma] = \left[G_0{}^{-1} - \frac{U^2}{4} G_0\right]. \tag{6.62}$$

From this Green's function we obtain the density of states possessing two peaks centred at $\omega = \pm U/2$ with width $2D$. In this case, from (6.60), $\Delta(i\omega_n)$ for small values of ω becomes

$$\Delta(i\omega_n) = G_0(i\omega_n)^{-1} - i\omega_n \simeq i\omega_n \frac{D^2}{U^2 - D^2}. \tag{6.63}$$

Thus we can confirm that $\Delta(i\omega_n)$ vanishes proportionally to $i\omega_n$. As a result we have obtained the following behaviour: $G \to i\omega$, $G_0 \to 1/i\omega$, $\Sigma \to 1/i\omega$ and $\Delta \sim i\omega$. The solution thus obtained represents the Mott insulator which is characterized by the divergent self-energy and the energy gap extending to a distance of U. When the system approaches the Mott insulator from the Fermi liquid, the quasi-particles approach the states just before the localization, and generally their momentum dependence becomes weak. As a result, the $d = \infty$ model can be considered not to lose the essential point in approaching the Mott insulator.

To conclude the method of approaching the Mott insulator from the Fermi liquid, when we increase the value of U in the half-filling case, $m^*/m = 1 - \partial\Sigma/\partial\omega|_{\omega=0}$ approaches an infinite value and the jump in occupation at the Fermi energy, $z = m/m^*$, tends to zero. The peak, which exists at the Fermi energy and keeps its height constant, disappears at a critical value $U = U_c$ in the limit of vanishing width, and the energy gap is created. During this process, the peaks corresponding to $\omega = \varepsilon_d$ and $\omega = \varepsilon_d + U$ develop gradually from the Fermi liquid state and transform into the two peaks situated above and below the Mott–Hubbard gap. We note again that in real systems the antiferromagnetic long-range order is realized at low temperatures owing to the super-exchange interaction.

6.5 One-dimensional Hubbard model

(a) Special nature of the one-dimensional system

The one-dimensional Hubbard model is special, since the energy conservation and momentum conservation are not independent. This system does not belong to the Fermi liquid, but to the Luttinger liquid, and does not possess any well-defined quasi-particle excitations. The excitations in this system are given by collective modes related to the charge and spin degrees. The jump z_k of the Fermi distribution n_k at $k = k_F$ tends to zero, corresponding to the disappearance of quasi-particle excitations. Actually, the behaviour of n_k near $k = k_F$ is continuous and given by the power law [11]

$$n_k = n_{k_F} - \text{const}|k - k_F|^{\theta}\,\text{sgn}(k - k_F), \tag{6.64}$$

where θ is given by Fig. 6.6. As long as U is not zero, θ tends to 1/8 as the filling approaches half-filled, $N_e/N = 1$.

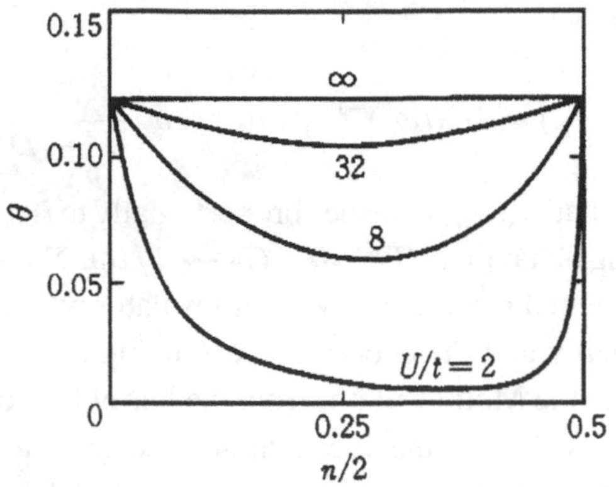

Fig. 6.6 The occupation number n_k near the Fermi point in the one-dimensional lattice is given by $|k - k_F|^\theta$. This figure shows the critical exponent θ as a function of U/t and the electron density n [11].

(b) Exact solution by Bethe ansatz

The one-dimensional Hubbard Hamiltonian has been solved on the basis of the Bethe ansatz by Lieb and Wu [12]. The Hamiltonian is given by

$$\mathcal{H} = -t \sum_{<i,j>} \sum_{\sigma} c_{i\sigma}^\dagger c_{j\sigma} + U \sum_i c_{i\uparrow}^\dagger c_{i\uparrow} c_{i\downarrow}^\dagger c_{i\downarrow}. \tag{6.65}$$

Here we put the nearest neighbour transfer matrix element $t = 1$. For the half-filled case $N/L = 1$, L being the number of lattice points, we obtain the ground state energy as

$$E = E\left(\frac{L}{2}, \frac{L}{2} : U\right) = -4L \int_0^\infty \frac{J_0(\omega) J_1(\omega) d\omega}{\omega[1 + \exp(\omega U/2)]}, \tag{6.66}$$

where J_0 and J_1 are the Bessel functions. As long as $U \neq 0$ the ground state of the half-filled Hubbard Hamiltonian is the Mott insulator. The expression of (6.66) is singular at $U = 0$ and cannot be expanded with respect to U.

The Wilson ratio $R_W = (\chi_s/\gamma)(\frac{2}{3}\pi k_B{}^2/(g\mu_B)^2)$ is 1 at $U = 0$ and becomes 2 discontinuously at $U \neq 0$. This is because the freedom of charge vanishes at nonzero U and γ is reduced to the half value corresponding to the lost degree of freedom.

The book written recently by Takahashi [13] explains the one-dimensional solvable models in detail.

6.6 Ferromagnetism of transition metals

The ferromagnetism of the transition metals has been studied by many people, such as Herring, Kanamori and Nagaoka. According to the recent theory of Okabe, the ferromagnetism of the transition metals is stabilized by Hund's rule coupling among electrons occupying the degenerate d-orbitals. It has been shown by photoemission measurements for Fe and Ni that there exists enough possibility for two electrons to meet on the same atoms and to be affected by the Hund's rule coupling. This experimental fact means that the effective correlation among d-electrons is not so strong owing to the screening effect by s- and p-electrons, surrounding d-electron pairs. From the above reason, we can accept the result given by the density functional approximation. The role of Hund's rule coupling in realizing the ferromagnetic ground state can be seen from a detailed examination of the Gutzwiller approximation, as done by Okabe [14].

6.7 Superconductivity in the Hubbard Hamiltonian

The Hubbard Hamiltonian gives rise to superconductivity from the electron correlations. There exist p-wave and the d-wave pairing states depending on the momentum dependence of interactions between quasi-particles in the Fermi liquid state.

This issue will be discussed in Chapter 9 in connection with high-temperature superconductivity.

References

[1] J. Hubbard, *Proc. Roy. Soc. (London) A* **276** (1963) 238; **277** (1964) 237; **281** (1964) 401.
[2] Y. Nagaoka, *Solid State Commun.* **3** (1965) 409; *Phys. Rev.* **147** (1966) 392.
[3] J. Kanamori, *Prog. Theor. Phys.* **30** (1963) 275.
[4] M. C. Gutzwiller, *Phys. Rev. Lett.* **10** (1963) 159; *Phys. Rev. A* **134** (1964) 923; **137** (1965) 1726.
[5] W. F. Brinkman and T. M. Rice, *Phys. Rev. B* **2** (1970) 4302.
[6] H. Yokoyama and H. Shiba, *J. Phys. Soc. Jpn.* **59** (1990) 3669.
[7] W. Metzner and D. Vollhardt, *Phys. Rev. Lett.* **62** (1989) 324.
[8] E. Müller-Hartmann, *Z. Phys. B* **74** (1989) 507; **76** (1989) 211.
[9] X. Y. Zhang, M. J. Rosenberg and G. Kotliar, *Phys. Rev. Lett.* **70** (1993) 1666.
[10] M. J. Rosenberg, X. Y. Zhang and G. Kotliar, *Phys. Rev. Lett.* **69** (1992) 1236.
[11] N. Kawakami and S. K. Yang, *Phys. Rev. Lett.* **65** (1990) 2039.
[12] E. H. Lieb and F. Y. Wu, *Phys. Rev. Lett.* **20** (1968) 1445.
[13] M. Takahashi, *Thermodynamics of One-dimensional Solvable Models* (Cambridge University Press, 1999).
[14] T. Okabe, *J. Phys. Soc. Jpn.* **65** (1996) 1056.

7
Fermi liquid theory of strongly correlated electron systems

Heavy fermion systems are explained on the basis of the Fermi liquid theory. The specific heat, magnetic susceptibility and electrical resistivity are discussed. Using Anderson's orthogonality theorem, we show that the Fermi liquid is nothing but a local spin singlet state at every site.

7.1 Heavy fermion systems

The Fermi liquid theory is independent of a model Hamiltonian, and can be applied to any system as long as the system remains a Fermi liquid. The theory tells us that even if the electron interaction becomes strong, physical quantities behave as those of the non-interacting Fermi gas. The difference between them with and without interaction is not qualitative but quantitative [1].

As seen at the Mott transition in Hubbard systems, the effective mass of the electron increases near the transition point. As a system realizing such a large effective mass, a Fermi liquid system called the heavy electron system or the heavy fermion system attracts general interest. The heavy fermion systems are composed of the rare earth metals such as Ce and Yb, and actinide atoms such as U. The heavy fermions are nothing but the quasi-particles in the Fermi liquid theory. As a quasi-particle in a strongly correlated electron system, the heavy fermion is an important issue to be studied in the development of the Fermi liquid theory. The heavy fermion realized in f-electron systems is one of the heavy quasi-particles appearing near the Mott transition.

Another important system among the strongly correlated electron systems is that of the cuprate high-temperature superconductors. This system shows different properties from those expected for the simple Fermi liquid. Examples of these different properties are the T-linear term of the electrical resistivity and the deviation from the Korringa relation, which means that $1/T_1T$ in a Fermi liquid is constant in the relaxation rate of the nuclear magnetic resonance. However, we show below that these anomalous properties can be explained by taking the two-dimensional antiferromagnetic fluctuations into account on the basis of the Fermi liquid theory.

Moreover, the Fermi liquid theory is important in understanding the strongly correlated electron systems for the following reasons. Since the heavy quasi-particles themselves play essential roles in the phase transitions such as superconductivity and magnetism, and make the long-range order, the Fermi liquid theory is indispensable in understanding the ordered states. A good example of this fact is the jump of the specific heat at the transition temperature; the jump is proportional to the mass of quasi-particles in the normal state above T_c. This means that heavy fermions themselves make the phase transition through the renormalized interaction among quasi-particles.

7.2 Heavy fermions

Now we explain the origin and physical properties of heavy fermions; they are seen in alloys and compounds that contain the rare earth metals such as Ce and Yb, and actinide atoms such as U [2].

The rare earth metals such as Gd and Tb usually show magnetic ordered states. These are ferromagnetism and screw structure, and originate through the RKKY (Rudermann–Kittel–Kasuya–Yosida) interaction. However, the systems containing the atoms that possess one f-electron $Ce(4f)^1$ or one f-hole $Yb\,(4f)^{13}$ do not show any magnetic order until very low temperatures and sometimes show Kondo like behaviour. The systems showing Kondo-like behaviour are called dense Kondo systems. In the periodic systems the resistivity shows T^2 behaviour after showing a peak with decreasing temperature. The reason that some heavy fermion systems do not show any magnetic order at low temperatures, but show Kondo-like behaviour, can be explained as follows.

The RKKY interaction (4.21) which determines the magnetic ordered state is proportional to the product of the localized spins S. $S = 7/2$ for Gd, while $S = 1/2$ for Ce. If the coupling constants of the RKKY interaction, J_{RKKY}, are assumed to be the same for Ce, Yb and Gd, their magnetic ordering temperatures T_c are almost determined by the values of S^2.

Since the ordering temperature T_c of Gd is 300K, those of Ce and Yb are estimated as 6K. The smallness of the localized spins for Ce and Yb is important in reducing the magnetic ordering temperature.

On the other hand, the alloys and compounds of the actinide atom U possess two or three f-electrons on a U site. The effective bandwidth of f-electrons in the U system seems to be wider than those of the Ce and Yb systems. The wider f-bandwidth of the U system may be important in suppressing the magnetic long-range orders as well as the orbital degeneracy. The following points are important in our understanding of the heavy fermions. The f-electron systems possessing f-electron number near an integer are the Fermi liquids near the insulating states. The f-electrons can be delocalized through the hybridization with conduction electrons. If there exists no hybridization between f and conduction electrons, the f-electron number becomes just an integer and the f-electrons are localized. Therefore the hybridization between f- and conduction electrons plays an important role in realizing heavy fermions. As a result, the heavy fermions give a large T^2 term in the resistivity.

7.3 Kondo temperature in crystalline fields

We calculate the Kondo temperature of a Ce impurity system using the scaling law introduced in Section 4.5 [3]. We consider the Coqblin–Schrieffer Hamiltonian which describes the exchange interaction between the conduction and f-electrons:

$$
\begin{aligned}
\mathcal{H} = & \sum_k \varepsilon_k \left(\sum_M c_{kM}{}^\dagger c_{kM} + \sum_m c_{km}{}^\dagger c_{km} \right) + \sum_M E_M a_M{}^\dagger a_M + \sum_m E_m a_m{}^\dagger a_m \\
& - \frac{J_0}{2N} \sum_{\substack{mm' \\ kk'}} c_{km}{}^\dagger c_{k'm'} a_{m'}{}^\dagger a_m - \frac{J_1}{2N} \sum_{\substack{MM' \\ kk'}} c_{kM}{}^\dagger c_{k'M'} a_{M'}{}^\dagger a_M \\
& - \frac{J_2}{2N} \sum_{\substack{Mm \\ kk'}} (c_{kM}{}^\dagger c_{k'm} a_m{}^\dagger a_M + c_{km}{}^\dagger c_{k'M} a_M{}^\dagger a_m). \qquad (7.1)
\end{aligned}
$$

Here we have put the origin at the centre of the Ce atom and expanded the wave-functions of conduction electrons in the spherical waves. We consider the case where levels of f-electrons split into higher levels M and lower levels m as shown in Fig. 7.1. We write the creation (annihilation) operator of the f-electron corresponding to each base-function M or m as $a_M^\dagger$ (a_M) or $a_m^\dagger$ (a_m), and write that of the conduction electron as $c_{kM}^\dagger$ (c_{kM}) or $c_{km}^\dagger$ (c_{km}), where M and m correspond to the wave-functions of the f-electron.

The energy levels E_M and E_m, which are determined by the spin–orbit coupling and the crystalline field, are determined so that their sum is zero:

$$\sum_M E_M + \sum_m E_m = 0. \qquad (7.2)$$

Moreover, keeping Ce in mind, we assume the f-electron number is unity:

$$\sum_M a_M{}^\dagger a_M + \sum_m a_m{}^\dagger a_m = 1. \qquad (7.3)$$

In (7.1), J_0, J_1 and J_2 are the exchange interactions within lower levels, within higher levels and between higher and lower levels, respectively.

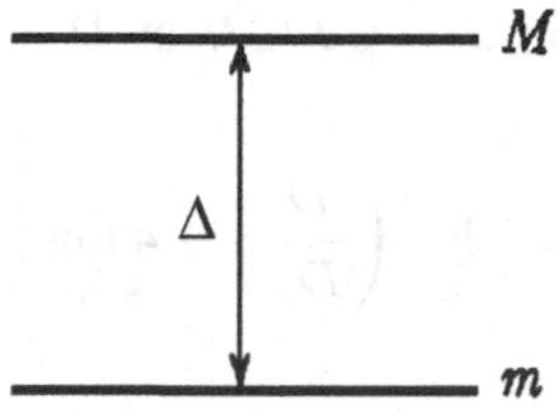

Fig. 7.1 Splitting of f-levels due to the crystal field. Here, levels split into two groups separated by Δ.

Using Poorman's scaling method derived in Chapter 4 and adopting the bandwidth D of conduction electrons as a scaling parameter, we obtain the following equations for the coupling constants $\tilde{J}_0$, $\tilde{J}_1$ and $\tilde{J}_2$:

$$\frac{d\tilde{J}_0}{dD} = \sum_m \frac{\tilde{J}_0{}^2}{D + E_m - \omega} + \sum_M \frac{\tilde{J}_2{}^2}{D + E_M - \omega}, \tag{7.4}$$

$$\frac{d\tilde{J}_1}{dD} = \sum_M \frac{\tilde{J}_1{}^2}{D + E_M - \omega} + \sum_m \frac{\tilde{J}_2{}^2}{D + E_m - \omega}, \tag{7.5}$$

$$\frac{d\tilde{J}_2}{dD} = \sum_M \frac{\tilde{J}_1\tilde{J}_2}{D + E_M - \omega} + \sum_m \frac{\tilde{J}_0\tilde{J}_2}{D + E_m - \omega}, \tag{7.6}$$

where ω is the energy of the electron system including crystalline field effects. Now let us consider the Ce ion in the cubic field. In this case the six states belonging to $j = 5/2$ split into the doublet Γ_7, $E_{\rm D} = -2\Delta/3$ and the quartet Γ_8, $E_{\rm Q} = \Delta/3$. Hereafter, we discuss separately the two cases when the crystal field $\Delta = E_{\rm Q} - E_{\rm D}$ is positive or negative.

(1) $\Delta > 0$: Γ_7 is the ground state, $\omega \simeq -2\Delta/3$.

$$\frac{d\tilde{J}_0}{dD} = \frac{2\tilde{J}_0{}^2}{D} + \frac{4\tilde{J}_2{}^2}{D + \Delta}, \tag{7.7}$$

$$\frac{d\tilde{J}_1}{dD} = \frac{4\tilde{J}_1{}^2}{D + \Delta} + \frac{2\tilde{J}_2{}^2}{D}, \tag{7.8}$$

$$\frac{d\tilde{J}_2}{dD} = \frac{4\tilde{J}_1\tilde{J}_2}{D + \Delta} + \frac{2\tilde{J}_0\tilde{J}_2}{D}. \tag{7.9}$$

If we put $J_0 = J_1 = J_2 = J$ and $\tilde{J}_0 = \tilde{J}_1 = \tilde{J}_2 = \tilde{J}$ for simplicity, we obtain

$$\frac{d\tilde{J}}{dD} = \frac{2\tilde{J}^2}{D} + \frac{4\tilde{J}^2}{D + \Delta}. \tag{7.10}$$

Assuming the initial condition $\tilde{J} = \rho J/2N$ at $D = D_0$, and solving (7.10), we obtain

$$-\frac{1}{\tilde{J}} + \frac{2N}{\rho J} = 2\log\left(\frac{D}{D_0}\right) + 4\log\left(\frac{D+\Delta}{D_0+\Delta}\right). \tag{7.11}$$

When $D_0 \gg \Delta$, $\tilde{J}$ is given by

$$\tilde{J} = \frac{\rho J}{2N}\left[1 + \frac{\rho|J|}{N}\log\left(\frac{D}{D_0}\right) + \frac{2\rho|J|}{N}\log\left(\frac{D+\Delta}{D_0}\right)\right]^{-1}. \tag{7.12}$$

Since the Kondo temperature T_K is given by the value of D at which $\tilde{J}$ diverges, for $k_B = 1$, T_K is given by

$$1 + \frac{\rho|J|}{N}\log\left(\frac{T_K}{D_0}\right) + \frac{2\rho|J|}{N}\log\left(\frac{T_K+\Delta}{D_0}\right) = 0, \tag{7.13}$$

$$T_K = \left(\frac{D_0}{T_K+\Delta}\right)^2 D_0 e^{-N/\rho|J|}. \tag{7.14}$$

If $T_K \ll \Delta$, we obtain

$$T_K = \left(\frac{D_0}{\Delta}\right)^2 D_0 e^{-N/\rho|J|} = \left(\frac{D_0}{\Delta}\right)^2 T_K{}^0, \tag{7.15}$$

where $T_K{}^0$ is the Kondo temperature for the case without the orbital degeneracy, that is $\Delta \simeq D_0$. The result of (7.15) is important. If we assume $D_0 = 10^4$ K and $\Delta = 10^2$ K as usual values, we obtain $(D_0/\Delta)^2 = 10^4$. By hybridizing higher levels situated around 100 K above the lowest levels, the binding energy of the ground state becomes large by a factor of 10^4.

(2) $\Delta < 0$: Γ_8 is the ground state, $\omega \simeq -|\Delta|/3$. Assuming $\tilde{J}_0 = \tilde{J}_1 = \tilde{J}_2 = J$, we obtain

$$\frac{d\tilde{J}}{dD} = \frac{4\tilde{J}^2}{D} + \frac{2\tilde{J}^2}{D+|\Delta|}, \tag{7.16}$$

$$\tilde{J} = \frac{\rho J}{2N}\left[1 + \frac{\rho|J|}{N}\log\left(\frac{D+|\Delta|}{D_0+|\Delta|}\right) + \frac{2\rho|J|}{N}\log\left(\frac{D}{D_0}\right)\right]^{-1}. \tag{7.17}$$

Similarly to case (1), T_K is given by

$$T_K = \left(\frac{D_0}{T_K+|\Delta|}\right)^{1/2} D_0 e^{-N/2\rho|J|}$$
$$\simeq \left(\frac{D_0}{|\Delta|}\right)^{1/2} D_0 e^{-N/2\rho|J|} \quad (T_K \ll |\Delta|). \tag{7.18}$$

Table 7.1 γ, χ *and A for various metals*

		γ^{-1} (mJ mol^{-1} K^2)	χ (memu mol^{-1})	A ($\mu\Omega$ cm K^{-2})
Paramag.	$CeCu_6$	1500	8.5–75.7	42–143
	$CeAl_3$	1600	36	35
Super.	$CeCu_2Si_2$	1000	12–16	11
	UBe_{13}	1100	15	–
	UPt_3	450	4.2–8.3	2.0
Mag.	U_2Zn_{17}	400	12	–
Normal	Pd	9.4	0.8	10^{-5}
	Ag	0.6	0.03	10^{-7}

In this case the exponent $\rho|J|$ of the exponential function is multiplied by a factor of 2 and the Kondo temperature becomes high. This is because the ground f-levels in the crystal field are four-fold degenerate and the number of channels combining the conduction electrons with the f-electrons becomes twice as many.

Including the above-mentioned cases and also the Yb system, we can obtain a general expression for T_K as

$$T_K = \left(\frac{D_0}{\Delta_1}\right)^{N_1/N_0} \left(\frac{D_0}{\Delta_2}\right)^{N_2/N_0} \cdots \left(\frac{D_0}{\Delta_m}\right)^{N_m/N_0} D_0 \exp\left[-\frac{2N}{\rho|J|}\frac{1}{N_0}\right], \tag{7.19}$$

where Δ_i and N_i are the energy difference between level i and the ground state and the degeneracy of level i, respectively. N_0 is the degeneracy of the ground state. As seen here, the degeneracy of f-levels and the effect of upper levels in the crystal field are important for the Kondo effect. Therefore we cannot ignore the degeneracy of f-levels when we discuss heavy fermions.

7.4 Fermi liquid theory on heavy fermion systems

Heavy fermion systems behave as a Fermi liquid with a strong enhancement due to the electron correlation. To see this fact, in Table 7.1 we show the T-linear coefficient γ of the specific heat, magnetic susceptibility χ and the coefficient A of T^2 terms of resistivity for typical electron systems. The values of γ and χ in heavy fermion systems are 10^2–10^3 times as large as those of free electrons, and A is 10^4–10^6 times as large compared with conventional metals. These large enhancement factors mean that the heavy fermions are the Fermi liquids close to the Mott transition. In the heavy fermion systems almost localized f-electrons and delocalized conduction electrons hybridize to deviate the f-electron number from an integer, and keep the metallic state against the strong electron correlation between f-electrons. In this way the existence of the hybridization is essentially important in the Fermi liquid behaviour of the heavy fermions.

The physics of heavy fermions can be described on the basis of the periodic Anderson Hamiltonian. The degeneracy of the f-orbital is essential in keeping the normal state against strong correlation effects. Here, for simplicity, we ignore the orbital degeneracy and assume that the correlation U between f-electrons is not so strong as to destroy the Fermi liquid state. Using the following simple model, we make clear the mechanism of the formation of heavy fermions:

$$\mathcal{H} = \mathcal{H}_0 + \mathcal{H}', \tag{7.20}$$

$$\begin{aligned}\mathcal{H}_0 = &\sum_{k\sigma} \varepsilon_k c_{k\sigma}{}^\dagger c_{k\sigma} + \sum_{k\sigma} E_k a_{k\sigma}{}^\dagger a_{k\sigma} \\ &+ \sum_{k\sigma} (V_k a_{k\sigma}{}^\dagger c_{k\sigma} + V_k{}^* c_{k\sigma}{}^\dagger a_{k\sigma}) - \frac{NU}{4} \langle n_0{}^f \rangle^2,\end{aligned} \tag{7.21}$$

$$\mathcal{H}' = \frac{U}{N} \sum_{kk'q} a_{k+q\uparrow}{}^\dagger a_{k'-q\downarrow}{}^\dagger a_{k'\downarrow} a_{k\uparrow}, \tag{7.22}$$

where $a_{k\sigma}{}^\dagger$ ($c_{k\sigma}{}^\dagger$) is the creation operator of the f- (conduction) electron with energy $E_k(\varepsilon_k)$, and σ denotes spin. The coulomb repulsion U between f-electrons on the same atom is the only many-body interaction in this Hamiltonian. The above two kinds of electrons, the f-electrons and the conduction electrons, hybridize through the mixing term V_k and form energy bands. The f-electron number averaged over the unperturbed state is denoted as $\langle n_0{}^f \rangle$.

To construct the Fermi liquid theory, we treat the electron interaction U as a perturbation for the state determined by $\mathcal{H}_0$ and introduce the self-energy $\Sigma_k(z)$ originating from the perturbation. Here $z = \omega - i\Gamma$ is complex. Now we can determine Green's functions for the f-electron and the conduction electron by the following equations:

$$(z\hat{1} - \hat{H})\hat{G} = \hat{1}, \tag{7.23}$$

$$z\hat{1} - \hat{H} = \begin{pmatrix} z - E_k - \Sigma_k(z) & -V_k \\ -V_k{}^* & z - \varepsilon_k \end{pmatrix}, \tag{7.24}$$

$$\hat{G} = \begin{pmatrix} G_{k\sigma}{}^f(z) & G_{k\sigma}{}^{fc}(z) \\ G_{k\sigma}{}^{cf}(z) & G_{k\sigma}{}^c(z) \end{pmatrix}, \tag{7.25}$$

where $\hat{1}$ is the unit matrix of two components. For example, the diagonal parts of

Green's functions, $G_{k\sigma}{}^{f}(z)$ and $G_{k\sigma}{}^{c}(z)$, are given by

$$G_{k\sigma}{}^{f}(z) = [z - E_{k\sigma} - \Sigma_{k\sigma}(z) - |V_k|^2/(z - \varepsilon_k)]^{-1}, \tag{7.26}$$

$$G_{k\sigma}{}^{c}(z) = [z - \varepsilon_{k\sigma} - |V_k|^2/(z - E_{k\sigma} - \Sigma_{k\sigma}(z))]^{-1}. \tag{7.27}$$

The energy of a quasi-particle is determined by the pole of Green's function, and is given by the solution of the equation

$$(z - E_{k\sigma} - \Sigma_{k\sigma}(z))(z - \varepsilon_{k\sigma}) - |V_k|^2 = 0. \tag{7.28}$$

(a) Electronic specific heat

Following Luttinger, we derive the T-linear coefficient of the specific heat ($\omega_+ = \omega + i\delta$):

$$\begin{aligned}
\gamma &= -\frac{\pi^2 k_B{}^2}{6\pi i} \sum_{k\sigma} \left\{ \frac{\partial}{\partial\omega} \log\left[\omega_+ + \mu - E_k - \Sigma_k{}^R(\omega_+) - \frac{|V_k|^2}{\omega_+ + \mu - \varepsilon_k}\right] - \text{C.C.} \right\}_{\omega=0} \\
&= \frac{2\pi^2 k_B{}^2}{3} \sum_k -\frac{1}{\pi}\text{Im}\left[\mu + i\delta - E_k - \Sigma_k{}^R(0) - \frac{|V_k|^2}{\mu + i\delta - \varepsilon_k}\right]^{-1} \\
&\quad \times \left(1 - \left.\frac{\partial \Sigma_k{}^R(\omega)}{\partial\omega}\right|_{\omega=0} + \frac{|V_k|^2}{(\mu - \varepsilon_k)^2}\right) \\
&= \frac{2\pi^2 k_B{}^2}{3} \left\{ \sum_k \rho_k{}^f(0)\tilde{\gamma}_k + \sum_k \rho_k{}^c(0) \right\},
\end{aligned} \tag{7.29}$$

$$\tilde{\gamma}_k = 1 - \left.\frac{\partial \Sigma_k(\omega)}{\partial\omega}\right|_{\omega=0}. \tag{7.30}$$

Here the density of states $\rho_k{}^f(\omega)$ for f-electrons and that for conduction electrons $\rho_k{}^c(\omega)$ are given by

$$\rho_k{}^f(\omega) = -\frac{1}{\pi}\text{Im}\left[\mu + \omega_+ - E_k - \Sigma_k{}^R(\omega) - \frac{|V_k|^2}{\mu + \omega_+ - \varepsilon_k}\right]^{-1}, \tag{7.31}$$

$$\rho_k{}^c(\omega) = -\frac{1}{\pi}\text{Im}\left[\mu + \omega_+ - \varepsilon_k - \frac{|V_k|^2}{\omega_+ + \mu - E_k - \Sigma_k{}^R(\omega)}\right]^{-1}. \tag{7.32}$$

Since the electronic specific heat at low temperatures is given by the thermal excitation of quasi-particles, the T-linear coefficient γ is also given by the density of states for quasi-particles on the Fermi surface:

$$\gamma = \frac{\pi^2 k_B{}^2}{3} \sum_{k\sigma} \delta(\mu - E_{k\sigma}{}^*) = \frac{2\pi^2 k_B{}^2}{3} \tilde{\gamma}. \tag{7.33}$$

Now let us introduce the renormalization factors $z_k{}^f$ for f-electrons and $z_k{}^c$ for conduction electrons. These are given by the residues of $G_k{}^f$ and $G_k{}^c$, respectively:

$$z_k{}^f = \left(\tilde{\gamma}_k + \frac{|V_k|^2}{(\mu - \varepsilon_k)^2}\right)^{-1}, \tag{7.34}$$

$$z_k{}^c = \frac{|V_k|^2}{(\mu - \varepsilon_k)^2} \Big/ \left(\tilde{\gamma}_k + \frac{|V_k|^2}{(\mu - \varepsilon_k)^2}\right). \tag{7.35}$$

Here the following relation holds:

$$\tilde{\gamma}_k z_k{}^f + z_k{}^c = 1. \tag{7.36}$$

From (7.36), (7.33) is given by

$$\begin{aligned} \gamma &= \frac{\pi^2 k_B{}^2}{3} \sum_{k\sigma} (\tilde{\gamma}_{k\sigma} z_{k\sigma}{}^f + z_{k\sigma}{}^c) \delta(\mu - E_{k\sigma}{}^*) \\ &= \frac{\pi^2 k_B{}^2}{3} \sum_{k\sigma} [\rho_{k\sigma}{}^f(0) \tilde{\gamma}_{k\sigma} + \rho_{k\sigma}{}^c(0)]. \end{aligned} \tag{7.37}$$

This agrees with (7.29). In our model, since the electron correlation works only among f-electrons, the contribution of f-electrons to the specific heat is enhanced by $\tilde{\gamma}_k$. From the above result, we conclude that the large coefficient of specific heat γ in heavy fermions originates from $\tilde{\gamma}_k$ because of the coulomb interaction among f-electrons. Actually, we can confirm this fact by calculating the second-order term of the self-energy. In this case, when the bandwidth of f-electrons becomes narrow and the density of states at Fermi energy becomes high in the unperturbed state, the enhancement factor becomes large through the coupling constant $\rho^f(0)U$.

If we assume $\tilde{\gamma}_k \gg 1$, we can derive the band of heavy fermions as follows. $\Sigma_k(\omega)$ can be expanded as

$$\Sigma_{k\sigma}(\omega) \simeq \Sigma_{k\sigma}(0) + \left.\frac{\partial \Sigma_{k\sigma}(\omega)}{\partial \omega}\right|_{\omega=0} \omega - i\Delta_k, \tag{7.38}$$

$$\Delta_k = -\mathrm{Im}\Sigma_{k\sigma}(\omega). \tag{7.39}$$

Substituting (7.38) into (7.28) and putting $\omega = E_k{}^* - i\Gamma_k{}^*$, we obtain

$$E_k{}^* = \frac{1}{\tilde{\gamma}_k}\left[E_{k\sigma} + \Sigma_{k\sigma}(0) + \frac{|V_k|^2}{E_k{}^* - \varepsilon_{k\sigma}}\right] = \tilde{E}_k + \frac{|\tilde{V}_k|^2}{E_k{}^* - \varepsilon_k}, \tag{7.40}$$

$$\tilde{E}_k = (E_{k\sigma} + \Sigma_{k\sigma}(0))/\tilde{\gamma}_k, \tag{7.41}$$

$$|\tilde{V}_k|^2 = |V_k|^2/\tilde{\gamma}_k. \tag{7.42}$$

Equation (7.40) represents the band of quasi-particles reduced by the factor $1/\tilde{\gamma}_k$. In other words, since the reduction of energy scale leads to the enhancement of density of states, the specific heat is enhanced by a factor $\tilde{\gamma}_k$. The energy width of quasi-particles $\Gamma_k{}^*$ represents the damping rate and is reduced as

$$\Gamma_k{}^* = z_k{}^f \Delta_k \simeq \Delta_k/\tilde{\gamma}_k. \tag{7.43}$$

(b) Magnetic susceptibility

The magnetization M is given by

$$M = \mu_B \sum_{k\sigma} \sigma\theta(\mu - E_{k\sigma}{}^*), \tag{7.44}$$

where we have assumed the Bohr magneton μ_B and $g = 2$ for both f- and conduction electrons. The function θ is the step function and $E_{k\sigma}{}^*$ is the energy of a quasi-particle in the presence of the magnetic field. The spin susceptibility χ_s is obtained from (7.44) as

$$\chi_s = \left.\frac{\partial M}{\partial H}\right|_{H=0} = \lim_{H\to 0} \mu_B \sum_{k\sigma} \sigma\delta(\mu - E_k{}^*)(-\partial E_{k\sigma}{}^*/\partial H)\Big|_{H=0}. \tag{7.45}$$

Using the eigenvalue equation, let us calculate $\partial E_{k\sigma}{}^*/\partial H$. Putting $H_\sigma = \sigma\mu_B H$, we obtain the eigenvalue equation (7.28) for $E_{k\sigma} = E_k - H_\sigma$ and $\varepsilon_{k\sigma} = \varepsilon_k - H_\sigma$:

$$[\omega + \mu - E_{k\sigma} - \Sigma_{k\sigma}(\omega)](\omega + \mu - \varepsilon_{k\sigma}) - |V_k|^2 = 0. \tag{7.46}$$

Keeping in mind that ω depends on H, we obtain from (7.46):

$$-\frac{\partial E_{k\sigma}{}^*}{\partial H} = -\mu_B\sigma[z_k{}^f \tilde{\chi}_s(\boldsymbol{k}) + z_k{}^c], \tag{7.47}$$

$$\tilde{\chi}_s(\boldsymbol{k}) = \tilde{\chi}_{\uparrow\uparrow}(\boldsymbol{k}) + \tilde{\chi}_{\uparrow\downarrow}(\boldsymbol{k}), \tag{7.48}$$

$$\tilde{\chi}_{\uparrow\uparrow}(\boldsymbol{k}) = 1 - \left.\frac{\partial \Sigma_{k\sigma}(0)}{\partial H_\sigma}\right|_{H=0}, \tag{7.49}$$

$$\tilde{\chi}_{\uparrow\downarrow}(\boldsymbol{k}) = \left.\frac{\partial \Sigma_{k\sigma}(0)}{\partial H_{-\sigma}}\right|_{H=0}. \tag{7.50}$$

Substituting (7.47) into (7.45), we obtain the susceptibility

$$\chi_s = 2\mu_B{}^2 \sum_k [\rho_k{}^f(0)\tilde{\chi}_s(\boldsymbol{k}) + \rho_k{}^c(0)]. \tag{7.51}$$

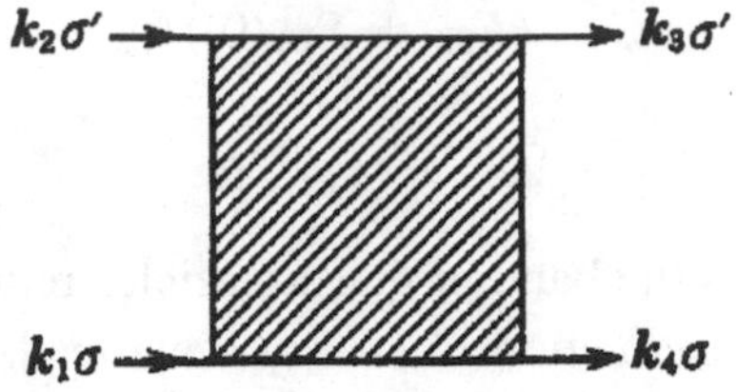

Fig. 7.2 Four-point vertex $\Gamma_{\sigma\sigma'}(\boldsymbol{k}_1, \boldsymbol{k}_2; \boldsymbol{k}_3, \boldsymbol{k}_4)$.

Since the electron interaction works only among f-electrons, the contribution of f-electrons to the susceptibility is enhanced by a factor $\tilde{\chi}_s$.

Now we use Ward's identity introduced in Chapter 5, which relates the self-energy to the vertex functions

$$\tilde{\gamma}_k = \tilde{\gamma}(\boldsymbol{k}) = \tilde{\chi}_{\uparrow\uparrow}(\boldsymbol{k}) + \sum_{k'} \rho_{k'}{}^f(0)\Gamma_{\sigma\sigma}(\boldsymbol{k}, \boldsymbol{k}'; \boldsymbol{k}', \boldsymbol{k}). \tag{7.52}$$

The vertex function $\Gamma_{\sigma\sigma}$ is defined by Fig. 7.2 and represents the renormalized interaction between f-electrons. Using (7.33) and (7.52), we can represent $\tilde{\gamma}$ as

$$\tilde{\gamma} = \chi_{\uparrow\uparrow} + \delta_{\uparrow\uparrow}, \tag{7.53}$$

$$\chi_{\uparrow\uparrow} = \sum_{k} \rho_k{}^f(0)\tilde{\chi}_{\uparrow\uparrow}(\boldsymbol{k}), \tag{7.54}$$

$$\delta_{\uparrow\uparrow} = \sum_{kq} \rho_k{}^f(0)\Gamma_{\uparrow\uparrow}{}^{\mathrm{A}}(\boldsymbol{k}, \boldsymbol{k}+\boldsymbol{q};\ \boldsymbol{k}+\boldsymbol{q}, \boldsymbol{k})\rho_{k+q}{}^f(0), \tag{7.55}$$

where we have introduced the anti-symmetrized vertex $\Gamma_{\uparrow\uparrow}{}^{\mathrm{A}}$ in place of $\Gamma_{\sigma\sigma}$. The vertex $\Gamma_{\uparrow\uparrow}{}^{\mathrm{A}}(\boldsymbol{k}, \boldsymbol{k}+\boldsymbol{q}; \boldsymbol{k}+\boldsymbol{q}, \boldsymbol{k})$ vanishes at $\boldsymbol{q} = 0$ by definition.

Similarly, we can write

$$\begin{aligned} \chi_{\uparrow\downarrow} &= \sum_{k} \rho_k{}^f(0)\tilde{\chi}_{\uparrow\downarrow}(\boldsymbol{k}) \\ &= \sum_{kq} \rho_k{}^f(0)\Gamma_{\uparrow\downarrow}(\boldsymbol{k}, \boldsymbol{k}+\boldsymbol{q};\ \boldsymbol{k}+\boldsymbol{q}, \boldsymbol{k})\rho_{k+q}{}^f(0). \end{aligned} \tag{7.56}$$

The charge susceptibility $\tilde{\chi}_c$ is given by

$$\begin{aligned} \tilde{\chi}_c &= \rho^c(0) + \chi_{\uparrow\uparrow} - \chi_{\uparrow\downarrow} \\ &= \sum_{k} \rho_k{}^c(0) + \sum_{k} \rho_k{}^f(0)[\tilde{\chi}_{\uparrow\uparrow}(\boldsymbol{k}) - \tilde{\chi}_{\uparrow\downarrow}(\boldsymbol{k})]. \end{aligned}$$

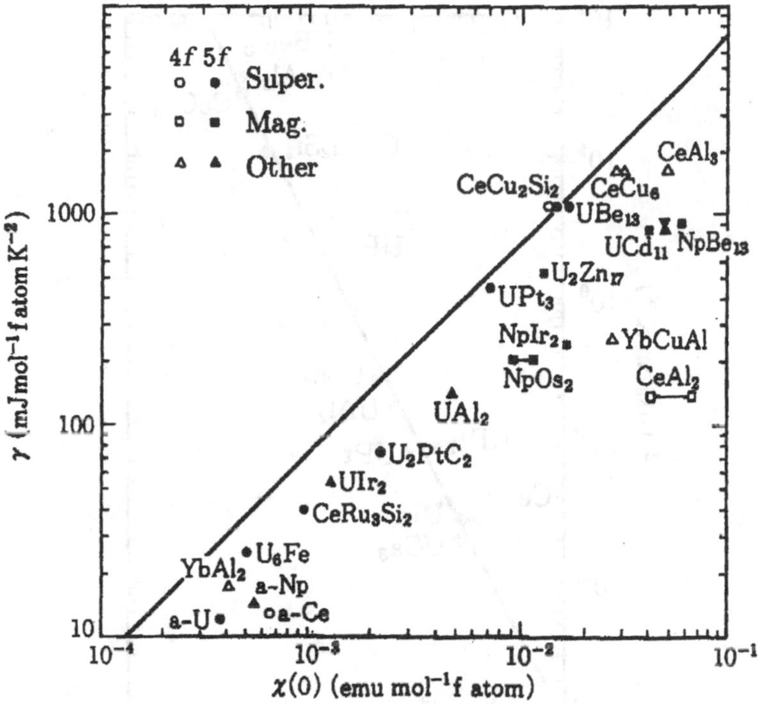

Fig. 7.3 The coefficient γ of specific heat and the susceptibility χ are proportional to each other [4].

When the coulomb repulsion U between f-electrons is large, the charge fluctuation of f-electrons is suppressed:

$$\chi_{\uparrow\uparrow} - \chi_{\uparrow\downarrow} = 0. \tag{7.57}$$

In this case the Wilson ratio R_W is given by

$$R_W = \tilde{\chi}_s/\tilde{\gamma} = \frac{(\chi_{\uparrow\uparrow} + \chi_{\uparrow\downarrow})}{(\chi_{\uparrow\uparrow} + \delta_{\uparrow\uparrow})} = \frac{2}{(1 + \delta_{\uparrow\uparrow}/\chi_{\uparrow\uparrow})}. \tag{7.58}$$

If we neglect $\delta_{\uparrow\uparrow}$, considering that $\Gamma_{\uparrow\uparrow}{}^{A}(\boldsymbol{k}, \boldsymbol{k}+\boldsymbol{q}; \boldsymbol{k}+\boldsymbol{q}, \boldsymbol{k}) = 0$ at $\boldsymbol{q} = 0$, we obtain $R_W = 2$.

As shown in Fig. 7.3, the specific heat coefficient γ and the susceptibility χ are almost proportional to each other in the experimental results. For $U = 0$, $R_W = 1$. Even when γ and χ are enhanced, R_W takes a value around unity. In the actual case, the estimation of the Wilson ratio is not easy owing to the degeneracy of f-orbitals.

(c) Resistivity

The electron interactions give the damping rate of quasi-particles, which is proportional to T^2. Although the T^2 term should be seen in the resistivity at low

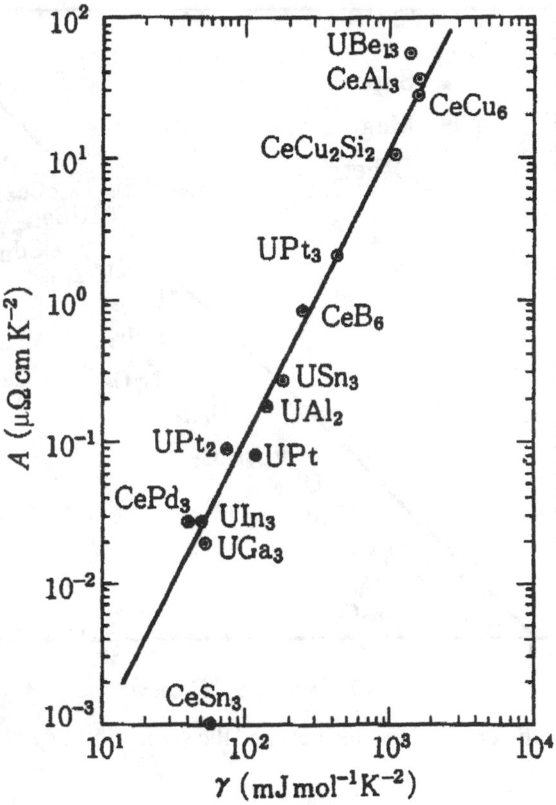

Fig. 7.4 The relation between γ and the coefficient A of the T^2 term in the resistivity. A is proportional to γ^2 [5].

temperatures, the value had been too small to be observed in experiments, until the heavy fermion was found. The coefficient A of the T^2 term in heavy fermion systems is from 10^4 times to 10^6 times as large as that in ordinary metals. As shown in Fig. 7.4, the coefficient A increases in proportion to γ^2. The T^2 term is also observed in the over-doped cuprate superconductors and the organic conductors. Now we discuss the basic problem related to the T^2 term of the resistivity.

When the electron interactions conserve the momentum, the electron collision does not contribute to the resistivity. Now we calculate the conductivity $\sigma_{\mu\nu}$ on the basis of the linear response theory, and show that in a free electron system the resistivity does not arise from the electron interaction. Here we use a periodic Anderson model in order to apply it to the heavy fermion system, but it is easy to discuss $\sigma_{\mu\nu}$ on the basis of the Hubbard model.

The current operator in the heavy fermion system is given by ($\hbar = 1$)

$$\hat{J} = \sum_{k\sigma} (v_k{}^f a_{k\sigma}{}^\dagger a_{k\sigma} + v_k{}^c c_{k\sigma}{}^\dagger c_{k\sigma} + \nabla_k V_k (a_{k\sigma}{}^\dagger c_{k\sigma} + c_{k\sigma}{}^\dagger a_{k\sigma})), \tag{7.59}$$

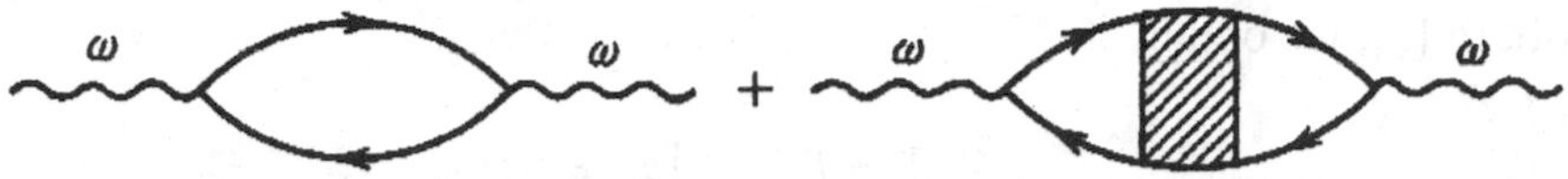

Fig. 7.5 Diagrams for the conductivity σ. The second diagram contains the vertex corrections.

where $v_k{}^f = \nabla_k E_k$ and $v_k{}^c = \nabla_k \varepsilon_k$. The conductivity $\sigma_{\mu\nu}$ is given by

$$\sigma_{\mu\nu} = \sum_{ij} \sigma_{\mu\nu}{}^{(ij)},$$

$$\sigma_{\mu\nu}{}^{(ij)} = e^2 \sum_{k\sigma,\, k'\sigma'} v_{k\mu}{}^{(i)} v_{k\nu}{}^{(j)} \lim_{\omega \to 0} \frac{1}{\omega} \text{Im}\, K_{k\sigma,k'\sigma'}{}^{(ij)}(\omega + i\delta), \tag{7.60}$$

where components i, j mean f- or conduction electrons. The two-particle Green's function $K_{k\sigma,k'\sigma'}{}^{(i,j)}(\omega + i\delta)$ is obtained by analytic continuation from the thermal Green's function $\tilde{K}_{k\sigma,k'\sigma'}(i\omega)$ $(k_B = 1)$.

The thermal two-particle Green's function is written as

$$\tilde{K}_{k\sigma,k'\sigma'}{}^{(ij)}(\omega_m) = \int_0^{1/T} d\tau\, e^{\omega_m \tau} \langle T_\tau [A_{k\sigma}{}^{(i)\dagger}(\tau) A_{k\sigma}{}^{(i)}(\tau) A_{k'\sigma'}{}^{(j)\dagger} A_{k'\sigma'}{}^{(j)}] \rangle, \tag{7.61}$$

$$A_{k\sigma}{}^{(i)}(\tau) = e^{(H-\mu N_i)\tau} A_{k\sigma}{}^{(i)} e^{-(H-\mu N_i)\tau},$$

where $A_{k\sigma}{}^{(c)} = c_{k\sigma}$, $A_{k\sigma}{}^{(f)} = a_{k\sigma}$, $\omega_m = 2m\pi i T$. The total conductivity can also be written in the quasi-particle picture as

$$\sigma_{\mu\nu} = e^2 \sum_{k\sigma,\, k'\sigma'} v_{k\mu}{}^* v_{k'\nu}{}^* \lim_{\omega \to 0} \frac{1}{\omega} \text{Im}\, K_{k\sigma,k'\sigma'}{}^*(\omega + i\delta), \tag{7.62}$$

where $K_{k\sigma,k'\sigma'}{}^*$ is the two-particle Green's function of quasi-particles. The velocity of quasi-particle $v_k{}^*$ is given by

$$v_k{}^* = \nabla_k E_k{}^* = z_k{}^f \tilde{v}_k{}^f + z_k{}^c v_k{}^c + z_k{}^f \frac{1}{\mu - \varepsilon_k} \nabla_k |V_k|^2, \tag{7.63}$$

$$\tilde{v}_k{}^f = \nabla_k (E_k + \Sigma_k(0)).$$

Now let us calculate (7.61). Taking into account the singularities of Green's functions and vertex functions and carrying out the analytic continuations done by

Éliashberg [6], we obtain

$$K_{k\sigma,k'\sigma'}{}^{(ij)}(\omega+i\delta) = -\frac{1}{4\pi i}\int_{-\infty}^{\infty} d\varepsilon\Big[\text{th}\frac{\varepsilon}{2T}K_1{}^{(ij)}(\varepsilon,\omega) + \left(\text{th}\frac{\varepsilon+\omega}{2T} - \text{th}\frac{\varepsilon}{2T}\right)K_2{}^{(ij)}(\varepsilon,\omega) - \text{th}\frac{\varepsilon+\omega}{2T}K_3{}^{(ij)}(\varepsilon,\omega)\Big], \tag{7.64}$$

$$K_l{}^{(ij)}(\varepsilon,\omega) = g_l{}^{(ijji)}(\varepsilon,\omega) + g_l{}^{(iffi)}(\varepsilon,\omega)\sum_{m=1}^{3}\frac{1}{4\pi i}\int d\varepsilon' T_{lm}(\varepsilon,\varepsilon':\omega) \times g_m{}^{(fjjf)}(\varepsilon',\omega). \tag{7.65}$$

Here g_l is given for $\omega > 0$ by

$$g_1{}^{(ijji)}(\varepsilon,\omega) = G^{R(ij)}(\varepsilon)G^{R(ji)}(\varepsilon+\omega), \tag{7.66a}$$

$$g_2{}^{(ijji)}(\varepsilon,\omega) = G^{A(ij)}(\varepsilon)G^{R(ji)}(\varepsilon+\omega), \tag{7.66b}$$

$$g_3{}^{(ijji)}(\varepsilon,\omega) = G^{A(ij)}(\varepsilon)G^{A(ji)}(\varepsilon+\omega), \tag{7.66c}$$

where R and A mean the retarded and the advanced Green's functions, respectively. Approximating Green's functions by the poles near the Fermi surface, we obtain

$$g_1{}^{(ijji)}(\varepsilon,\omega) \simeq \{G^{R(ij)}(\varepsilon)\}^2 = z_k{}^i z_k{}^j(\varepsilon - E_k{}^* + i\delta)^{-2}, \tag{7.67a}$$

$$g_2{}^{(ijji)}(\varepsilon,\omega) = G^{A(ij)}(\varepsilon)G^{R(ji)}(\varepsilon+\omega) \simeq 2\pi i z_k{}^i z_k{}^j \delta(\varepsilon - E_k{}^*)/(\omega + 2i\Gamma_k{}^*), \tag{7.67b}$$

$$g_3{}^{(ijji)}(\varepsilon,\omega) = \{g_1{}^{(ijji)}(\varepsilon,\omega)\}^*. \tag{7.67c}$$

The four-point vertex T_{lm} in (7.65) contributes to the renormalization of the velocity, except for T_{22} possessing g_2 sections on both sides. At $T = 0$, the vertex correction $\Lambda_{k\sigma}{}^0(0)$ is given by

$$\Lambda_{k\sigma}{}^0(0) = \sum_{k'\sigma'}\int\frac{d\omega'}{2\pi i}\Gamma_{\sigma\sigma'}(\boldsymbol{k},\boldsymbol{k}')[G_{k'}{}^f(\omega')]^2\left[v_{k'}{}^f + \frac{|V_k|^2}{(\omega'+\mu-\varepsilon_k)^2}v_k{}^c + \frac{\partial|V_k|^2/\partial\boldsymbol{k}'}{(\omega'+\mu-\varepsilon_{k'})}\right], \tag{7.68}$$

where $\Gamma_{\sigma\sigma'}(\boldsymbol{k},\boldsymbol{k}') = \Gamma_{\sigma\sigma'}(\boldsymbol{k},\boldsymbol{k}';\boldsymbol{k}',\boldsymbol{k})$.

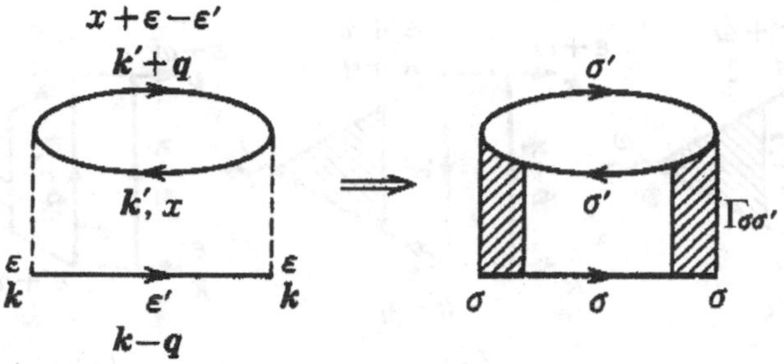

Fig. 7.6 Self-energy diagrams giving the T^2 term in its imaginary part. Solid lines and broken lines represent the electron lines and interaction lines, respectively. The left diagram is the second-order term in U and the right one is the general diagram giving the T^2 term in the imaginary part.

On the other hand, the momentum derivative of self-energy is given by

$$\begin{aligned}\frac{\partial \Sigma_{k\sigma}(0)}{\partial \boldsymbol{k}} &= \sum_{k'\sigma'} \int \Gamma_{\sigma\sigma'}(\boldsymbol{k},\boldsymbol{k}') \lim_{q\to 0} \frac{1}{\boldsymbol{q}} [G_{k'+q}{}^f(\omega') - G_{k'}{}^f(\omega')] \frac{d\omega'}{2\pi i} \\ &= \sum_{k'\sigma'} \int \frac{d\omega'}{2\pi i} \Gamma_{\sigma\sigma'}(\boldsymbol{k},\boldsymbol{k}')[G_{k'}{}^f(\omega')]^2 \Bigg[v_{k'}{}^f + \frac{|V_{k'}|^2}{(\omega'+\mu-\varepsilon_{k'})^2} v_{k'}{}^c \\ &\quad + \frac{\partial |V_{k'}|^2/\partial \boldsymbol{k}'}{(\omega'+\mu-\varepsilon_{k'})} \Bigg] - \sum_{k'\sigma'} \Gamma_{\sigma\sigma'}(\boldsymbol{k},\boldsymbol{k}') z_{k'}{}^f \delta(\mu - E_{k'}{}^*) v_{k'}{}^*. \qquad (7.69)\end{aligned}$$

Using the above results, we obtain the electron current at $T = 0$:

$$\begin{aligned}\boldsymbol{j}_k &= z_k{}^f \left(v_k{}^f + \Lambda_k{}^0(0) + \frac{1}{\mu - \varepsilon_k} \nabla_k |V_k|^2 \right) + z_k{}^c v_k{}^c \\ &= v_k{}^* + \sum_{k'\sigma'} f_{\sigma\sigma'}(\boldsymbol{k},\boldsymbol{k}') \delta(\mu - E_{k'}{}^*) v_{k'}{}^*. \qquad (7.70)\end{aligned}$$

The second term of (7.70) represents the back-flow term, and the interaction between quasi-particles is given by

$$f_{\sigma\sigma'}(\boldsymbol{k},\boldsymbol{k}') = z_k{}^f \Gamma_{\sigma\sigma'}(\boldsymbol{k},\boldsymbol{k}';\boldsymbol{k}',\boldsymbol{k}) z_{k'}{}^f. \qquad (7.71)$$

At a finite temperature in the hydrodynamic regime the back-flow term is replaced by the second term of the following equation:

$$\begin{aligned}\sigma_{\mu\nu}(\omega) &= \frac{i}{2} e^2 \sum_k j_{k\mu} \frac{1}{2T} \frac{\mathrm{ch}^{-2}(E_k{}^*/2T)}{\omega + 2i\Gamma_k{}^*} j_{k\nu} \\ &\quad + \frac{1}{2} e^2 \sum_{k,k'} j_{k\mu} z_k{}^f \frac{(1/2T)\mathrm{ch}^{-2}(E_k{}^*/2T) T_{22}(\boldsymbol{k},\boldsymbol{k}';\omega)}{(\omega + 2i\Gamma_k{}^*)(\omega + 2i\Gamma_{k'}{}^*)} z_{k'}{}^f j_{k'\nu}. \qquad (7.72)\end{aligned}$$

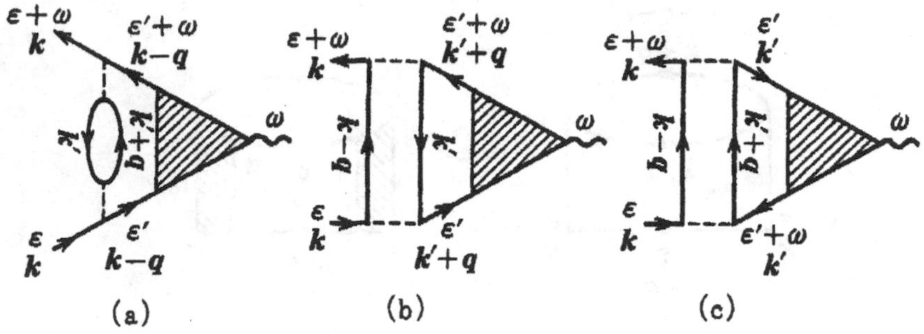

Fig. 7.7 The second-order diagrams for the vertex corrections necessary to conserve the total momentum. Solid and broken lines denote the electron and interaction lines, respectively. These diagrams can be obtained by attaching the vertex correction to each of three electron lines in Fig. 7.6.

In this case ($T \neq 0$), we can put $j_k = v_k{}^*$ and consider hereafter the second term of (7.72).

Now we discuss the vertex correction T_{22}. In Fig. 7.7, we show the vertex corrections corresponding to the second-order term of self-energy, $\Sigma_{k\sigma}{}^{(2)}(\varepsilon)$. These three diagrams give the T^2 terms in the same way as the T^2 term of self-energy. Here the T_{22} term appears in the product with the g_2 section, and the T^2 term in T_{22} cancels out with the factor $1/2i\Gamma_k{}^*$ which arises from the g_2 section. As a result, the vertex correction becomes of order unity. We therefore cannot neglect the vertex correction due to the T_{22}, even at low temperatures. After some calculations, the vertex correction $\boldsymbol{\Lambda}_k$ is determined by the following equations:

$$\boldsymbol{\Lambda}_k{}^{(a)}(\varepsilon) \simeq U^2 \sum_{k'q} \pi \rho_{k-q}(0)\rho_{k'+q}(0)\rho_{k'}(0)\frac{\varepsilon^2 + (\pi T)^2}{2\Delta_{k-q}(\varepsilon)}\boldsymbol{\Lambda}_{k-q}(\varepsilon), \tag{7.73a}$$

$$\boldsymbol{\Lambda}_k{}^{(b)}(\varepsilon) \simeq U^2 \sum_{k'q} \pi \rho_{k-q}(0)\rho_{k'+q}(0)\rho_{k'}(0)\frac{\varepsilon^2 + (\pi T)^2}{2\Delta_{k'+q}(\varepsilon)}\boldsymbol{\Lambda}_{k'+q}(\varepsilon), \tag{7.73b}$$

$$\boldsymbol{\Lambda}_k{}^{(c)}(\varepsilon) \simeq -U^2 \sum_{k'q} \pi \rho_{k-q}(0)\rho_{k'}(0)\rho_{k'+q}(0)\frac{(\pi T)^2 + \varepsilon^2}{2\Delta_{k'}(-\varepsilon)}\boldsymbol{\Lambda}_{k'}(-\varepsilon). \tag{7.73c}$$

If we do not confine ourselves to the U^2 term but include higher order terms, the vertex correction is given by the diagrams shown in Fig. 7.8. These diagrams correspond to the general diagram of the self-energy in Fig. 7.6. As a result, we

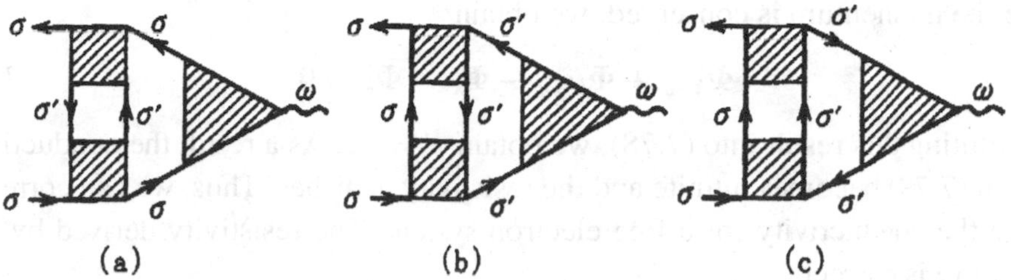

Fig. 7.8 General diagrams of the vertex correction necessary to recover the momentum conservation.

obtain the self-consistent equations for the vertex correction $\boldsymbol{\Lambda}_k(\varepsilon)$:

$$\begin{aligned}\boldsymbol{\Lambda}_k(\varepsilon) &= \boldsymbol{j}_k + \boldsymbol{\Lambda}_k{}^{(a)}(\varepsilon) + \boldsymbol{\Lambda}_k{}^{(b)}(\varepsilon) + \boldsymbol{\Lambda}_k{}^{(c)}(\varepsilon)\\ &= \boldsymbol{j}_k + \sum_{k'q} \Delta_0(\boldsymbol{k}, \boldsymbol{k}'; \boldsymbol{k}' + \boldsymbol{q}, \boldsymbol{k} - \boldsymbol{q}) \left\{ \frac{\boldsymbol{\Lambda}_{k-q}(\varepsilon)}{2\Delta_{k-q}(\varepsilon)} + \frac{\boldsymbol{\Lambda}_{k'+q}(\varepsilon)}{2\Delta_{k'+q}(\varepsilon)} \right.\\ &\quad \left. - \frac{\boldsymbol{\Lambda}_{k'}(-\varepsilon)}{2\Delta_{k'}(-\varepsilon)} \right\}, \end{aligned} \tag{7.74}$$

where

$$\begin{aligned}\Delta_0(\boldsymbol{k}, \boldsymbol{k}'; \boldsymbol{k}' + \boldsymbol{q}, \boldsymbol{k} - \boldsymbol{q}) &= \pi \rho_{k-q}{}^f(0) \rho_{k'+q}{}^f(0) \rho_{k'}{}^f(0)[(\pi T)^2 + \varepsilon^2]\\ &\times \left[\Gamma_{\uparrow\downarrow}{}^2(\boldsymbol{k}, \boldsymbol{k}'; \boldsymbol{k}' + \boldsymbol{q}, \boldsymbol{k} - \boldsymbol{q}) + \frac{1}{2}(\Gamma_{\uparrow\uparrow}{}^A(\boldsymbol{k}, \boldsymbol{k}'; \boldsymbol{k}' + \boldsymbol{q}, \boldsymbol{k} - \boldsymbol{q}))^2 \right],\end{aligned} \tag{7.75}$$

$$\Delta_k = \frac{1}{2} \sum_{k'q} \Delta_0(\boldsymbol{k}, \boldsymbol{k}'; \boldsymbol{k}' + \boldsymbol{q}, \boldsymbol{k} - \boldsymbol{q}). \tag{7.76}$$

If we put

$$\boldsymbol{\Phi}_k(\varepsilon) = \boldsymbol{\Lambda}_k(\varepsilon)/2\Delta_k(\varepsilon) = \boldsymbol{\Phi}_k(-\varepsilon), \tag{7.77}$$

(7.74) is written as

$$\boldsymbol{j}_k + \sum_{k'q} \Delta_0(\boldsymbol{k}, \boldsymbol{k}'; \boldsymbol{k}' + \boldsymbol{q}, \boldsymbol{k} - \boldsymbol{q})[\boldsymbol{\Phi}_{k-q} + \boldsymbol{\Phi}_{k'+q} - \boldsymbol{\Phi}_{k'} - \boldsymbol{\Phi}_k] = 0. \tag{7.78}$$

Finally, the conductivity (7.72) is given by

$$\sigma_{\mu\nu}(0) = e^2 \sum_k j_{k\mu} \left(-\frac{\partial f(x)}{\partial x} \right)_{x=E_k{}^*} \frac{\boldsymbol{\Lambda}_{k\nu}}{2\Gamma_k{}^*}. \tag{7.79}$$

Here, if we assume a free electron system,

$$\boldsymbol{\Phi}_k = \boldsymbol{k}F, \tag{7.80}$$

since the momentum is conserved, we obtain

$$\Phi_{k-q} + \Phi_{k'+q} - \Phi_{k'} - \Phi_k = 0. \tag{7.81}$$

Substituting this result into (7.78), we obtain $F = \infty$. As a result the conductivity $\sigma(0)$ in (7.79) becomes infinite and the resistivity vanishes. Thus, we can correctly derive the conductivity for a free electron system. The resistivity derived by this procedure is correct.

Now we consider actual electron systems in a lattice. Taking the Umklapp scattering into account, we assume

$$\Phi_{k-q} + \Phi_{k'+q} - \Phi_{k'} - \Phi_k = \sum_i \boldsymbol{K}_i F, \tag{7.82}$$

where $\boldsymbol{K}_i$ is a reciprocal lattice vector. Substituting (7.82) into (7.78), we obtain

$$\Phi_k = \frac{\boldsymbol{k}}{2\Delta_k} \frac{j_k \cdot \boldsymbol{k}}{\sum_i \boldsymbol{K}_i \cdot \boldsymbol{k}}, \qquad j_k \propto \boldsymbol{k}. \tag{7.83}$$

Using this result, we obtain the conductivity as

$$\sigma_{\mu\nu}(0) = e^2 \sum_k \delta(\mu - E_k{}^*) j_{k\mu} \frac{1}{2\Gamma_k{}^*} \frac{\boldsymbol{k}^2}{\sum_i \boldsymbol{K}_i \cdot \boldsymbol{k}} j_{k\nu}. \tag{7.84}$$

This equation gives the resistivity proportional to T^2 at low temperatures. Once a Fermi surface is given, we can correctly estimate the T^2 term of the resistivity by calculating the Umklapp scattering process on the basis of (7.78) and (7.79):

$$\sigma_{\mu\nu} = e^2 \sum_k \delta(\mu - E_k{}^*) j_{k\mu} \frac{1}{2\Gamma_k{}^*} \frac{1}{C_k} j_{k\nu}. \tag{7.85}$$

Here C_k is a numerical factor determined by the Umklapp scattering. The current of quasi-particles $\boldsymbol{j}_k = \boldsymbol{v}_k{}^*$ is reduced by the renormalization factor $z_k{}^f$ the same as $\Gamma_k{}^*$ and $E_k{}^*$. The density of states $\rho_k{}^*(0) = \delta(\mu - E_k{}^*)$ is enhanced by $1/z_k{}^f$ owing to the renormalization of $E_k{}^*$. From these renormalizations, the renormalization factors in (7.85) cancel each other out and the conductivity (7.85) can be written only by the physical quantities of the bare f-electrons. As a result the resistivity is determined by the imaginary part Δ_k of the self-energy for f-electrons, which is given by the T^2 term:

$$\Delta_k \simeq \frac{4(\pi T)^2}{3} \sum_{k'q} \pi \rho_{k-q}{}^f(0) \rho_{k'}{}^f(0) \rho_{k'+q}{}^f(0) \Big[\Gamma_{\uparrow\downarrow}{}^2(\boldsymbol{k}, \boldsymbol{k}'; \boldsymbol{k}' + \boldsymbol{q}, \boldsymbol{k} - \boldsymbol{q}) + \frac{1}{2} (\Gamma_{\uparrow\uparrow}{}^{\mathrm{A}}(\boldsymbol{k}, \boldsymbol{k}'; \boldsymbol{k}' + \boldsymbol{q}, \boldsymbol{k} - \boldsymbol{q}))^2 \Big]. \tag{7.86}$$

On the other hand, from (7.37), (7.52) and (7.57) the specific heat coefficient γ is given by

$$\gamma = \frac{2\pi^2 k_{\rm B}{}^2}{3} \sum_{kq} \rho_k{}^f(0)[\Gamma_{\uparrow\downarrow}(\boldsymbol{k}, \boldsymbol{k}+\boldsymbol{q};\ \boldsymbol{k}+\boldsymbol{q}, \boldsymbol{k})$$
$$+\Gamma_{\uparrow\uparrow}{}^{\rm A}(\boldsymbol{k}, \boldsymbol{k}+\boldsymbol{q};\ \boldsymbol{k}+\boldsymbol{q}, \boldsymbol{k})]\rho_{k+q}{}^f(0), \tag{7.87}$$

where we have assumed that the charge fluctuation of f-electrons is suppressed.

Comparing (7.86) giving the T^2 term of the resistivity with (7.87) giving γ, we can show that when the spin fluctuation is localized in the real space and the momentum dependence of $\Gamma_{\sigma\sigma'}$ can be ignored, the Kadowaki–Woods relaltion holds:

$$A \propto \gamma^2. \tag{7.88}$$

The coefficient of (7.88), from (7.84), (7.86) and (7.87), is given by $A/\gamma^2 = 1/v^2\rho^2 \propto 1/k_{\rm F}{}^4$ and independent of the bare mass of electrons. This relation, therefore, holds not only for f-electrons but also for general Fermi liquids.

7.5 Spin fluctuation and Fermi liquid

The periodic Anderson Hamiltonian reduces to the Kondo lattice where the localized spins and conduction electrons interact with each other, when the mixing term $|V_k|$ is small compared with U and $\mu - E^f$. The cuprate superconductors are also the systems possessing strong spin fluctuations. Here let us discuss the relation between the spin fluctuations and the Fermi liquid.

Let us consider an arbitrary operator $\hat{O}$ and two ground states $|{\rm i}\rangle$ and $|{\rm f}\rangle$. Anderson's orthogonality theorem says that for the matrix element $\langle {\rm f}|\hat{O}|{\rm i}\rangle$ not to vanish, it requires that the local electron number in the matrix element should be conserved.

(a) Anderson Hamiltonian

The Anderson Hamiltonian discussed in Chapter 5 is written as

$$\mathcal{H}_{\rm A} = \sum_{k\sigma} \varepsilon_k n_{k\sigma} + E_d \sum_{\sigma} n_{d\sigma} + \frac{1}{\sqrt{N}} \sum_{k\sigma} (V_k d_\sigma{}^\dagger c_{k\sigma} + V_k{}^* c_{k\sigma}{}^\dagger d_\sigma) + U n_{d\uparrow} n_{d\downarrow}. \tag{7.89}$$

This model reduces to the s–d exchange Hamiltonian, when $\mu - E_d, U \gg \Delta = \pi\rho|V|^2$. In this case the s–d exchange interaction J is given by

$$\frac{J_z}{2N} = \frac{J_\perp}{2N} = -|V|^2 \left(\frac{1}{|\mu - E_d|} + \frac{1}{E_d + U - \mu} \right). \tag{7.90}$$

Here we have represented the localized orbitals as d-electrons. In the rare earth metals, the localized orbitals are f-electrons.

The ground state of the Anderson Hamiltonian Ψ_g is generally given by

$$\Psi_g = A_0\varphi_0 + A_\uparrow d_\uparrow{}^\dagger\varphi_\uparrow + A_\downarrow d_\downarrow{}^\dagger\varphi_\downarrow + A_2 d_\uparrow{}^\dagger d_\downarrow{}^\dagger\varphi_2, \tag{7.91}$$

where A_i is the coefficient of each state. The states φ_0 and φ_2 are the wave-functions of conduction electrons corresponding to no d-electron and two d-electron states, respectively. Although it is difficult to calculate φ_i and A_i, we can discuss the electronic state using the orthogonality theorem.

The essential term to gain the energy in the Anderson Hamiltonian is the third term mixing the d-electrons and conduction electrons. If the expectation value of the mixing term with respect to the ground state Ψ_g vanishes, d-electrons and conduction electrons cannot couple coherently and the localized spin and conduction electrons fluctuate independently. To avoid this, the following expectation value in the ground state should be finite:

$$\begin{aligned}
\langle\Psi_g|\mathcal{H}_{\mathrm{mix}}|\Psi_g\rangle &= \frac{V}{\sqrt{N}}\sum_{k\sigma}\langle\Psi_g|d_\sigma{}^\dagger c_{k\sigma} + c_{k\sigma}{}^\dagger d_\sigma|\Psi_g\rangle \\
&= V\sum_\sigma\left\{\left\langle A_0\varphi_0\middle|\frac{1}{\sqrt{N}}\sum_k c_{k\sigma}{}^\dagger\middle|A_\sigma\varphi_\sigma\right\rangle\right. \\
&\quad + \left\langle A_2\varphi_2\middle|\frac{1}{\sqrt{N}}\sum_k c_{k\sigma}\middle|A_{-\sigma}\varphi_{-\sigma}\right\rangle + \left\langle A_\sigma\varphi_\sigma\middle|\frac{1}{\sqrt{N}}\sum_k c_{k\sigma}\middle|A_0\varphi_0\right\rangle \\
&\quad \left. + \left\langle A_{-\sigma}\varphi_{-\sigma}\middle|\frac{1}{\sqrt{N}}\sum_k c_{k\sigma}{}^\dagger\middle|A_2\varphi_2\right\rangle\right\}.
\end{aligned} \tag{7.92}$$

Since $c_{0\sigma} = (1/\sqrt{N})\sum_k c_{k\sigma}$ annihilates one electron, Anderson's orthogonality theorem gives the following relations among the local numbers of conduction electrons corresponding to each component of Ψ_g:

$$n_{0\sigma}{}^{\mathrm{c}} = n_{\sigma\sigma}{}^{\mathrm{c}} + 1, \tag{7.93}$$

$$n_{0-\sigma}{}^{\mathrm{c}} = n_{\sigma-\sigma}{}^{\mathrm{c}} = n_{-\sigma\sigma}{}^{\mathrm{c}} = n_{0\sigma}{}^{\mathrm{c}}, \tag{7.94}$$

$$n_{2\sigma}{}^{\mathrm{c}} = n_{-\sigma\sigma}{}^{\mathrm{c}} - 1, \tag{7.95}$$

$$n_{2-\sigma}{}^{\mathrm{c}} = n_{-\sigma-\sigma}{}^{\mathrm{c}} = n_{\sigma\sigma}{}^{\mathrm{c}}, \tag{7.96}$$

where ${n_{i\sigma}}^{\rm c}$ denotes the local conduction electron number with spin σ for the d-electron state i in (7.91). These are the necessary conditions for the ground state. As a result,

$$ {n_{\sigma\sigma}}^{\rm c} = {n_{-\sigma\sigma}}^{\rm c} - 1. \tag{7.97} $$

This is the same equation as (4.56):

$$ {n_{\sigma\sigma}}^{\rm c} = -\frac{1}{2}, \qquad {n_{\sigma-\sigma}}^{\rm c} = \frac{1}{2}, \tag{7.98} $$

$$ {n_{0\sigma}}^{\rm c} = \frac{1}{2}, \qquad {n_{2\sigma}}^{\rm c} = -\frac{1}{2}. \tag{7.99} $$

This result means that when including d-electrons, every component possesses a half up-spin electron and a half down-spin electron around the impurity. This is the same as that of the resonance orbital for the Anderson Hamiltonian with $\mu - E_d = U = 0$.

(b) Periodic Anderson Hamiltonian

Using f-orbitals, we can write the periodic Anderson Hamiltonian as

$$ \begin{aligned} \mathcal{H}_{\rm PA} = &\sum_{k\sigma} \varepsilon_k n_{k\sigma} + E^f \sum_{i\sigma} {n_{i\sigma}}^f + U \sum_i {n_{i\uparrow}}^f {n_{i\downarrow}}^f \\ &+ \frac{1}{\sqrt{N}} \sum_{k\sigma} \left\{ V_k {f_{i\sigma}}^\dagger c_{k\sigma} e^{i\boldsymbol{k}\cdot\boldsymbol{R}_i} + {V_k}^* {c_{k\sigma}}^\dagger f_{i\sigma} e^{-i\boldsymbol{k}\cdot\boldsymbol{R}_i} \right\}. \end{aligned} \tag{7.100} $$

Here, to simplify the argument, we assume a separable case $V_{\boldsymbol{k}} = V f(\boldsymbol{k})$, and obtain

$$ \frac{1}{\sqrt{N}} \sum_{\boldsymbol{k}} f(\boldsymbol{k}) e^{i\boldsymbol{k}\cdot\boldsymbol{R}_i} c_{\boldsymbol{k}\sigma} \equiv c_{i\sigma}. \tag{7.101} $$

The operator $c_{i\sigma}$ (${c_{i\sigma}}^\dagger$) is the annihilation (creation) operator of the conduction electron at site i. For the periodic Anderson Hamiltonian we can apply the same argument as case (a) with respect to an arbitrary lattice point of the f-electron. Writing a lattice point chosen arbitrarily as site 0, we expand the ground state as

$$ \Psi_{\rm g} = A_0 \varphi_0 + A_\uparrow {f_{0\uparrow}}^\dagger \varphi_\uparrow + A_\downarrow {f_{0\downarrow}}^\dagger \varphi_\downarrow + A_2 {f_{0\uparrow}}^\dagger {f_{0\downarrow}}^\dagger \varphi_2. \tag{7.102} $$

The mixing term is rewritten using (7.101) as

$$ \mathcal{H}_{\rm mix} = V \sum_{i\sigma} ({f_{i\sigma}}^\dagger c_{i\sigma} + {c_{i\sigma}}^\dagger f_{i\sigma}). \tag{7.103} $$

For $\Psi_{\rm g}$ given by (7.102) to be the ground state, it is necessary for the matrix element $\langle \Psi_{\rm g} | \mathcal{H}_{\rm mix} | \Psi_{\rm g} \rangle$ not to vanish. The necessary condition can be obtained by the same argument as the Anderson Hamiltonian. As a result, when the f-electron at arbitrary site 0 possesses σ spin, the local electron number possessing σ' spin, $n_{\sigma\sigma'}$, is given by

$$n_{\sigma\sigma} = -\frac{1}{2}, \tag{7.104}$$

$$n_{\sigma-\sigma} = \frac{1}{2}. \tag{7.105}$$

As is seen in the Friedel oscillation, the spatial change of the electron distribution varies depending on each system. The local number of electrons denoted as $n_{\sigma\sigma'}$ means the deviation from the uniform distribution, irrespective of spatial extension.

What is the difference between the periodic and impurity Anderson Hamiltonians? In the periodic case every site possesses f-electrons and the neighbouring local electron number $n_{\sigma\sigma'}$ includes f-electrons at other sites than site 0. That is electrons, which construct the singlet state with the central f-electron we have chosen, include not only conduction electrons but also f-electrons at other sites. The ratio of components and spatial extension of the distribution depend on the system parameters such as U and V. However, one important thing is that as far as the ground state is concerned, an f-electron and electrons surrounding it always compose the singlet state. If we include the f-electron at the chosen site in the local electron number, the spin of the f-electron should always cancel out with neighbouring electron spins and in a sufficiently wide region the spin distribution is spatially uniform. This uniform distribution holds for each component of the chosen f-spin, which is true without taking the average over all the components. This fact means that when the up-spin of the f-electron is fixed there remains down-spin in the neighbour of the f-spin. The distribution of the down-spin is not uniform and local in space. In conclusion, under the necessary condition of local spin conservation, coherent mixing as in the Fermi liquid can be maintained. When coherent mixing is destroyed, the separation between localized f-electrons and conduction electrons occurs similarly to Mott insulators. Although the Hamiltonian conserves the total spins as a whole, in the ground state spins should be locally conserved everywhere as in the Fermi liquid. This condition is more strict than that required by the Hamiltonian. In the periodic systems we can shift the chemical potential and reduce by one electron at each site. By this procedure the electron distribution changes from Fig. 7.9(a) to (b). As seen in Fig. 7.9(b), an f-electron always moves accompanying a hole with the same spin around it. As a result spins distribute uniformly as a whole.

We have stressed the important role of the mixing term in maintaining the Fermi liquid. On the other hand, what is the role of electron correlation in the periodic

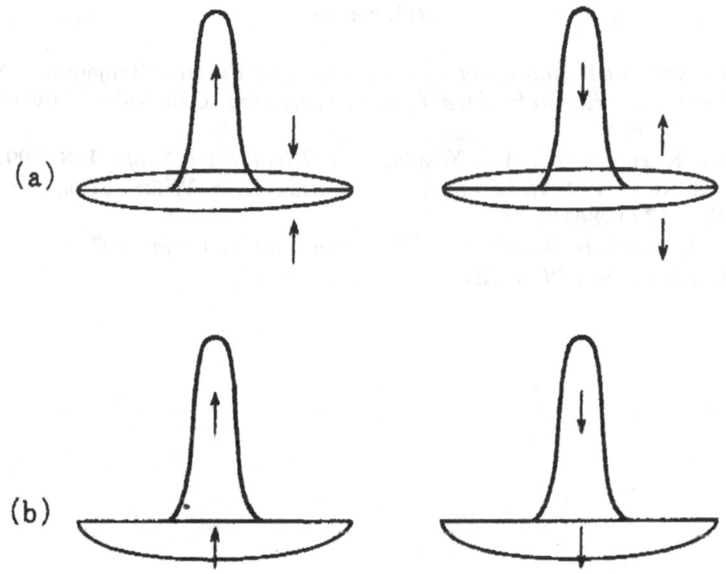

Fig. 7.9 (a) The local electron distribution associated with localized up-spin and down-spin f-electrons. (b) The electron distribution obtained from (a) by reducing uniformly half up- and down-spin electrons. The f-electron and surrounding hole conserve spins locally.

Anderson Hamiltonian? The electron correlation reduces the width of the band constructed by the mixing term and enhances the electron mass. This effect results from the fact that the electrons move so as to avoid each other to reduce the coulomb repulsive energy. Although the electron correlation is important in enhancing the electron mass, the energy gain originates from the mixing term. As a result, the vanishing of the mixing term separates the Mott insulator from the Fermi liquid.

(c) *d–p* Hamiltonian and Hubbard Hamiltonian

The argument for the periodic Anderson Hamiltonian can be directly applied to the d–p Hamiltonian, by replacing f-electrons and conduction electrons with d-electrons and p-electrons, respectively. As a result, in the ground state of the normal state, d-electrons construct the local spin singlet states coupling with surrounding d- and p-electrons. In the Hubbard Hamiltonian, each electron composes a local spin singlet state by coupling with the neighbouring electrons in the ground state.

Hence, the strongly correlated Fermi liquid state is nothing but the resonating valence bond (RVB) state in metals. From this point of view, it seems to be a simple and correct strategy to study the effect of electron correlation on the basis of the Fermi liquid. As a result it is natural to consider that high-temperature superconductivity originates from the electron correlation between d-electrons.

References

[1] P. W. Anderson, *Basic Notions of Condensed Matter Physics* (Benjamin, 1984).
[2] A. C. Hewson, *The Kondo Problem to Heavy Fermions* (Cambridge University Press, 1993).
[3] K. Yamada, K. Hanzawa and K. Yosida, *Prog. Theor. Phys. Suppl.* **108** (1992).
[4] P. A. Lee, T. M. Rice, J. W. Serene, L. J. Sham and L. W. Wilkins, *Comm. Condens. Matter Phys.* **12** (1986) 99.
[5] K. Kadowaki and S. B. Woods, *Solid State Commun.* **58** (1986) 507.
[6] G. M. Éliashberg, *Sov. Phys. JETP* **14** (1962) 886.

8

Transport theory based on Fermi liquid theory

Starting with the linear response theory, we derive the transport equations for the conductivity, Hall coefficient and cyclotron resonance. In the derivation, the importance of the vertex correction is stressed in recovering the momentum conservation. In the lattice system the Umklapp scattering plays an essential role in transport phenomena.

8.1 Conductivity

The transport equation for a degenerate system of Fermi particles was derived by Éliashberg [1]. Let us introduce the general formula of the conductivity starting with the linear response theory [2] (Appendix D). By using the result in Appendix E derived by Éliashberg [1], the conductivity $\sigma_{\mu\nu}(\omega)$ is given by

$$\sigma_{\mu\nu}(\omega) = \frac{ie^2}{2\Omega}\left\{ \sum_{k} v_{k\mu}{}^* \frac{1}{2T} \frac{\cosh^{-2}[(\varepsilon_k{}^* - \mu)/2T]}{\omega + 2i\gamma_k{}^*} v_{k\nu}{}^* \right.$$
$$\left. + \frac{1}{2}\sum_{k,k'} z_k v^*_{k\mu} \frac{1}{2T} \frac{\cosh^{-2}[(\varepsilon_k{}^* - \mu)/2T] T_{22}(k, k'; \omega)}{(\omega + 2i\gamma_k{}^*)(\omega + 2i\gamma_{k'}{}^*)} z_{k'} v_{k'\nu}{}^* \right\}, \tag{8.1}$$

where $\varepsilon_k{}^*$, $v_k{}^*$ and z_k are the energy, the velocity and the renormalization factor of quasi-particles, respectively. In this chapter we put the damping rate of quasi-particles $\gamma_k{}^*$ as

$$\gamma_k{}^* = -z_k \operatorname{Im} \Sigma_k(0). \tag{8.2}$$

In the hydrodynamic region where $\omega\tau = \omega/\gamma_k{}^* \ll 1$, the electrical conductivity is given by

$$\sigma_{\mu\nu}(0) = \sigma_{\mu\nu} = \frac{e^2}{\Omega}\sum_k \frac{1}{2\gamma_k{}^*}\left(-\frac{\partial f}{\partial \varepsilon}\right)_{\varepsilon=\varepsilon_k{}^*} v_{k\mu}{}^* \Lambda_{k\nu}, \tag{8.3}$$

$$\Lambda_{k\nu} = v_{k\nu}{}^* + \sum_{k'} \frac{z_k T_{22}(\boldsymbol{k},\boldsymbol{k}') z_{k'}}{2i\gamma_{k'}{}^*} v_{k'\nu}{}^*. \tag{8.4}$$

Here we have used the relation

$$\frac{1}{4T}\cosh^{-2}\left(\frac{\varepsilon_k{}^* - \mu}{2T}\right) = \left(-\frac{\partial f}{\partial \varepsilon}\right)_{\varepsilon=\varepsilon_k{}^*}. \tag{8.5}$$

For the static conductivity the imaginary part of T_{22} contributes to the vertex correction, as discussed in the previous chapter. Without any Umklapp scattering the electric current is conserved even in the presence of electron–electron scattering. In this case we have no resistivity even at finite temperatures. To derive this result from (8.3), we have to take the vertex correction into account not to violate the Ward identity [3].

In actual lattice systems, we can correctly obtain the coefficient of the T^2 term of resistivity, by taking the Umklapp scattering into account [3–5].

8.2 Optical conductivity

In the collisionless regime $\gamma_k{}^*/\omega \ll 1$, where $\gamma_k{}^* \propto T^2$, the optical conductivity is obtained from (8.1). For our purpose we need only the real part of $T_{22}(\boldsymbol{k},\boldsymbol{k}';\omega)$, which is given by

$$\mathrm{Re}\, T_{22}(\boldsymbol{k},\boldsymbol{k}';\omega) = \left\{\tanh\left(\frac{\varepsilon_{k'}{}^* - \mu + \omega}{2T}\right) - \tanh\left(\frac{\varepsilon_{k'}{}^* - \mu}{2T}\right)\right\} \times \mathrm{Re}\,\Gamma(\boldsymbol{k},\boldsymbol{k}';\omega). \tag{8.6}$$

From (8.1) and (8.6) we obtain

$$\begin{aligned}\sigma_{\mu\nu}(\omega) &= \frac{ie^2}{\omega + i\delta}\frac{1}{\Omega}\sum_k \Big\{ v_{k\mu}{}^* v_{k\nu}{}^* \delta(\mu - \varepsilon_k{}^*) \\ &\quad + \sum_{k'} f(\boldsymbol{k},\boldsymbol{k}') v_{k\mu}{}^* v_{k'\nu}{}^* \delta(\mu - \varepsilon_k{}^*)\delta(\mu - \varepsilon_{k'}{}^*)\Big\} \\ &= \frac{ie^2}{\omega + i\delta}\frac{1}{\Omega}\sum_k v_{k\mu}{}^* j_{k\nu}{}^* \delta(\mu - \varepsilon_k{}^*), \end{aligned} \tag{8.7}$$

where

$$f(\boldsymbol{k}, \boldsymbol{k}') = z_{\boldsymbol{k}} z_{\boldsymbol{k}'} \Gamma^{\omega}(\boldsymbol{k}, \boldsymbol{k}'), \tag{8.8}$$

and

$$j_{\boldsymbol{k}\nu}{}^* = v_{\boldsymbol{k}\nu}{}^* + \sum_{\boldsymbol{k}'} f(\boldsymbol{k}, \boldsymbol{k}') v_{\boldsymbol{k}'\nu}{}^* \delta(\mu - \varepsilon_{\boldsymbol{k}'}{}^*). \tag{8.9}$$

The second term of (8.9) represents the back-flow due to the quasi-particle interaction. Equation (8.7) can be obtained also by the following derivation [4, 5]. Here we write $k = (\boldsymbol{k}, \omega)$ for simplicity. To obtain the linear response to the vector potential $\boldsymbol{A}$, the total current operator is defined by

$$\hat{J}_\mu = \hat{v}_\mu - e \sum_\nu \hat{\varepsilon}''_{\mu\nu} A_\nu, \tag{8.10}$$

where

$$\hat{v}_\mu \equiv \sum_{\boldsymbol{k}} v_{\boldsymbol{k}\mu} c_{\boldsymbol{k}}{}^\dagger c_{\boldsymbol{k}}, \qquad v_{\boldsymbol{k}\mu} = \frac{\partial \varepsilon_{\boldsymbol{k}}}{\partial k_\mu} \tag{8.11}$$

and

$$\hat{\varepsilon}''_{\mu\nu} = \sum_{\boldsymbol{k}} \frac{\partial^2 \varepsilon_{\boldsymbol{k}}}{\partial k_\mu \partial k_\nu} c_{\boldsymbol{k}}{}^\dagger c_{\boldsymbol{k}}. \tag{8.12}$$

The optical conductivity is given by

$$\sigma_{\mu\nu}(\omega) = \frac{e^2}{\omega + i\delta} \left(i K_{\mu\nu}(\omega + i\delta) + \frac{1}{\Omega} \int \frac{d^4 k}{(2\pi)^4} \frac{\partial^2 \varepsilon_{\boldsymbol{k}}}{\partial k_\mu \partial k_\nu} G(k) e^{i\omega 0} \right), \tag{8.13}$$

where $K_{\mu\nu}(\omega)$ is defined by

$$K_{\mu\nu}(\omega) = \frac{i}{\Omega} \int_{-\infty}^{\infty} \langle 0 | \mathrm{T} \hat{v}_\mu(t) \hat{v}_\nu | 0 \rangle \exp(i\omega t) dt, \tag{8.14}$$

and corresponds to the uniform limit $\boldsymbol{k} \to 0$ of the current correlation function $K_{\mu\nu}(k)$. In terms of the vertex function $\Lambda(k', k)$, $K_{\mu\nu}(k)$ is given by

$$K_{\mu\nu}(k) = -\frac{i}{\Omega} \int \frac{d^4 k'}{(2\pi)^4} v_{\boldsymbol{k}\mu} G(k' + k/2) G(k' - k/2) \Lambda_\nu(k', k). \tag{8.15}$$

The vertex function satisfies the following equation:

$$\Lambda_\mu(k', k) = v_{\boldsymbol{k}'\mu} - i \int \frac{d^4 k''}{(2\pi)^4} \Gamma^{(0)}(k', k'') \times G(k'' + k/2) G(k'' - k/2) \Lambda_\mu(k'', k), \tag{8.16}$$

where $\Gamma^{(0)}(k, k')$ is the irreducible four-point vertex function. As is well known, in the limit $k = (\boldsymbol{k}, \omega) \to 0$, the product of the Green's functions $G(k' + k/2)G(k' - k/2)$ behaves irregularly:

$$G(k' + k/2)G(k' - k/2) \\ = G(k')^2 + 2\pi i z_{k'}{}^2 \frac{\boldsymbol{k} \cdot \boldsymbol{v}_{k'}{}^*}{\omega - \boldsymbol{k} \cdot \boldsymbol{v}_{k'}{}^*} \delta(\mu - \varepsilon_{k'}{}^*)\delta(\omega'). \tag{8.17}$$

In particular, between the ω-limit ($k/\omega = 0$) and the k-limit ($k/\omega = \infty$) of (8.17) we obtain the relation

$$\left\{G(k')^2\right\}^{\omega} - \left\{G(k')^2\right\}^{k} = 2\pi i z_{k'}{}^2 \delta(\mu - \varepsilon_{k'}{}^*)\delta(\omega'). \tag{8.18}$$

On the other hand, the derivative of the self-energy is given by

$$\frac{\partial \Sigma(k, \omega)}{\partial k_\mu} = -i \int \frac{d^4k'}{(2\pi)^4} \Gamma^{(0)}(k, k') \frac{\partial}{\partial k'_\mu} G(k', \omega') \\ = -i \int \frac{d^4k'}{(2\pi)^4} \Gamma^{(0)}(k, k') \{G(k')^2\}^k \left(v_{k'\mu} + \frac{\partial \Sigma(k', \omega')}{\partial k'_\mu} \right). \tag{8.19}$$

Comparing this equation with

$$\Lambda_\mu{}^k(k') = v_{k'\mu} - i \int \frac{d^4k''}{(2\pi)^4} \Gamma^{(0)}(k', k'') \{G(k'')^2\}^k \Lambda_\mu{}^k(k''), \tag{8.20}$$

which is derived from (8.16), we obtain the relation

$$\Lambda_\mu{}^k(k') = v_{k'\mu} + \frac{\partial \Sigma(k', \omega')}{\partial k'_\mu}, \tag{8.21}$$

hence

$$v_{k'\mu}{}^* = z_{k'} \Lambda_\mu{}^k(k'). \tag{8.22}$$

Using the above results, we obtain the k-limit of $K_{\mu\nu}(k)$, as

$$K_{\mu\nu}{}^k = -\frac{i}{\Omega} \int \frac{d^4k'}{(2\pi)^4} v_{k'\mu} \{G(k')^2\}^k \Lambda_\nu{}^k(k') \\ = -\frac{i}{\Omega} \int \frac{d^4k'}{(2\pi)^4} v_{k'\mu} \frac{\partial}{\partial k'_\nu} G(k') \\ = \frac{i}{\Omega} \int \frac{d^4k'}{(2\pi)^4} \frac{\partial v_{k'\mu}}{\partial k'_\nu} G(k'). \tag{8.23}$$

Substituting (8.23) into the second term of (8.13), we obtain

$$\sigma_{\mu\nu}(\omega) = \frac{ie^2}{\omega + i\delta}(K_{\mu\nu}(\omega + i\delta) - K_{\mu\nu}{}^k). \tag{8.24}$$

Thus, the effective mass m' in the Drude weight, defined by

$$\operatorname{Re}\sigma_{\mu\nu}(\omega) = \pi e^2 \left(\frac{n}{m'}\right)_{\mu\nu} \delta(\omega) + \sigma_{\text{inc}}, \tag{8.25}$$

is given by

$$\left(\frac{n}{m'}\right)_{\mu\nu} = K_{\mu\nu}{}^{\omega} - K_{\mu\nu}{}^{k}. \tag{8.26}$$

To estimate (8.26), from (8.16) we obtain

$$\Lambda_\mu{}^\omega(\boldsymbol{k}) = \Lambda_\mu{}^k(\boldsymbol{k}) - i\int \frac{d^4k'}{(2\pi)^4}\Gamma^\omega(\boldsymbol{k},\boldsymbol{k}')[\{G(k')^2\}^\omega - \{G(k')^2\}^k]\Lambda_\mu{}^k(k'). \tag{8.27}$$

Thus we obtain

$$z_{\boldsymbol{k}}\Lambda_\mu{}^\omega(\boldsymbol{k}) = v_{\boldsymbol{k}\mu}{}^* + \sum_{k'} f(\boldsymbol{k},\boldsymbol{k}')v_{\boldsymbol{k}'\mu}{}^*\delta(\mu - \varepsilon_{\boldsymbol{k}'}{}^*) \equiv j_{\boldsymbol{k}\mu}{}^*, \tag{8.28}$$

where $f(\boldsymbol{k},\boldsymbol{k}') \equiv z_{\boldsymbol{k}} z_{\boldsymbol{k}'}\Gamma^\omega(k,k')$.

Since we have

$$K_{\mu\nu}{}^\omega = K_{\mu\nu}{}^k + \frac{1}{\Omega}\int \frac{d^4k'}{(2\pi)^4}\Lambda_\mu{}^k(k')(\{G(k')^2\}^\omega - \{G(k')^2\}^k)\Lambda_\nu{}^\omega(k'), \tag{8.29}$$

we finally obtain

$$\left(\frac{n}{m'}\right)_{\mu\nu} = \frac{1}{\Omega}\sum_{k} v_{\boldsymbol{k}\mu}{}^* j_{\boldsymbol{k}\nu}{}^*\delta(\mu - \varepsilon_{\boldsymbol{k}}{}^*). \tag{8.30}$$

Equations (8.28) and (8.30) agree with (8.9) and (8.7), respectively. In strongly correlated electron systems the vertex correction reduces the Drude weight.

In Galilean invariant systems, $\boldsymbol{j}_{\boldsymbol{k}}{}^* = \boldsymbol{v}_{\boldsymbol{k}} = \boldsymbol{k}/m$. When $\partial v_{\boldsymbol{k}\mu}/\partial k_\nu = \delta_{\mu\nu}/m$, from (8.23) we obtain

$$K_{\mu\nu}{}^k = \frac{\delta_{\mu\nu}}{m}\frac{i}{\Omega}\int \frac{d^4k'}{(2\pi)^4}G(k') = -\frac{n}{m}\delta_{\mu\nu}. \tag{8.31}$$

In this case the sum rule

$$\int_{-\infty}^{\infty} \operatorname{Re}\sigma_{\mu\nu}(\omega)d\omega = \frac{\pi e^2}{\Omega}\langle\hat{\varepsilon}''_{\mu\nu}\rangle = \frac{\pi n e^2}{m}\delta_{\mu\nu} \tag{8.32}$$

is satisfied with the Drude weight, hence $K_{\mu\nu}{}^\omega = 0$.

In the general case, the lost weight in the coherent part is transferred to the incoherent part:

$$\left(\frac{n}{m}\right)_{\mu\nu} - \left(\frac{n}{m'}\right)_{\mu\nu} = -\frac{1}{\pi}\int_{-\infty}^{\infty} \frac{\operatorname{Im} K_{\mu\nu}(\omega')}{\omega'} d\omega' \geq 0, \tag{8.33}$$

where we have defined the effective mass in the total weight:

$$\left(\frac{n}{m}\right)_{\mu\nu} = \frac{1}{\Omega}\langle \hat{\varepsilon}''_{\mu\nu} \rangle.$$

8.3 Hall conductivity

A general expression for the Hall conductivity was derived from the Kubo formula by Kohno and Yamada [6]:

$$\sigma_{\mu\nu} = \frac{e^3}{c} H \sum_{\boldsymbol{k}} \left[\Lambda_{\boldsymbol{k}\mu} \frac{\partial \Lambda_{\boldsymbol{k}\nu}}{\partial k_\nu} - \frac{\partial \Lambda_{\boldsymbol{k}\mu}}{\partial k_\nu} \Lambda_{\boldsymbol{k}\nu} \right] v_{\boldsymbol{k}\mu}{}^* \frac{1}{(2\gamma_{\boldsymbol{k}}{}^*)^2} \left(-\frac{\partial f}{\partial \varepsilon} \right)_{\varepsilon = \varepsilon_{\boldsymbol{k}}{}^*}, \tag{8.34}$$

where the current $\Lambda_{\boldsymbol{k}\nu}$ is given by (8.4) and includes the vertex correction T_{22}. Equation (8.34) is exact as far as the most divergent terms with respect to $(\gamma_{\boldsymbol{k}}{}^*)^{-1}$ are concerned. The damping rate $\gamma_{\boldsymbol{k}}{}^*$ is the smallest scale in a Fermi liquid. When a magnetic field is applied along the z-axis, we can put $\mu = x$ and $\nu = y$.

On the other hand, the Boltzmann equation gives

$$\sigma_{\mu\nu} = \frac{e^3}{c} H\tau^2 \sum_{\boldsymbol{k}} \left[v_{\boldsymbol{k}\mu}{}^* \frac{\partial v_{\boldsymbol{k}\nu}{}^*}{\partial k_\nu} - \frac{\partial v_{\boldsymbol{k}\mu}{}^*}{\partial k_\nu} v_{\boldsymbol{k}\nu}{}^* \right] v_{\boldsymbol{k}\mu}{}^* \left(-\frac{\partial f}{\partial \varepsilon} \right)_{\varepsilon = \varepsilon_{\boldsymbol{k}}{}^*}, \tag{8.35}$$

where τ is the relaxation time. If $\boldsymbol{\Lambda}_{\boldsymbol{k}}$ is proportional to $\boldsymbol{v}_{\boldsymbol{k}}{}^*$, i.e.,

$$\Lambda_{\boldsymbol{k}\mu} = v_{\boldsymbol{k}\mu}{}^* \alpha(\boldsymbol{k}), \tag{8.36}$$

we can define a transport relaxation time τ_{tr} by

$$\tau_{\mathrm{tr}} = \frac{\alpha(\boldsymbol{k})}{2\gamma_{\boldsymbol{k}}{}^*}. \tag{8.37}$$

If (8.36) and (8.37) hold, (8.34) reduces to (8.35).

The Hall coefficient is given by

$$R = \frac{-\sigma_{yx}}{\sigma_{xx}\sigma_{yy} - \sigma_{xy}\sigma_{yx}} \frac{1}{H}. \tag{8.38}$$

Here we assume the weak field limit, where $\omega_c\tau \ll 1$, ω_c being the cyclotron frequency and τ the electron mean free time. When we retain only the terms up to the first order in $\boldsymbol{H}$, in the prefactor of (8.38), we obtain

$$R = \frac{{\sigma_{xy}}^{(1)}}{{\sigma_{xx}}^{(0)}{\sigma_{yy}}^{(0)}}\frac{1}{H}. \tag{8.39}$$

Here we write the term $\boldsymbol{H}^m$ as ${\sigma_{\mu\nu}}^{(m)}$.

Thus the Hall coefficient R takes a constant value independent of H and temperature T in the ordinary Fermi liquid. It should be noted that the current $\Lambda_{k\nu}$ including the vertex correction gives finite contributions through (8.34). In some systems such as the cuprate superconductors, antiferromagnetic spin fluctuations are strong. In these cases the resistivity shows the T-linear term in the temperature dependence. Corresponding to this behaviour the Hall coefficient depends on temperature through the momentum derivative of $\Lambda_{k\nu}$ in (8.34). In Chapter 9, we discuss the Hall coefficient in high-T_c superconductors. In this case the vertex correction plays an essential role in explaining the temperature dependence of the Hall coefficient.

8.4 Cyclotron resonance

In strongly correlated electron systems, Kohn's theorem [7] concerning the cyclotron resonance is well known and says that the frequency of the cyclotron resonance is not affected by the electron–electron interaction, when the mutual electron interactions conserve the total momentum, as seen in free electron gas. However, the momentum conservation is violated by the Umklapp process in the usual lattice systems. Now let us consider the effect of the electron correlation on the cyclotron resonance in lattice systems [8].

For the cyclotron resonance we consider the following correlation function:

$$K_+{}^R(\omega) = -i\int_{-\infty}^{\infty}\theta(t)\langle[j_-(t), j_+(0)]\rangle e^{i\omega t}dt, \tag{8.40}$$

where $j_\pm \equiv j_x \pm \mathrm{i} j_y$ raises the Landau quantum number by unity. Equation (8.40) can be rewritten as

$$K_+{}^R(\omega) = \sum_n \frac{2\omega_{n0}|\langle\psi_n|j_+|\psi_0\rangle|^2}{(\omega + i\delta)^2 - {\omega_{n0}}^2}, \tag{8.41}$$

where δ is a positive infinitesimal. The state $|\psi_n\rangle$ is an exact eigenstate of the many-electron system; $|\psi_0\rangle$ is the ground state. $\omega_{n0} = E_n - E_0$ ($\hbar = 1$), E_n being the energy of an eigenstate denoted as n.

Now, writing $p = (\boldsymbol{p}, \omega)$ and $k = (\boldsymbol{q}, \omega)$, we put

$$\left.\begin{aligned} g_1(p;k) &= G^R(p+k)G^R(p), \\ g_2(p;k) &= G^R(p+k)G^A(p), \\ g_3(p;k) &= G^A(p+k)G^A(p). \end{aligned}\right\} \tag{8.42}$$

The singular term g_2 is written in the Fermi liquid theory as

$$g_2(p;k) = \frac{2\pi i z_{\boldsymbol{p}}{}^2 \delta(\varepsilon - \varepsilon_{\boldsymbol{p}}{}^*)}{\omega - \boldsymbol{v}^* \cdot \boldsymbol{q} + 2i\gamma_{\boldsymbol{p}}{}^*}, \tag{8.43}$$

where $\boldsymbol{v}^* = \partial \varepsilon_{\boldsymbol{p}}{}^*/\partial \boldsymbol{p}$ and $\gamma_{\boldsymbol{p}}{}^*$ is the damping rate of the quasi-particle. The irreducible vertex correction is written as

$$\begin{aligned} T^{(0)}(p, p'; k) &\equiv T_{22}{}^{(0)}(p, p'; k) \\ &= \left(\tanh\frac{\varepsilon' + \omega}{2T} - \tanh\frac{\varepsilon'}{2T}\right)\Gamma^k(p, p'; k) + iT'(p, p'; k), \end{aligned} \tag{8.44}$$

where $\Gamma^k(p, p')$ is a real quantity and independent of ε, ε' and k for small values of ε and ε'. Here we use the relation

$$\tanh[(\varepsilon + \omega)/2T] - \tanh(\varepsilon/2T) \simeq 2\omega\delta(\varepsilon), \tag{8.45}$$

and the imaginary part $iT'(p, p'; k)$ is ignored in the discussion on the cyclotron resonance (see Section 8.2).

Thus the total vertex correction connected to g_2 sections on both sides is written as

$$T(p, p'; k) \cong \left(\tanh\frac{\varepsilon' + \omega}{2T} - \tanh\frac{\varepsilon'}{2T}\right)\Gamma(p, p'; k), \tag{8.46}$$

where $\Gamma_{\sigma\sigma'}(p, p'; k)$ satisfies the following equation:

$$\begin{aligned} \Gamma_{\sigma\sigma'}(\boldsymbol{p}, \boldsymbol{p}'; k) &= \Gamma_{\sigma\sigma'}{}^k(\boldsymbol{p}, \boldsymbol{p}') + \sum_{\boldsymbol{p}'', \sigma''} \Gamma_{\sigma\sigma''}{}^k(\boldsymbol{p}, \boldsymbol{p}'') \\ &\times \frac{\omega}{\omega + i\delta - \boldsymbol{v}''^* \cdot \boldsymbol{q}} z_{\boldsymbol{p}''}{}^2 \delta(\mu - \varepsilon_{\boldsymbol{p}''}{}^*)\Gamma_{\sigma''\sigma'}(\boldsymbol{p}'', \boldsymbol{p}'; k). \end{aligned} \tag{8.47}$$

Using the relation between Γ^k and Γ^ω, we obtain

$$\begin{aligned} \Gamma_{\sigma\sigma'}(\boldsymbol{p}, \boldsymbol{p}'; k) &= \Gamma_{\sigma\sigma'}{}^\omega(\boldsymbol{p}, \boldsymbol{p}') + \sum_{\boldsymbol{p}'', \sigma''} \Gamma_{\sigma\sigma''}{}^\omega(\boldsymbol{p}, \boldsymbol{p}'') \frac{\boldsymbol{v}''^* \cdot \boldsymbol{q}}{\omega + i\delta - \boldsymbol{v}''^* \cdot \boldsymbol{q}} \\ &\times z_{\boldsymbol{p}''}{}^2 \delta(\mu - \varepsilon_{\boldsymbol{p}''}{}^*)\Gamma_{\sigma''\sigma'}(\boldsymbol{p}'', \boldsymbol{p}'; k). \end{aligned} \tag{8.48}$$

Now we define the following functions:

$$g = \omega z_{\boldsymbol{p}}{}^2 \delta(\mu - \varepsilon_{\boldsymbol{p}}{}^*)/(\omega + i\delta - \boldsymbol{v}^* \cdot \boldsymbol{q}), \tag{8.49}$$

$$\tilde{g} = (\boldsymbol{v}^* \cdot \boldsymbol{q}/\omega) g. \tag{8.50}$$

Equations (8.47) and (8.48) can be written as

$$\Gamma = \Gamma^k + \Gamma^k g \Gamma = \Gamma^k + \Gamma g \Gamma^k, \tag{8.51}$$

$$\Gamma = \Gamma^\omega + \Gamma^\omega \tilde{g} \Gamma = \Gamma^\omega + \Gamma \tilde{g} \Gamma^\omega, \tag{8.52}$$

$$\Gamma^k = \Gamma^\omega - \Gamma^\omega (g - \tilde{g}) \Gamma^k. \tag{8.53}$$

Now we consider a situation in which an oscillating electric field (microwave) is applied perpendicularly to the uniform magnetic field. In considering the cyclotron resonance, the Landau quantization of quasi-particle states should be taken into account. In this case we replace the g_2 section as

$$g_2(p; n \to n+1, \omega) \cong \frac{2\pi i z_p{}^2 \delta(\varepsilon - \varepsilon_p{}^*)}{\omega + i\delta - \omega_c{}^*(p_z)}, \tag{8.54}$$

where $\omega_c{}^*(p_z)$ is the cyclotron frequency $eH/m^*(p_z)c$. The cyclotron mass of quasi-particles is defined by

$$m^*(p_z) = \frac{1}{2H} \frac{\partial S(\varepsilon, p_z)}{\partial \varepsilon}\bigg|_{\varepsilon=\mu}, \tag{8.55}$$

$S(\varepsilon, p_z)$ being the area of the cross-section of the constant-energy surface by the plane p_z = constant. The notation $n \to n+1$ means a transition to the Landau level with higher quantum number by unity. That is, the particle–hole pair has a different Landau quantum number by unity. This transition is induced by an external oscillating electric field, which couples to the quasi-particle current density.

If we neglect the magnetic field dependencies except those existing in the singularity of the g_2 section, the vertex functions satisfy the following equation:

$$\begin{aligned} \Gamma_{\sigma\sigma'}(\boldsymbol{p}, \boldsymbol{p}'; n \to n+1, \omega) &= \Gamma_{\sigma\sigma'}{}^{\omega}(\boldsymbol{p}, \boldsymbol{p}') \\ &+ \sum_{\boldsymbol{p}'', \sigma''} \Gamma_{\sigma\sigma''}{}^{\omega}(\boldsymbol{p}, \boldsymbol{p}'') \frac{\omega_c{}^*(p''{}_z)}{\omega + i\delta - \omega_c{}^*(p''{}_z)} z_{p''}{}^2 \delta(\mu - \varepsilon_{p''}{}^*) \\ &\times \Gamma_{\sigma''\sigma'}(\boldsymbol{p}'', \boldsymbol{p}'; n \to n+1, \omega). \end{aligned} \tag{8.56}$$

If interactions between electrons are unimportant, the orbits at which the cyclotron frequency $\omega_c{}^*(p_z)$ has its extreme values with respect to p_z are likely to determine the resonance frequencies. The resonance condition for interacting electron systems is expressed as an integral equation which is homogeneous with respect to a vector $\boldsymbol{\Lambda_p}$:

$$\begin{aligned} \sum_{p} \frac{z_p{}^2 \delta(\mu - \varepsilon_p)}{\omega - \omega_c{}^*(p_z)} Q(p) \cdot \bigg\{ &(\omega - \omega_c{}^*(p_z)) \boldsymbol{\Lambda_p} \\ &- \sum_{p', \sigma'} \Gamma_{\sigma\sigma'}{}^{\omega}(\boldsymbol{p}, \boldsymbol{p}') z_{p'}{}^2 \delta(\mu - \varepsilon_{p'}) \omega_c{}^*(p'_z) \boldsymbol{\Lambda_{p'}} \bigg\} = 0, \end{aligned} \tag{8.57}$$

where $\boldsymbol{Q}(\boldsymbol{p})$ is the renormalized vertex coupled to the external electric field and connected to a g_2 section on the other side. Since we have assumed that the electric field is in the plane perpendicular to the uniform magnetic field, $\boldsymbol{Q}(\boldsymbol{p})$ is a two-dimensional vector on this plane, and $\boldsymbol{\Lambda}_{\boldsymbol{p}}$ is also some two-dimensional nonzero vector. In an isotropic electron gas $\omega_c{}^*(p_z)$ is independent of p_z, i.e., $\omega_c{}^* = eH/m^*c$ with an isotropic effective mass m^*, and the resonance condition reduces to the following much simpler equation:

$$\omega = \left\{ 1 + \sum_{p',\sigma'} \Gamma_{\sigma\sigma'}{}^{\omega}(\boldsymbol{p}, \boldsymbol{p}') z_{p_{\mathrm{F}}}{}^2 \delta(\mu - \varepsilon_{p'})(\boldsymbol{p} \cdot \boldsymbol{p}') \right\} \omega_c{}^*. \tag{8.58}$$

The same equation as (8.58) with the corresponding phenomenological interaction function of the quasi-particles can be derived from the quasi-classical Boltzmann transport equation on the basis of the phenomenological Landau Fermi liquid theory:

$$\omega = \left(1 + \frac{1}{3} F_1{}^s \right) \omega_c{}^*, \tag{8.59}$$

where $F_1{}^s$ is the first harmonic of the spin-symmetric sum of the function $z_{p_{\mathrm{F}}}{}^2 \Gamma_{\sigma\sigma'}{}^{\omega}(\hat{\boldsymbol{p}}, \hat{\boldsymbol{p}}') N(\mu)$, $N(\mu)$ being the density of states of quasi-particles at the Fermi energy.

In a Galilean invariant isotropic system the effective mass of the quasi-particles is given by

$$\frac{m^*}{m} = 1 + \frac{F_1{}^s}{3}. \tag{8.60}$$

The term $F_1{}^s/3$ comes from the back-flow. As a result the resonance frequency is given by $\omega_c = eH/mc$ (m is the bare electron mass) and contains no effect of electron–electron interactions. Thus Kohn's theorem in Galilean invariant systems has been proved on the basis of the Fermi liquid theory. For Bloch electrons on a lattice, on the other hand, Kohn's theorem no longer holds owing to the Umklapp processes. In both cases, it is important to treat the vertex correction properly, and in Galilean invariant systems the momentum conservation is recovered by the back-flow term. Although we have given a general formula to determine the cyclotron resonance frequency, it is difficult to predict the value of the resonance frequency in real systems.

In order to discuss the effect of lattices on the cyclotron resonance frequency, Kanki and Yamada have made some calculations for the Hubbard model on the square lattice [8]. They have found that near the half-filling case in this model, Umklapp processes make the back-flow a backward flow in contrast to a forward flow in the Galilean invariant systems. In this case a kind of effective mass in the electron transport is enhanced for a part of the Fermi surface even from the thermodynamic effective mass of the quasi-particles.

In general we expect that Umklapp processes tend to reduce the cyclotron resonance frequency from that of non-interacting electrons, in contrast to Kohn's theorem [7].

8.5 Magnetoresistance

Now we apply the Fermi liquid theory to the calculation of the magnetoresistance under the magnetic field along the z-axis.

When we discuss the transport phenomena, we use the Boltzmann equation in which we usually assume a constant relaxation time. The relaxation time is approximated by the lifetime of quasi-particles:

$$1/\tau_{k} = -2\,\mathrm{Im}\,\Sigma_{k}{}^{\mathrm{R}}(0). \tag{8.61}$$

By the relaxation time approximation (RTA), the magnetoconductivity $\Delta\sigma_{xx}{}^{\mathrm{RTA}}$ is given by

$$\Delta\sigma_{xx}{}^{\mathrm{RTA}} = -H^2\frac{e^4}{4}\int_{\mathrm{FS}}\frac{dS_{k}}{|\boldsymbol{v}_{k}|}\{(\boldsymbol{v}_{k}\times\boldsymbol{e}_z)\cdot\boldsymbol{\nabla}(2\tau_{k}v_{kx})\}^2\,(2\tau_{k}), \tag{8.62}$$

where $\boldsymbol{e}_z$ is the unit vector along the magnetic field. In the above equation we have a term which arises from the $\boldsymbol{k}$-derivative of τ_{k}, in contrast to the expression for σ_{xy} in (8.35).

Based on the Kubo formula, we can reformulate the expression for $\Delta\sigma_{xx}$ so that it does not violate the momentum conservation and gives the τ^3 term correctly. The derivation has been done by Kontani [9], although it is not easy.

The final result obtained by Kontani is the following:

$$\Delta\sigma_{\mu\nu} = -H^2\cdot\frac{e^4}{4}\oint_{\mathrm{FS}}\frac{dS_{k}}{|\boldsymbol{v}_{k}|}\frac{1}{\gamma_{k}}d_{\mu}(\boldsymbol{k})D_{\nu}(\boldsymbol{k}), \tag{8.63}$$

$$d_{\mu}(\boldsymbol{k}) = \left(v_x(\boldsymbol{k})\frac{\partial}{\partial k_y} - v_y(\boldsymbol{k})\frac{\partial}{\partial k_x}\right)\cdot\left(\frac{\Lambda_{\mu}(\boldsymbol{k})}{\gamma_{k}}\right), \tag{8.64}$$

$$\begin{aligned} D_{\mu}(\boldsymbol{k}) &= \sum_{k'}\int\frac{d\varepsilon'}{4\pi i}T_{22}(\boldsymbol{k}0|\boldsymbol{k}'\varepsilon')|G_{k'}(\varepsilon')|^2 d_{\mu}(\boldsymbol{k}') + d_{\mu}(\boldsymbol{k}) \\ &= \oint_{\mathrm{FS}}\frac{dS_{k'}}{|\boldsymbol{v}_{k'}|}\int\frac{d\varepsilon'}{4i}T_{22}(\boldsymbol{k}0|\boldsymbol{k}'0)\frac{1}{\gamma_{k'}}d_{\mu}(\boldsymbol{k}') + d_{\mu}(\boldsymbol{k}), \end{aligned} \tag{8.65}$$

where $\oint_{\mathrm{FS}}dS_{k}$ represents the two-dimensional integration on the Fermi surface. Equation (8.63) is of order $\gamma_{k}{}^{-3}\propto\tau_{k}{}^3$, as expected. Both $D_{\mu}(\boldsymbol{k})$ and $d_{\mu}(\boldsymbol{k})$ are real, since T_{22} is purely imaginary for $\varepsilon = 0$. If we neglect the vertex correction arising from T_{22} in (8.65), we obtain the result (8.62) given by the relaxation time approximation.

If the fourfold symmetry around the z-axis ($\boldsymbol{e}_z \parallel \boldsymbol{B}$) is assumed, we obtain

$$\Delta\sigma_{xx} = -H^2 \cdot \frac{e^4}{4} \oint_{\mathrm{FS}} \frac{dS_{\boldsymbol{k}}}{|\boldsymbol{v}_{\boldsymbol{k}}|\gamma_{\boldsymbol{k}}} \boldsymbol{d}_\perp(k) \cdot \boldsymbol{D}_\perp(k)/2, \tag{8.66}$$

where $\boldsymbol{d}_\perp = (d_x, d_y)$ and $\boldsymbol{D}_\perp = (D_x, D_y)$.

The Nernst coefficient and magnetoresistance in high-T_c superconductors are well explained by Kontani [10].

8.6 Concluding remarks

In this chapter we have shown the important roles of vertex correction. In particular, the Umklapp scattering originating from a crystal potential plays an important role in strongly correlated electron systems. For example, in Section 8.2 we have shown that the Drude weight is much reduced by the quasi-particle interaction. Relating to this fact, the London constant equal to the inverse square of the magnetic penetration length λ at $T = 0$ is given by [11]:

$$\Lambda_{\mu\nu} = 1/\lambda_{\mu\nu}{}^2 = e^2 \int_{\mathrm{FS}} \frac{dS_{\boldsymbol{k}}}{4\pi^3|v^*(\boldsymbol{k})|} v_\mu{}^* j_\nu{}^*(\boldsymbol{k}), \tag{8.67}$$

where j_ν^* is given by (8.9) and reduced by the Umklapp scattering. Thus, the London constant Λ becomes small in the strongly correlated systems. This fact is essential to the understanding of the small value of Λ in the under-doped cuprates. The small value of Λ in cuprates can be attributed to neither small superfluid density n_s nor large effective electron mass m^*, although the so-called Uemura plot is often explained as small n_s.

References

[1] G. M. Éliashberg, *Sov. Phys. JETP* **14** (1962) 886.
[2] R. Kubo, *J. Phys. Soc. Jpn.* **12** (1957) 570.
[3] K. Yamada and K. Yosida, *Prog. Theor. Phys.* **76** (1986) 621.
[4] T. Okabe, *J. Phys. Soc. Jpn.* **67** (1998) 2792.
[5] T. Okabe, *J. Phys. Soc. Jpn.* **67** (1998) 4178.
[6] H. Kohno and K. Yamada, *Prog. Theor. Phys.* **80** (1988) 623.
[7] W. Kohn, *Phys. Rev.* **123** (1961) 1242.
[8] K. Kanki and K. Yamada, *J. Phys. Soc. Jpn.* **66** (1997) 1103.
[9] H. Kontani, *Phys. Rev. B* **64** (2001) 054413; *J. Phys. Soc. Jpn.* **70** (2001) 1873.
[10] H. Kontani, *Phys. Rev. Lett.* **89** (2002) 237003.
[11] T. Jujo, *J. Phys. Soc. Jpn.* **70** (2001) 1349; **71** (2002) 888.

9

Superconductivity in strongly correlated electron systems

In various strongly correlated electron systems, anisotropic superconductivities originate from the coulomb repulsion. Here we explain the mechanism on the basis of the Fermi liquid theory.

9.1 Cuprate high-temperature superconductors

9.1.1 Model Hamiltonian

In the cuprate systems, in which the superconducting critical temperature T_c reaches up to 150 K, the main part in the essential role of realizing the superconductivity is played by the copper-oxide plane, CuO_2. This plane changes from an insulating state into a metallic state by the doping of holes or electrons. The crystal structure of the CuO_2 plane and the phase diagram are shown in Figs. 9.1 and 9.2, respectively. To describe the CuO_2 plane the d–p Hamiltonian is used:

$$\mathcal{H} = \varepsilon_d \sum_{i,\sigma} n_{i\sigma}{}^d + \sum_{k,\sigma} \varepsilon_p(\boldsymbol{k}) p_{k\sigma}{}^\dagger p_{k\sigma} + \sum_{k\sigma} (v_k d_{k\sigma}{}^\dagger p_{k\sigma} + v_k{}^* p_{k\sigma}{}^\dagger d_{k\sigma}) + U \sum_i n_{i\uparrow} n_{i\downarrow}, \tag{9.1}$$

$$v_k{}^2 = 2t^2(2 - \cos k_x a - \cos k_y a). \tag{9.2}$$

Here d-electrons with energy ε_d and p-electrons $\varepsilon_p(\boldsymbol{k})$ hybridize through the transfer matrix v_k between the d- and p-orbitals situated at neighbouring Cu and O sites, respectively. The distance between Cu and O sites is denoted as $a/2$. These CuO_2 planes are connected by the doping layers stacked along the c-axis. In the d–p Hamiltonian only the coulomb repulsion U is the many-body interaction. As a result the magnetism and the superconductivity arise from the coulomb repulsion U. The Dyson–Gor'kov equation for the d–p Hamiltonian is written only by the

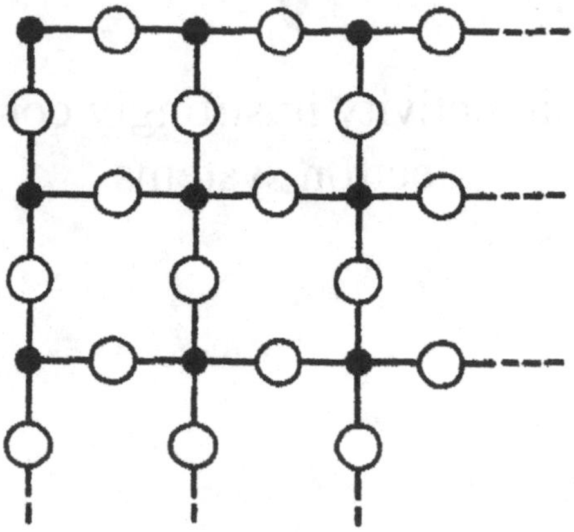

Fig. 9.1 CuO_2 plane in the cuprate superconductors. Open circles and closed circles represent O atoms and Cu atoms, respectively.

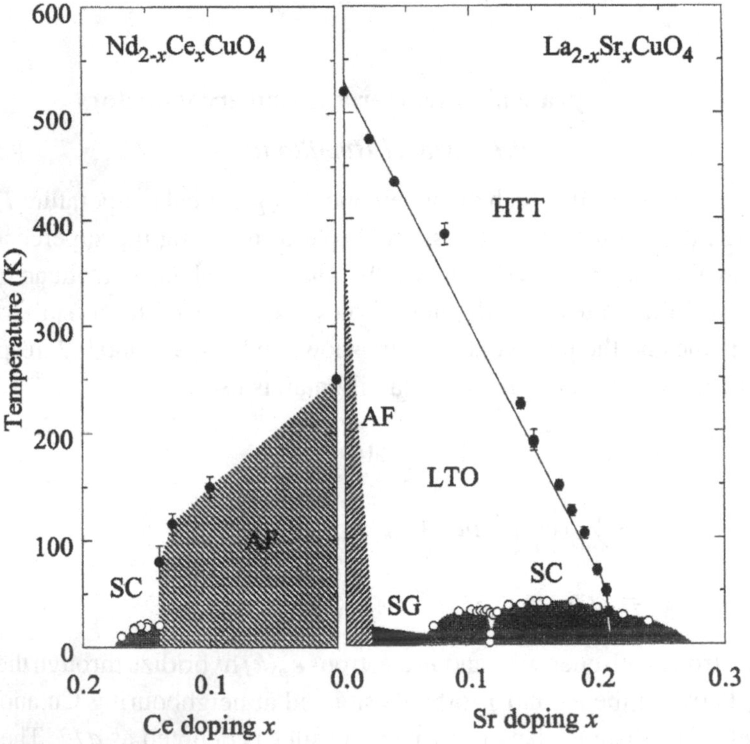

Fig. 9.2 Phase diagram of the cuprate superconductor as functions of temperature T and doping x. $La_{2-x}Sr_xCuO_4$ and $Nd_{2-x}Ce_xCuO_4$ are hole doped and electron doped systems, respectively. This figure was provided by T. Uefuji.

d-electron Green's functions, and the p-electron Green's functions are included implicitly in the d-electron Green's functions.

When we consider the superconducting state, we can use the Hubbard Hamiltonian possessing nearest and next nearest neighbour transfer integrals t and t'. The transfer integral t' effectively represents the transfer integral between the p-orbitals situated at the neighbouring oxide sites:

$$\mathcal{H} = \sum_{k,\sigma} \varepsilon_k a_{k\sigma}{}^\dagger a_{k\sigma} + \frac{U}{N} \sum_{k,k'q} a_{k-q\uparrow}{}^\dagger a_{k'+q\downarrow}{}^\dagger a_{k'\downarrow} a_{k\uparrow}, \tag{9.3}$$

$$\varepsilon_k = -2t(\cos k_x a + \cos k_y a) + 4t' \cos k_x a \cos k_y a - \mu. \tag{9.4}$$

Here we have included the chemical potential μ. Hereafter we put $2t = 1$ and $t' = 0.1 \sim 0.4t$.

As is shown in Fig. 9.2, the critical temperature T_c possesses a peak around $\delta \simeq 0.15$. The region around the peak is called the optimally doped region. The region doped more than the optimally doped region is called the over-doped region, and the region less than that is called the under-doped region.

9.1.2 Resistivity

We discuss the normal state above T_c on the basis of the Fermi liquid theory. As shown in Fig. 9.3, the electrical resistivity shows a temperature dependence

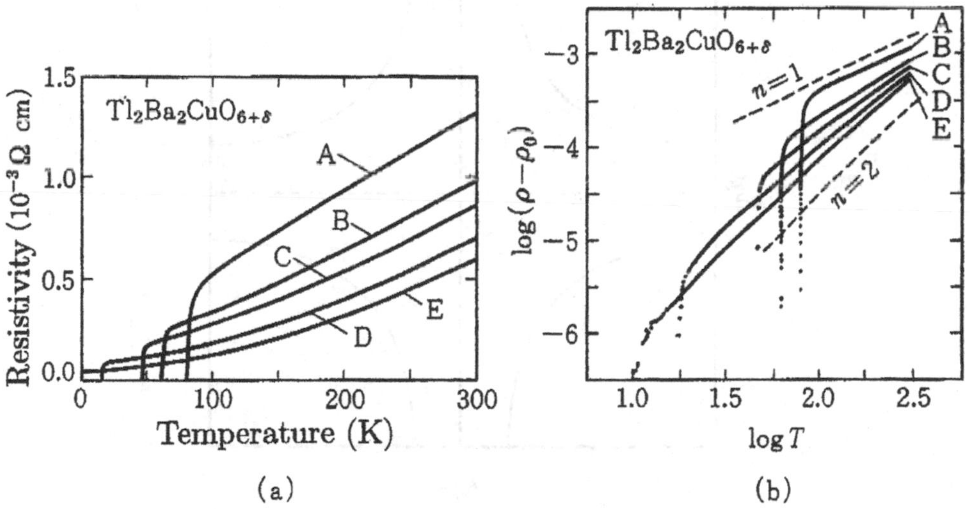

Fig. 9.3 Resistivity of the cuprate superconductor $Tl_2Ba_2CuO_{6+\delta}$. The hole doping increases from A to E. The temperature dependence of the resistivity changes from a T^2 term to a T-linear term with decreasing doping [1].

from T^2 to T-linear behaviour with decreasing doping [1]. This behaviour can be explained by the electron–electron scattering due to the antiferromagnetic spin fluctuations [2].

The dynamical susceptibility in high-T_c cuprates is given approximately by

$$\chi(\boldsymbol{q}, \omega) = \chi_Q/(1 + \xi^2(\boldsymbol{q} - \boldsymbol{Q})^2 - i\omega/\omega_s), \qquad (9.5)$$
$$\boldsymbol{Q} = (\pi, \pi),$$

where ξ and ω_s are correlation length and frequency of antiferromagnetic spin fluctuation, respectively; χ_Q is the staggered susceptibility at $\omega = 0$. Here the following relations are realized in high-T_c cuprates: $\chi_Q \propto \xi^2$, $\omega_s \propto \xi^{-2}$ and $\xi^2 \propto T^{-1}$. We adopt the magnetic interaction, which is given by

$$\begin{aligned}\Gamma(\boldsymbol{p}, \varepsilon_n; \boldsymbol{p}', \varepsilon_{n'}; \boldsymbol{q}, \omega_m) &= \Gamma(\boldsymbol{q}, \omega_m) \\ &= g^2\chi(\boldsymbol{q}, \omega_m) \\ &= g^2\chi_Q/(1 + \xi^2(\boldsymbol{q} - \boldsymbol{Q})^2 + |\omega_m|/\omega_s). \qquad (9.6)\end{aligned}$$

Here g is the coupling energy. The Fermi surface is given by $\varepsilon_k = 0$, and is shown in Fig. 9.4. The electron energy ε_k is assumed to be given by (9.4) with $t = 0.5$ eV and

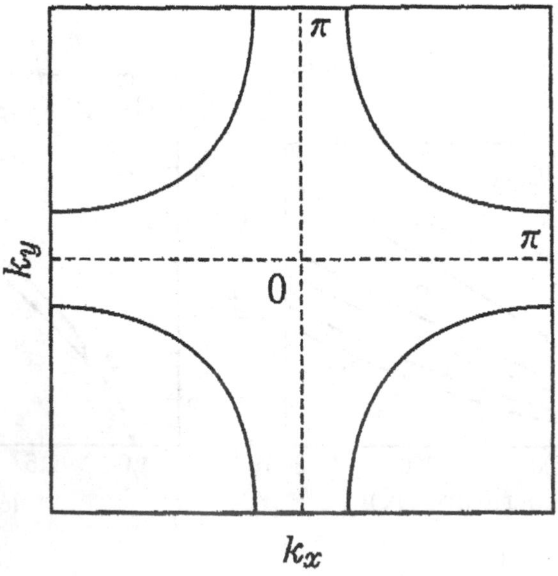

Fig. 9.4 The Fermi surface of cuprates is shown in the first Brillouin zone $(\pm\pi, \pm\pi)$. The parts near $(\pi, 0)$ and near $(\pi/2, \pi/2)$ are called hot spots and cold spots, respectively.

$t' = 0.45t$ to reproduce the shape of the hole-like Fermi surface. The self-energy is calculated as

$$\Sigma(\boldsymbol{k}, \varepsilon_n) = g^2 T \sum_{n',\boldsymbol{q}} G(\boldsymbol{k}-\boldsymbol{q}, \varepsilon_{n'})\chi(\boldsymbol{q}, \varepsilon_n - \varepsilon_{n'}). \tag{9.7}$$

By the analytic continuation, we obtain

$$\operatorname{Im}\Sigma^{\mathrm{R}}(\boldsymbol{k}, \varepsilon) = -g^2 \sum_{\boldsymbol{q}} \int_{-\infty}^{\infty} \frac{d\varepsilon'}{2\pi} \left[\coth\left(\frac{\varepsilon'-\varepsilon}{2T}\right) - \tanh\left(\frac{\varepsilon'}{2T}\right)\right] \operatorname{Im} G^{\mathrm{R}}(\boldsymbol{k}-\boldsymbol{q}, \varepsilon') \times \operatorname{Im}\chi^{\mathrm{R}}(\boldsymbol{q}, \varepsilon - \varepsilon'). \tag{9.8}$$

Here we assume that the temperature T, the frequency ε and the energy scale of spin fluctuation ω_{s} are all much smaller than the bandwidth D; that is, $T, \varepsilon, \omega_{\mathrm{s}} \ll D$. Under this condition we obtain

$$\operatorname{Im}\Sigma^{\mathrm{R}}(\boldsymbol{k}, \varepsilon) = -\frac{g^2}{2}\chi_Q \omega_{\mathrm{s}} \int_{\mathrm{FS}} \frac{d\boldsymbol{q}}{(2\pi)^2 |v_{\boldsymbol{k}-\boldsymbol{q}}|} \left\{ \log\left[1 + \frac{\varepsilon^2}{\omega_{\boldsymbol{q}}{}^2}\right] + \frac{4T}{\omega_{\boldsymbol{q}}} \tan^{-1}\left[\frac{\pi^2}{4}\frac{\pi}{\omega_{\boldsymbol{q}}}\right]\right\}, \tag{9.9}$$

$$\omega_{\boldsymbol{q}} = \omega_{\mathrm{s}}(1 + \xi^2(\boldsymbol{q} - \boldsymbol{Q})^2). \tag{9.10}$$

In (9.9), the integration is carried out on the Fermi surface $\varepsilon_{\boldsymbol{k}-\boldsymbol{q}} = 0$. The real part of the self-energy transforms the Fermi surface and is given at $T = 0$ as

$$\begin{aligned}\operatorname{Re}\Sigma^{\mathrm{R}}(\boldsymbol{k}, \varepsilon) &= -g^2 \sum_{\boldsymbol{q}} \int_{-\infty}^{\infty} \frac{d\varepsilon'}{2\pi} \left[\coth\left(\frac{\varepsilon'-\varepsilon}{2T}\right) \operatorname{Re} G^{\mathrm{R}}(\boldsymbol{k}-\boldsymbol{q}, \varepsilon')\operatorname{Im}\chi^{\mathrm{R}}(\boldsymbol{q}, \varepsilon-\varepsilon') \right.\\ &\quad \left. - \tanh\left(\frac{\varepsilon'}{2T}\right)\operatorname{Im} G^{\mathrm{R}}(\boldsymbol{k}-\boldsymbol{q}, \varepsilon')\operatorname{Re}\chi^{\mathrm{R}}(\boldsymbol{q}, \varepsilon-\varepsilon')\right]\\ &= -\frac{\alpha\omega_{\mathrm{s}} g^2}{2\pi}\sum_{\boldsymbol{q}} \frac{1}{\omega_{\boldsymbol{q}}{}^2 + (\varepsilon_{\boldsymbol{k}-\boldsymbol{q}} - \varepsilon)^2}\left[2(\varepsilon_{\boldsymbol{k}-\boldsymbol{q}} - \varepsilon)\log\left(\frac{\varepsilon_{\boldsymbol{k}-\boldsymbol{q}} - \varepsilon}{\omega_{\boldsymbol{q}}}\right) + \pi\omega_{\boldsymbol{q}}\theta(\varepsilon_{\boldsymbol{k}-\boldsymbol{q}})\right],\end{aligned} \tag{9.11}$$

where $\theta(x) = x/|x|$. The transformation of the Fermi surface due to the spin fluctuation is shown in Fig. 9.5. The antiferromagnetic spin fluctuation transforms the Fermi surface to approach the antiferromagnetic Brillouin zone. The conductivity $\sigma_{\mu\nu}$ is given by

$$\sigma_{\mu\nu} = e^2 \int \frac{d\boldsymbol{k}}{(2\pi)^2} \left\{ \frac{1}{2\operatorname{Im}\Sigma^{\mathrm{R}}(\boldsymbol{k}, \varepsilon_{\boldsymbol{k}})} \left(\frac{df}{d\varepsilon}\right)_{\varepsilon=\varepsilon_{\boldsymbol{k}}} \right\} v_{\boldsymbol{k}\mu}{}^* J_{\boldsymbol{k}\nu}, \tag{9.12}$$

$$J_{\boldsymbol{k}\nu} = v_{\boldsymbol{k}\nu}{}^* + \int \frac{d\boldsymbol{k}'}{(2\pi)^2} \frac{z_{\boldsymbol{k}} T_{22}(\boldsymbol{k}-\boldsymbol{k}', 0) z_{\boldsymbol{k}'}}{4i(-\operatorname{Im}\Sigma^{\mathrm{R}}(\boldsymbol{k}', \varepsilon_{\boldsymbol{k}}))} v'_{\boldsymbol{k}\nu}{}^*, \tag{9.13}$$

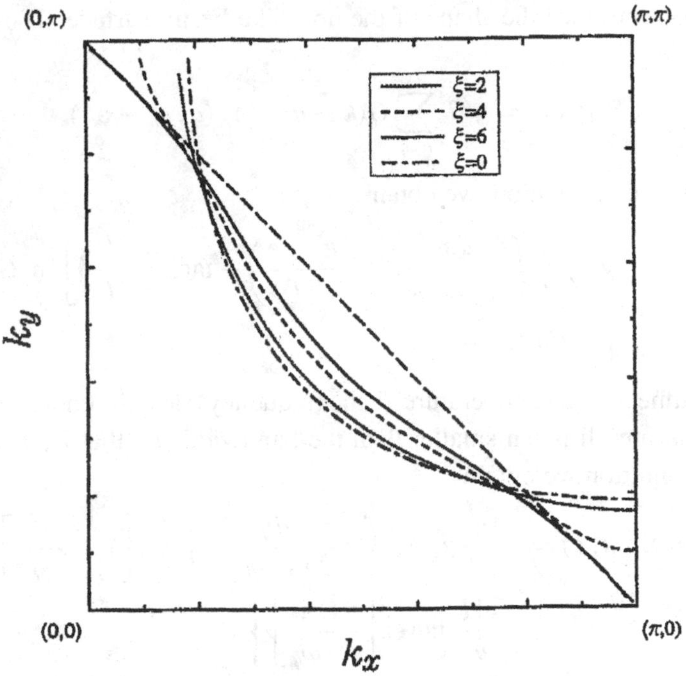

Fig. 9.5 The transformed Fermi surface for $\xi = 0$ (dot-dashed line), $\xi = 2$ (dotted line), $\xi = 4$ (dashed line) and $\xi = 6$ (solid line), respectively. $\omega_s = 5\,\text{meV}$. As ξ increases, the Fermi surface approaches the antiferromagnetic Brillouin zone.

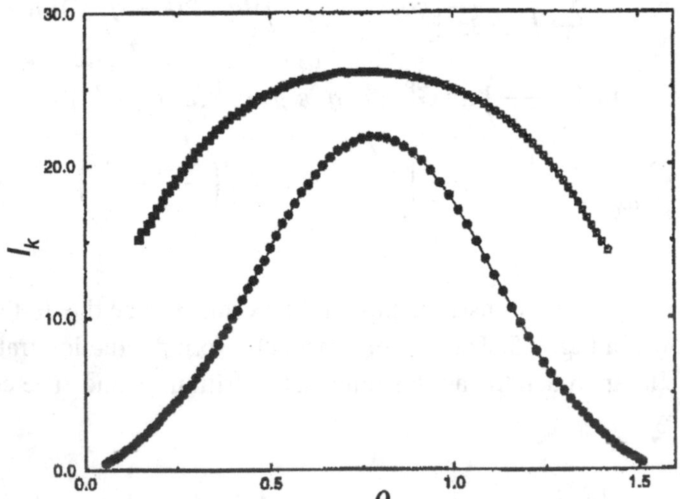

Fig. 9.6 The momentum dependence of the mean free path. The lower and upper curves correspond to under-doped and over-doped cuprates, respectively. Here $\theta = \tan^{-1}(k_y/k_x)$. The under-doped case shows strong anisotropy.

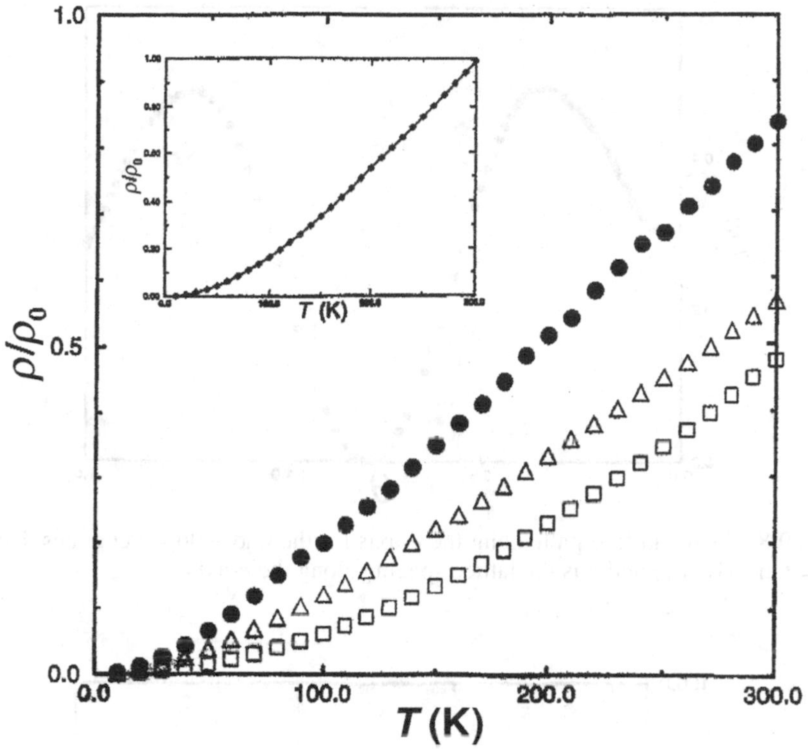

Fig. 9.7 The temperature dependence of the in-plane resistivity. Three curves from upper to lower correspond to under-doped, optimally doped and over-doped cuprates, respectively. The inset shows the in-plane resistivity for under-doped cuprate without the transformation of the Fermi surface.

where T_{22} is the vertex correction. Thus we obtain the mean free path shown in Fig. 9.6 and show the temperature dependence of the in-plane resistivity in Fig. 9.7. In the under-doped case the mean free path at the hot spots near $(\pi, 0)$ or $(0, \pi)$ becomes short, as shown in Fig. 9.6.

Now we consider the c-axis resistivity. For the cuprates the momentum dependence of $t_\perp$ along the c-axis is important in understanding the temperature dependence of ρ_c. It is obtained by the band calculation [3] as

$$t_\perp \propto \left(\frac{\cos k_x a - \cos k_y a}{2} \right)^2 . \tag{9.14}$$

In this case the mean free path is given by Fig. 9.8 and the c-axis resistivity is given by Fig. 9.9. As is seen, the c-axis resistivity becomes insulating in the under-doped region at low temperatures. This is because the c-axis transport is determined by the hot spot region owing to (9.14) and the mean free path near the hot spots becomes short in the under-doped systems.

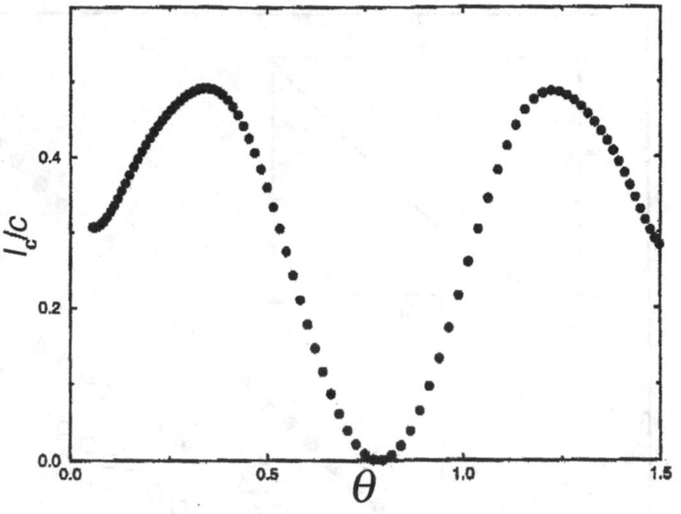

Fig. 9.8 The mean free path along the c-axis for the under-doped cuprates. Here $\theta = \tan^{-1}(k_y/k_x)$ and c is the lattice spacing along the c-axis.

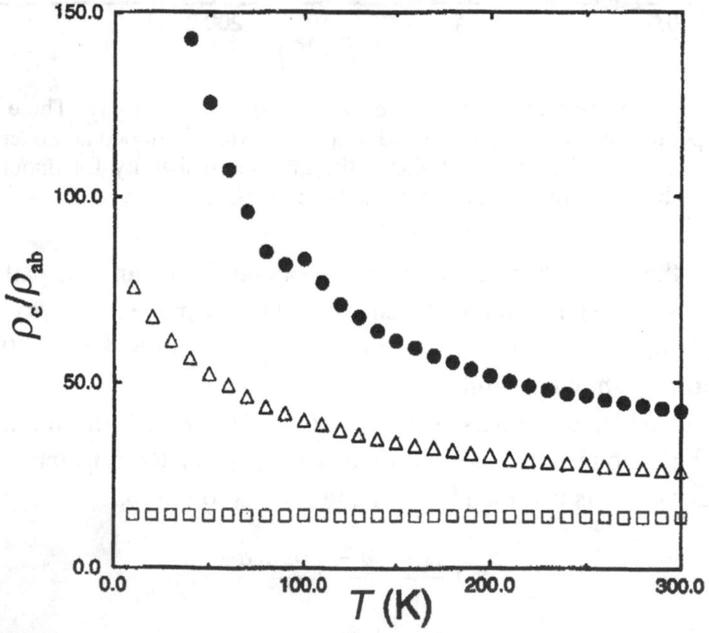

Fig. 9.9 The resistivity along the c-axis. From upper to lower the curves correspond to under-doped, optimally doped and over-doped cuprates, respectively. The discontinuity in the upper curve is due to a numerical error.

9.1.3 Hall coefficient

Now we discuss the Hall conductivity. According to the Fermi liquid theory derived by Kohno and Yamada [4], the Hall conductivity is given by

$$\sigma_{xy} = 2\frac{e^3}{c}H\int \frac{d\boldsymbol{k}}{(2\pi)^2}\left[J_{kx}\frac{\partial J_{ky}}{\partial k_y} - \frac{\partial J_{kx}}{\partial k_y}J_{ky}\right]v_x{}^*\frac{z_k{}^2}{2(\gamma_k{}^*)^2}\left(-\frac{\partial f}{\partial \varepsilon}\right)_{\varepsilon=\varepsilon_k}$$

$$= -2\frac{e^3}{2c}H\int \frac{d\boldsymbol{k}}{(2\pi)^3}\left(\boldsymbol{J}_k \times \frac{\partial \boldsymbol{J}_k}{\partial k_\parallel}\right)_z |v_\perp{}^*|\frac{z_k{}^2}{(2\gamma_k{}^*)^2}\left(-\frac{\partial f}{\partial \varepsilon}\right)_{\varepsilon=\varepsilon_k}. \quad (9.15)$$

In the second expression $k_\parallel$ is tangential to the curve of the intersection of a constant energy surface with x–y plane perpendicular to the magnetic field (i.e., $\boldsymbol{k}_\parallel$ is parallel to $\boldsymbol{e}_z \times \boldsymbol{v}_k$) and $k_\perp$ is perpendicular to a constant energy curve in the plane. Here $J_{k\mu}$, which was written as $\Lambda_{k\mu}$ in the previous chapter, includes the vertex correction and is given by

$$J_{k\mu} = v_{k\mu}{}^* + \int_{\infty}^{\infty}\frac{d\varepsilon'}{4\pi i}\int\frac{d\boldsymbol{k}'}{(2\pi)^3}z_k T_{22}{}^{(0)}(k\varepsilon;k'\varepsilon')\frac{2\pi i z_{k'}\delta(\varepsilon'-\varepsilon_{k'}{}^*)}{2i\gamma_{k'}{}^*}J_{k'\mu}. \quad (9.16)$$

If we approximate $-\partial f/\partial\varepsilon = \delta(\mu - \varepsilon_k{}^*)$, the integration over $k_\perp$ cancels with the factor $|v_\perp{}^*|$ and the Hall conductivity (9.15) is reduced to the form [5]

$$\sigma_{xy} = -2\frac{e^3}{2c}H\int_{FS}\frac{dk_z dk_\parallel}{(2\pi)^3}\left(\boldsymbol{J}\times\frac{\partial \boldsymbol{J}}{\partial k_\parallel}\right)_z\frac{z_k{}^2}{(2\gamma_k{}^*)^2}$$

$$= -\frac{e^3}{c}H\int_{FS}\frac{dk_z dk_\parallel}{(2\pi)^3}|\boldsymbol{J}|^2\frac{\partial\theta_J}{\partial k_\parallel}|\boldsymbol{v}^*|\frac{z_k{}^2}{(2\gamma_k{}^*)^2}, \quad (9.17)$$

where $\theta_J = \tan^{-1}(J_y/J_x)$. In an anisotropic system on a lattice, the vertex correction arising from the quasi-particle interaction (9.16) cannot be reduced to a momentum-independent factor. The current $\boldsymbol{J}$ is not in the direction of $\boldsymbol{v}^*$ unless the momentum is at symmetric points or at the Brillouin zone boundary. As a result the vertex correction affects the Hall coefficient significantly in strongly correlated electron systems.

Now we discuss the effect of vertex correction on the Hall coefficient using the actual model appropriate to the cuprate system [5]. We consider the nearly anti-ferromagnetic Fermi liquid model. The effective interaction $V_{\text{eff}}(\boldsymbol{q},\omega)$ is given by

$$V_{\text{eff}}(\boldsymbol{q},\omega) = g^2\chi(\boldsymbol{q},\omega), \quad (9.18)$$

where $\chi(\boldsymbol{q},\omega)$ is given by (9.5). Using (9.18) and (9.5), we approximate the imaginary part of the self-energy by the lowest-order term with respect to the exchange of spin fluctuations as

$$\operatorname{Im}\Sigma^{\mathrm{R}}(\boldsymbol{k},\varepsilon=0)=g^2\int\frac{d^2k'}{(2\pi)^2}\frac{\operatorname{Im}\chi^{\mathrm{R}}(\boldsymbol{k}-\boldsymbol{k}',\mu-\varepsilon_{\boldsymbol{k}}{}^*)}{\sinh\left(\dfrac{\varepsilon_{\boldsymbol{k}}{}^*-\mu}{T}\right)}$$
$$\simeq -g^2\int\frac{d^2q}{(2\pi)^2}\chi_Q\omega_{\mathrm{s}}\frac{(\pi T)^2}{\omega_{\boldsymbol{q}}(2\omega_{\boldsymbol{q}}+\pi T)}\rho_{\boldsymbol{k}-\boldsymbol{q}}(0). \qquad (9.19)$$

This corresponds to (9.9). Here $\omega_{\boldsymbol{q}}=\omega_{\mathrm{s}}(1+\xi^2(\boldsymbol{q}-\boldsymbol{Q})^2)$ and $\rho_{\boldsymbol{k}-\boldsymbol{q}}(0)$ is the spectral density at μ. Using (9.18), we obtain the vertex correction $T_{22}{}^{(0)}$ in (9.16) as

$$T_{22}{}^{(0)}(\boldsymbol{k}\varepsilon=0,\boldsymbol{k}'\varepsilon')=-4ig^2\frac{\operatorname{Im}\chi(\boldsymbol{k}-\boldsymbol{k}',\varepsilon')}{\sinh(\varepsilon'/T)}. \qquad (9.20)$$

Thus, the integral kernel in (9.16) possesses a sharp peak around $\boldsymbol{k}-\boldsymbol{k}'=\pm\boldsymbol{Q}$ and combines quasi-particles near the hot spots.

We show the solution of the integral equation (9.16) schematically in Fig. 9.10. We can see that vectors $\boldsymbol{J}$ and $\boldsymbol{v}$ are not parallel but take different directions. As is seen from (9.17), when the momentum variable moves along the Fermi surface, the sign of the contribution to the Hall conductivity is determined by the momentum derivative of the rotational angle θ_J. If we neglect the vertex correction, the velocity of the quasi-particle $\boldsymbol{v}^*$ is perpendicular to the Fermi surface

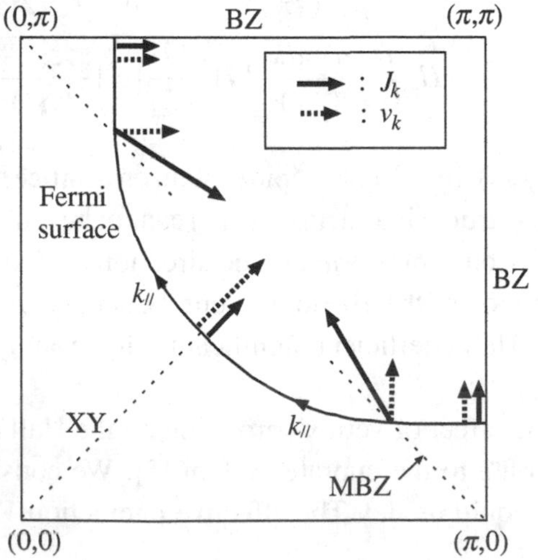

Fig. 9.10 Schematic behaviour of $\boldsymbol{J}_{\boldsymbol{k}}$ and $\boldsymbol{v}_{\boldsymbol{k}}$. $\boldsymbol{J}_{\boldsymbol{k}}$ is not perpendicular to the Fermi surface. $(d\theta_J/dk_\parallel)<0$ on the XY line and $(d\theta_J/dk_\parallel)>0$ on the Brillouin zone boundary.

and the Hall conductivity is determined by the curvature of the Fermi surface, $K = \partial\theta_v/\partial k_\parallel$. When we take the vertex correction into account in the presence of antiferromagnetic fluctuations, it gives a contribution with opposite sign to the curvature, when $\mathrm{sgn}(d\theta_J/dk_\parallel) = -\mathrm{sgn}(d\theta_v/dk_\parallel)$. Approximately with respect to the boundary given by the antiferromagnetic Brillouin zone (MBZ), the inside part gives a positive contribution to the Hall conductivity and the outside part gives a negative contribution, although the curvature of the Fermi surface is hole-like.

The results for the Hall coefficient are shown in Fig. 9.11. For the electron doped system the Hall coefficient becomes negative, since the outside part of the MBZ increases owing to the electron doping, although the Fermi surface is still hole-like everywhere [5].

As shown by the above argument, in the region where the antiferromagnetic correlation develops, the Hall coefficient increases in proportion to ξ^2, ξ being the antiferromagnetic correlation length. Thus we can understand that the Hall coefficient increases with the development of the antiferromagnetic correlation at low temperatures.

The contribution of the vertex correction to the electrical resistivity ρ increases the resistivity since the correction reduces the current $\boldsymbol{J}$ to a smaller value than $\boldsymbol{v}^*$ around the cold spots. This is shown in Fig. 9.12.

In the same way we can show that the magnetoresistance $\Delta\rho/\rho$ is approximately proportional to $\xi^4\rho^{-2}$ in the presence of the AF fluctuations using the expression (8.66) [6]. The coefficient ξ^4 violates Kohler's rule drastically in nearly antiferromagnetic Fermi liquids such as cuprates.

9.1.4 Nuclear spin lattice relaxation rate

The nuclear spin relaxation rate is given by the linear response theory as

$$\frac{1}{T_1} = k_\mathrm{B}T(g_N\mu_N)^2\sum_{\boldsymbol{q}}|A(\boldsymbol{q})|^2\left[\frac{1}{\omega}\mathrm{Im}\,\chi^{+-}(\boldsymbol{q},\omega+i\delta)\right]_{\omega=0}, \tag{9.21}$$

where $A(\boldsymbol{q})$ is the hyperfine coupling constant, and μ_N and g_N are the nuclear Bohr magneton and the g-value, respectively. The transverse susceptibility $\chi^{+-}(\boldsymbol{q},\omega + i\delta)$ is defined by d-electron spins as

$$\chi^{+-}(\boldsymbol{q},\omega+i\delta) = i\int_0^\infty dt e^{i(\omega+i\delta)t}\langle[S_{\boldsymbol{q}}{}^\dagger(t), S_{\boldsymbol{q}}{}^-(0)]\rangle, \tag{9.22}$$

$$S_{\boldsymbol{q}}{}^\dagger = \sum_{\boldsymbol{k}} d_{\boldsymbol{k}\uparrow}{}^\dagger d_{\boldsymbol{k}+\boldsymbol{q}\downarrow}. \tag{9.23}$$

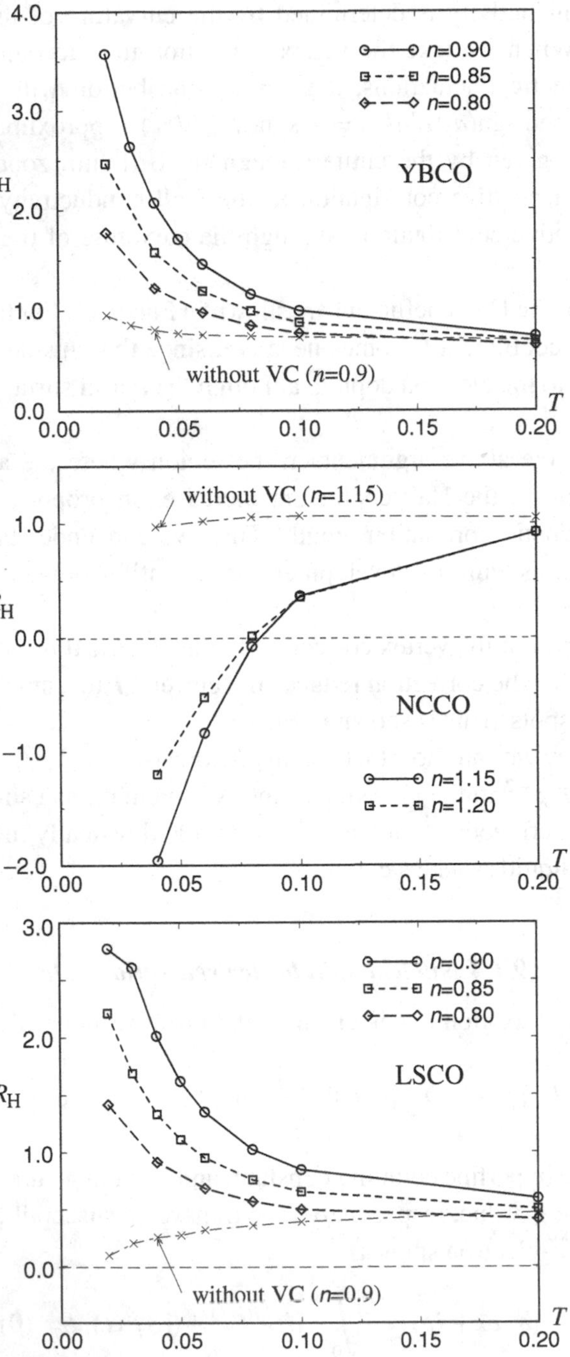

Fig. 9.11 Temperature dependence of the Hall coefficient R_H [5]. Here note that $1/|ne| \simeq 1.5 \times 10^{-3}\,\mathrm{cm^3\,C^{-1}}$.

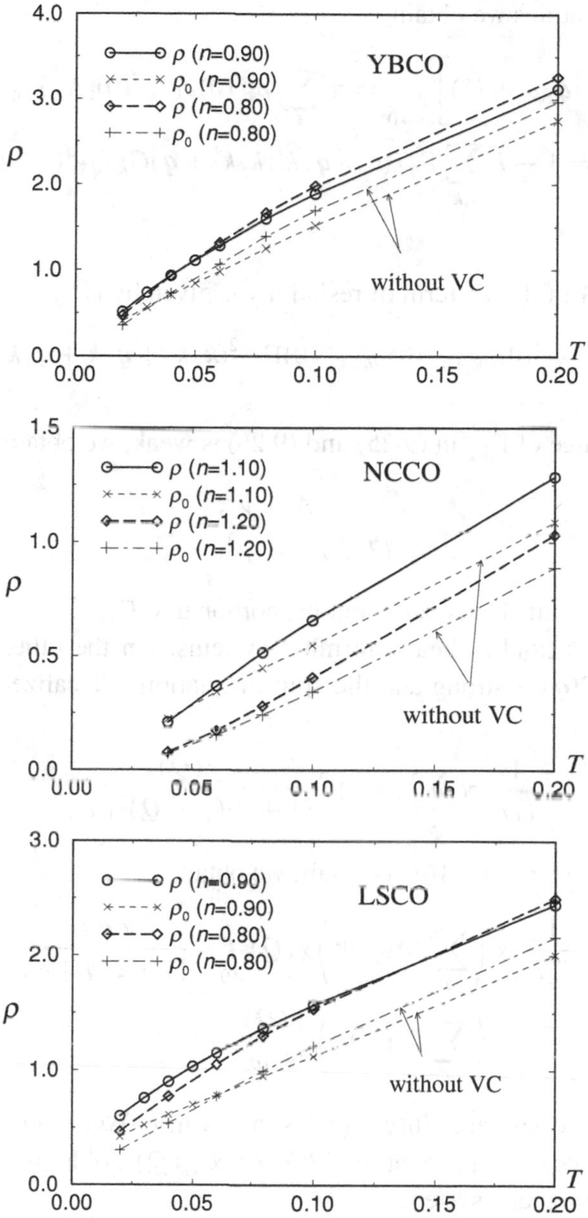

Fig. 9.12 Temperature dependence of the resistivity ρ [5]. Here, $\rho = 1.0$ corresponds to $\sim 4 \times 10^{-4}\ \Omega\,\text{cm}$.

At low temperatures we obtain

$$\left[\frac{1}{\omega}\mathrm{Im}\,\chi^{+-}(\boldsymbol{q},\omega+i\delta)\right]_{\omega\to 0}=\pi\sum_{\boldsymbol{k}}\rho_{\boldsymbol{k}}{}^{d}(0)\rho_{\boldsymbol{k}+\boldsymbol{q}}{}^{d}(0)[\Lambda_{\boldsymbol{k},\boldsymbol{k}+\boldsymbol{q}}(0)]^{2},\tag{9.24}$$

$$\Lambda_{\boldsymbol{k},\boldsymbol{k}+\boldsymbol{q}}(0)=1-T\sum_{n,\boldsymbol{k}'}\Gamma_{\uparrow\downarrow}(\boldsymbol{k}+\boldsymbol{q},\boldsymbol{k}';\boldsymbol{k},\boldsymbol{k}'+\boldsymbol{q})G_{\boldsymbol{k}+\boldsymbol{q}\uparrow}{}^{d}(i\varepsilon_n)G_{\boldsymbol{k}'\downarrow}{}^{d}(i\varepsilon_n).\tag{9.25}$$

The coefficient A of the T^2 term of resistivity is given by [7]

$$A\propto\sum_{\boldsymbol{k},\boldsymbol{q},\boldsymbol{k}'}\rho_{\boldsymbol{k}'}{}^{d}(0)\rho_{\boldsymbol{k}'+\boldsymbol{q}}{}^{d}(0)\rho_{\boldsymbol{k}+\boldsymbol{q}}{}^{d}(0)\Gamma_{\uparrow\downarrow}{}^{2}(\boldsymbol{k},\boldsymbol{k}'+\boldsymbol{q};\boldsymbol{k}+\boldsymbol{q},\boldsymbol{k}').\tag{9.26}$$

If the $\boldsymbol{q}$ dependence of $\Gamma_{\uparrow\downarrow}$ in (9.25) and (9.26) is weak, we obtain

$$A\propto\gamma^2,\tag{9.27}$$

$$(T_1T)^{-1}\propto\gamma^2,\tag{9.28}$$

where γ is the specific heat coefficient proportional to $\Gamma_{\uparrow\downarrow}$.

These relations hold in heavy fermion systems. On the other hand, if the $\boldsymbol{q}$ dependence of $\Gamma(\boldsymbol{q})$ is strong and the spin fluctuation is localized around $\boldsymbol{Q}$, we obtain

$$\frac{1}{T_1T}\propto\sum_{\boldsymbol{q}}|A(\boldsymbol{q})|^2\frac{\chi(\boldsymbol{Q})}{[1+\xi^2(\boldsymbol{q}-\boldsymbol{Q})^2]^2\omega_{\mathrm{s}}}.\tag{9.29}$$

When the $\boldsymbol{q}$ dependence of $A(\boldsymbol{q})$ is small, we obtain

$$\frac{1}{T_1T}\propto\left\langle\sum_{\boldsymbol{q}}|A(\boldsymbol{q}')|^2\right\rangle\chi(\boldsymbol{Q})\int_0^\infty\frac{q\,dq}{[1+\xi^2q^2]^2\omega_{\mathrm{s}}}$$

$$=\left\langle\sum_{\boldsymbol{q}}|A(\boldsymbol{q}')|^2\right\rangle\frac{\chi(\boldsymbol{Q})}{2\xi^2\omega_{\mathrm{s}}}.\tag{9.30}$$

Here the staggered susceptibility $\chi(\boldsymbol{Q})$ shows Curie–Weiss temperature dependence. Since $\omega_{\mathrm{s}}\propto\xi^{-2}$, the relation $1/(T_1T)\propto\chi(\boldsymbol{Q})$ holds in two-dimensional nearly antiferromagnetic systems.

9.2 Anisotropic superconductivity due to coulomb repulsion

One of the important issues in strongly correlated electron systems is to develop our understanding of the superconductivity in real systems and describe it in a unified form. The conventional superconductivity arises from the attractive interaction mediated by electron–phonon interactions. On the other hand, the so-called unconventional superconductivity has been observed in strongly correlated electron systems. These systems are the cuprates, the organic conductors BEDT–TTF, Sr_2RuO_4 and heavy fermions. In these systems the nuclear magnetic relaxation rate does not show any coherence peak inherent in the s-wave pairing near T_c, and shows power law behaviour such as T^3 at low temperatures below T_c. The electrical resistivity generally shows T^2-like behaviour above T_c. This fact means that the electron–electron scattering dominates the electron–phonon scattering up to a rather high temperature. Because of the strong on-site repulsion, the symmetry of their pairing states is not the s-wave symmetry but an anisotropic symmetry such as the p- and d-wave symmetry. From the measurement of the Knight shift, a p-wave triplet pairing is suggested for Sr_2RuO_4 [8–10]. The momentum dependence of the interaction between quasi-particles on the Fermi surface determines the symmetry of pairs [11]. It is possible that the p- or d-wave symmetry part of the effective interactions due to the coulomb repulsion is attractive, although the s-wave part is repulsive. The mechanism of the superconductivity is based on the pairing scenario which coincides with the BCS theory. The attractive force and the pairing symmetry are different from conventional ones. From the theoretical point of view the superconductivity in strongly correlated electron systems can be discussed on the basis of the Dyson–Gor'kov equation [12], which corresponds to the Éliashberg equation based on the vertex function arising from the electron–electron interactions in place of the usual electron–phonon interactions. We consider the following Hubbard Hamiltonian:

$$\mathcal{H} = \sum_{k,\sigma} \varepsilon_k c_{k\sigma}{}^{\dagger} c_{k\sigma} + \frac{U}{N} \sum_{q,k,k'} c_{k+q\uparrow}{}^{\dagger} c_{k'-q\downarrow}{}^{\dagger} c_{k'\downarrow} c_{k\uparrow}, \tag{9.31}$$

where U and ε_k are the on-site coulomb interaction and the band energy, respectively. From this Hamiltonian the superconductivity is realized by the coulomb repulsion U. This process can be discussed on the basis of the following Dyson–Gor'kov equation:

$$G(k) = G^{(0)}(k) + G^{(0)}(k)\Sigma^{(1)}(k)G(k) + G^{(0)}(k)\Sigma^{(2)}(k)F^{\dagger}(k), \tag{9.32}$$

$$F^{\dagger}(k) = G^{(0)}(-k)\Sigma^{(1)}(-k)F^{\dagger}(k) + G^{(0)}(-k)\Sigma^{(2)}(-k)G(k), \tag{9.33}$$

$$F(k) = G^{(0)}(k)\Sigma^{(1)}(k)F(k) + G^{(0)}(k)\Sigma^{(2)}(k)G(-k). \tag{9.34}$$

These equations are shown in Fig. 9.13. The argument k denotes $\boldsymbol{k}$ and ω. The normal and anomalous Green's functions G and F are introduced. Similarly the normal and anomalous self-energies $\Sigma^{(1)}$ and $\Sigma^{(2)}$ are introduced and can be calculated when the interaction vertex functions are given. In order to calculate the vertex function, the fluctuation exchange approximation [13] and the perturbation expansion [14] are used, depending on the strength of the interaction U.

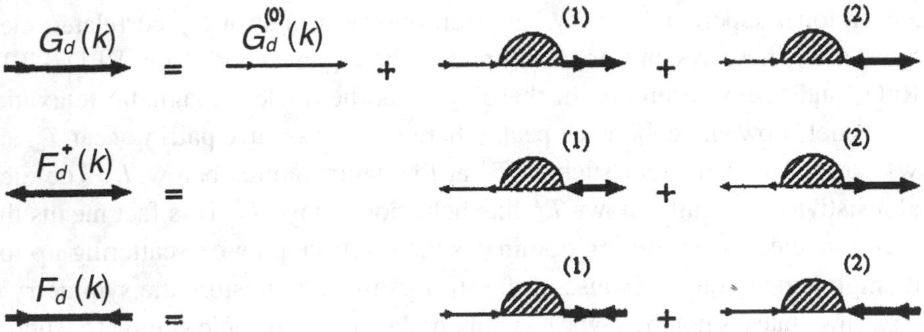

Fig. 9.13 The Dyson–Gor'kov equations are shown by diagrams. Thin and thick lines denote, respectively, unperturbed and perturbed Green's functions. Suffices (1) and (2) denote, respectively, normal and anomalous self-energy parts.

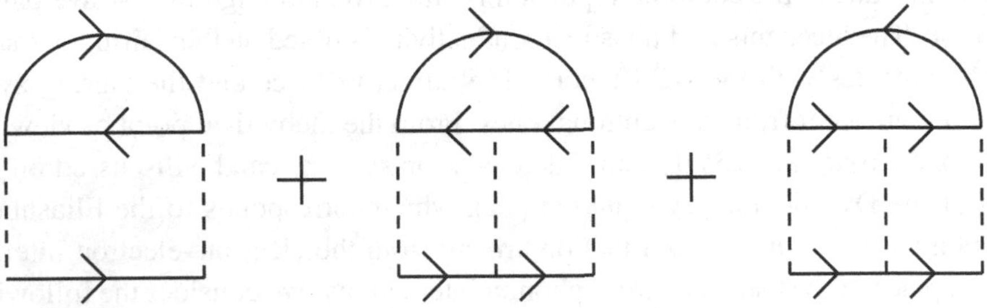

$\Sigma_2^{(1)}$ $\Sigma_{3a}^{(1)}$ $\Sigma_{3b}^{(1)}$

Fig. 9.14 Diagrams for the normal self-energy within the third-order perturbation. The solid and broken lines denote the normal Green's function and the interaction U, respectively.

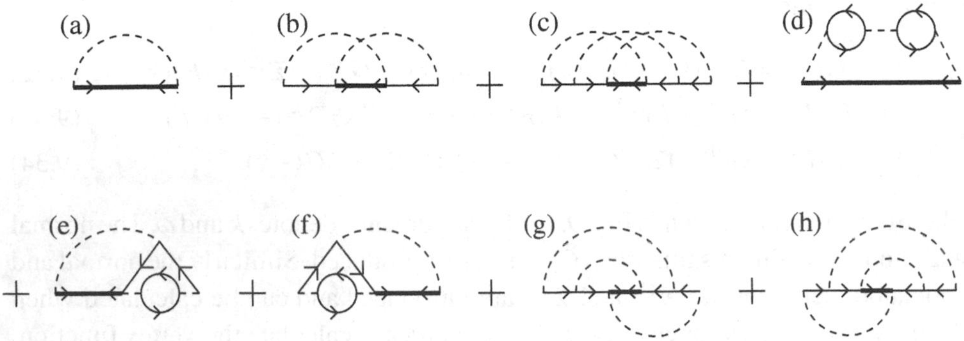

Fig. 9.15 Diagrams for the anomalous self-energy within the third-order perturbation. The thick line denotes the anomalous Green's function.

When we calculate the critical temperature T_c, the Dyson–Gor'kov equation is linearized with respect to $\Sigma^{(2)}$ and reduced to

$$G(k)^{-1} = G^{(0)-1} - \Sigma^{(1)}(k), \tag{9.35}$$

$$F^{\dagger}(k) = G(-k)\Sigma^{(2)}(-k)G(k), \tag{9.36}$$

$$F(k) = G(k)\Sigma^{(2)}(k)G(-k). \tag{9.37}$$

In the following subsections we discuss the nature of the superconductivity realized in the cuprates, the organic conductors and Sr_2RuO_4.

9.2.1 Superconductivity in over-doped cuprates

With increasing doped holes the cuprate system becomes a better metal and the electron correlation in the system becomes weaker. The resistivity shows T^2 temperature dependence as an ideal Fermi liquid. In the over-doped system we can adopt the perturbation theory with respect to the coulomb repulsion U between d-electrons. Hotta calculated the vertex part up to the third-order term in U and substituted it into the Dyson–Gor'kov equation [14]. By solving the integral equation, he obtained a d-wave superconducting state with a high critical temperature. In Figs. 9.14 and 9.15 we show the diagrams of normal and anomalous self-energy up to the third-order terms, respectively. The obtained critical temperature T_c is shown in Fig. 9.16. The calculation based on the third-order perturbation in U gives rather

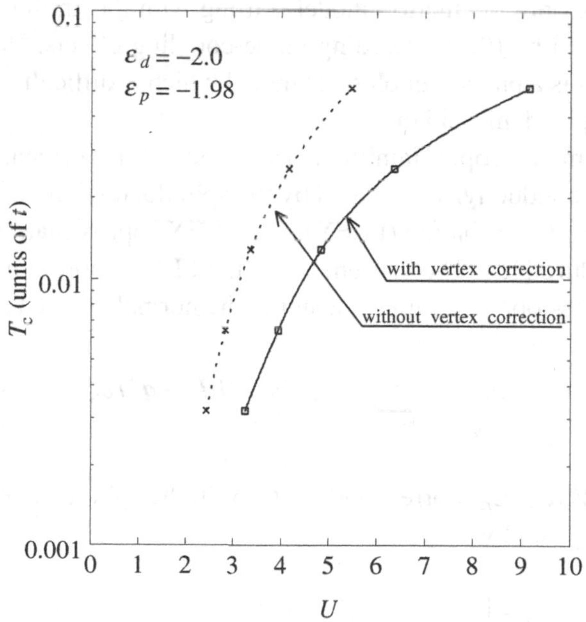

Fig. 9.16 Calculated results for T_c with and without the vertex corrections.

a high critical temperature T_c compared with that based on the random phase approximation (RPA), which is not shown here. The RPA calculation underestimates T_c owing to the overestimate of the damping rate of quasi-particles. The reason is the following. In the perturbation calculation of the normal self-energy the third-order terms, $\Sigma_{3a}{}^{(1)}$ and $\Sigma_{3b}{}^{(1)}$, almost cancel each other out when the system is near the half-filled case. This cancellation reduces the damping rate of quasi-particles. On the other hand, the RPA calculation takes only the particle–hole term $\Sigma_{3a}{}^{(1)}$ into account and overestimates the damping rate. The overestimated damping rate results in a low T_c. Thus we conclude that the correct calculation naturally leads to high-T_c superconductivity, which originates from the coulomb repulsion.

The doping dependence of T_c is shown in Fig. 9.17. The T_c increases rapidly with decreasing doping δ in the over-doped region, in good agreement with the experimental results for the Tl system. In the d–p Hamiltonian the effective transfer matrix between the neighbouring d-orbitals is given by $t^2/(\varepsilon_p - \varepsilon_d)$, t being the transfer integral between neighbouring d- and p-orbitals. Thus, the expansion parameter with respect to U is given by $U(\varepsilon_p - \varepsilon_d)/t^2$. As a result, the perturbation calculation is appropriate to the Tl system where ε_p and ε_d are nearly equal.

9.2.2 Superconductivity in optimally doped cuprates

The superconductivity in the optimally doped cuprates is well described by the antiferromagnetic spin fluctuation model starting with the antiferromagnetic spin susceptibility given by (9.5), including mode-coupling effects. However the spin fluctuation model is a phenomenological model which is difficult to derive directly from a microscopic Hamiltonian.

To start with a microscopic Hamiltonian and calculate the critical temperature for the d-wave superconductivity mediated by the spin fluctuations we adopt the fluctuation exchange approximation (FLEX). The FLEX approximation consists of the following procedure [13]. The self-energy in the FLEX approximation $\Sigma_F(\boldsymbol{k}, i\varepsilon_n)$ is given by the one-loop diagram exchanging the normal vertex $V_n(\boldsymbol{q}, i\omega_n)$,

$$\Sigma_F(\boldsymbol{k}, i\varepsilon_n) = T \sum_{\boldsymbol{q}, i\omega_n} V_n(\boldsymbol{q}, i\omega_n) G(\boldsymbol{k} - \boldsymbol{q}, i\varepsilon_n - i\omega_n). \tag{9.38}$$

The interaction $V_n(\boldsymbol{q}, i\omega_n)$ corresponds to (9.6) in the spin fluctuation model. The normal vertex is given by

$$V_n(\boldsymbol{q}, i\omega_n) = U^2 \left[\frac{3}{2}\chi_s(\boldsymbol{q}, i\omega_n) + \frac{1}{2}\chi_c(\boldsymbol{q}, i\omega_n) - \chi_0(\boldsymbol{q}, i\omega_n) \right], \tag{9.39}$$

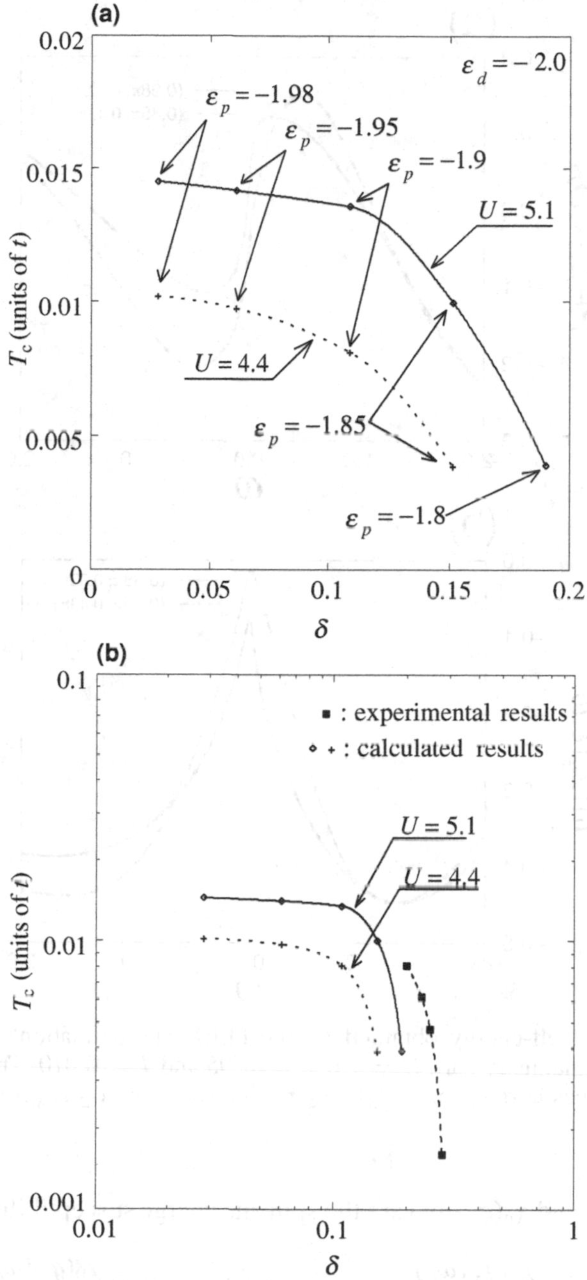

Fig. 9.17 (a) T_c as a function of doping δ for the Tl system. (b) The experimental results of Kubo *et al.* [1] are given by solid squares.

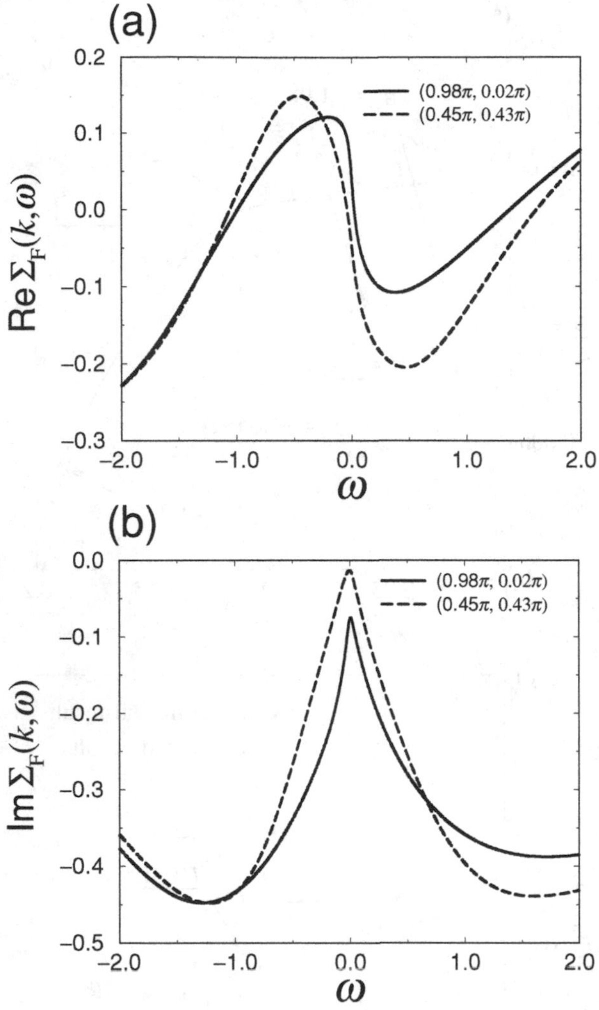

Fig. 9.18 The self-energy obtained by the FLEX approximation. (a) The real part. (b) The imaginary part. $U = 1.6$, $\delta = 0.095$ and $T = 0.010$. The solid lines and broken lines correspond to $(\frac{63}{64}\pi, \frac{1}{64}\pi)$ (hot spot) and $(\frac{29}{64}\pi, \frac{27}{64}\pi)$ (cold spot), respectively.

where $\chi_s(\boldsymbol{q}, i\omega_n)$ and $\chi_c(\boldsymbol{q}, i\omega_n)$ are the spin and charge susceptibility, respectively:

$$\chi_s(\boldsymbol{q}, i\omega_n) = \frac{\chi_0(\boldsymbol{q}, i\omega_n)}{1 - U\chi_0(\boldsymbol{q}, i\omega_n)}, \qquad \chi_c(\boldsymbol{q}, i\omega_n) = \frac{\chi_0(\boldsymbol{q}, i\omega_n)}{1 + U\chi_0(\boldsymbol{q}, i\omega_n)}. \tag{9.40}$$

Here $\chi_0(\boldsymbol{q}, i\omega_n)$ is the irreducible susceptibility:

$$\chi_0(\boldsymbol{q}, i\omega_n) = -T\sum_{\boldsymbol{k}, i\varepsilon_n} G(\boldsymbol{k}, i\varepsilon_n)G(\boldsymbol{k}+\boldsymbol{q}, i\varepsilon_n + i\omega_n), \tag{9.41}$$

where $G(\boldsymbol{k}, i\varepsilon_n)$ is the dressed Green function $G(\boldsymbol{k}, i\varepsilon_n) = (i\varepsilon_n - \varepsilon_{\boldsymbol{k}} - \Sigma_{\mathrm{F}}(\boldsymbol{k}, i\varepsilon_n))^{-1}$. The first-order term in $V_{\mathrm{n}}(\boldsymbol{q}, i\omega_n)$ with respect to U is omitted and treated as a shift of chemical potential. In the FLEX approximation, the self-energy and the spin susceptibility are determined self-consistently. The spin susceptibility $\chi_{\mathrm{s}}(\boldsymbol{q}, i\omega_n)$ given by the FLEX approximation is enhanced near the antiferromagnetic wave-vector $\boldsymbol{Q} = (\pi, \pi)$ for the Fermi surface of cuprates shown in Fig. 9.4. The antiferromagnetic spin fluctuations described by $\chi_{\mathrm{s}}(\boldsymbol{q}, i\omega_n)$ play an essential role in the FLEX approximation.

The superconducting critical temperature T_{c} is determined as the temperature below which the linearized Dyson–Gor'kov equation possesses a nontrivial solution. The Éliashberg equation which determines T_{c} is given by the following eigenvalue equation:

$$\lambda\phi(\boldsymbol{k}, i\varepsilon_n) = -T \sum_{\boldsymbol{p}, i\varepsilon_m} V_{\mathrm{a}}(\boldsymbol{k} - \boldsymbol{p}, i\varepsilon_n - i\varepsilon_m)|G(\boldsymbol{p}, i\varepsilon_m)|^2 \phi(\boldsymbol{p}, i\varepsilon_m). \tag{9.42}$$

Here $V_{\mathrm{a}}(\boldsymbol{q}, i\omega_n)$ is the anomalous vertex for the singlet channel and is given by the FLEX approximation as

$$V_{\mathrm{a}}(\boldsymbol{q}, i\omega_n) = U^2 \left[\frac{3}{2}\chi_{\mathrm{s}}(\boldsymbol{q}, i\omega_n) - \frac{1}{2}\chi_{\mathrm{c}}(\boldsymbol{q}, i\omega_n) \right] + U. \tag{9.43}$$

The critical temperature T_{c} is determined as the temperature where the maximum eigenvalue $\lambda_{\max}$ reaches unity. The eigenfunction $\phi_{\max}(\boldsymbol{p}, i\varepsilon_n)$ corresponding to the eigenvalue $\lambda_{\max}$ is the wave-function of the Cooper pairs.

Now we show the results obtained by the numerical calculation for the Hubbard model where $\varepsilon_{\boldsymbol{k}} = -2t(\cos k_x + \cos k_y) + 4t' \cos k_x \cos k_y - \mu$ is assumed ($t = 1/2, t' = t/4$) [15]. The self-energy is shown in Fig. 9.18(a) and (b). The negative slope of Re $\Sigma_{\mathrm{F}}(\boldsymbol{k}, \omega)$ and the small absolute value of Im $\Sigma_{\mathrm{F}}(\boldsymbol{k}, \omega)$ near the Fermi energy show the Fermi liquid behaviour. The single particle spectral weight and the density of states are shown in Fig. 9.19(a) and (b). The superconducting critical temperature is shown in Fig. 9.20(a) and (b) as a function of δ and U, respectively. In the FLEX approximation the critical temperature T_{c} increases with decreasing doping δ, and/or with increasing U. This behaviour agrees with that seen in the real systems possessing carrier number corresponding to the over-doped and the optimally doped cuprates.

9.2.3 Superconductivity in under-doped cuprates

The new physical phenomena in the under-doped cuprates are the pseudogap phenomena. The pseudogap originates from the superconducting fluctuation which

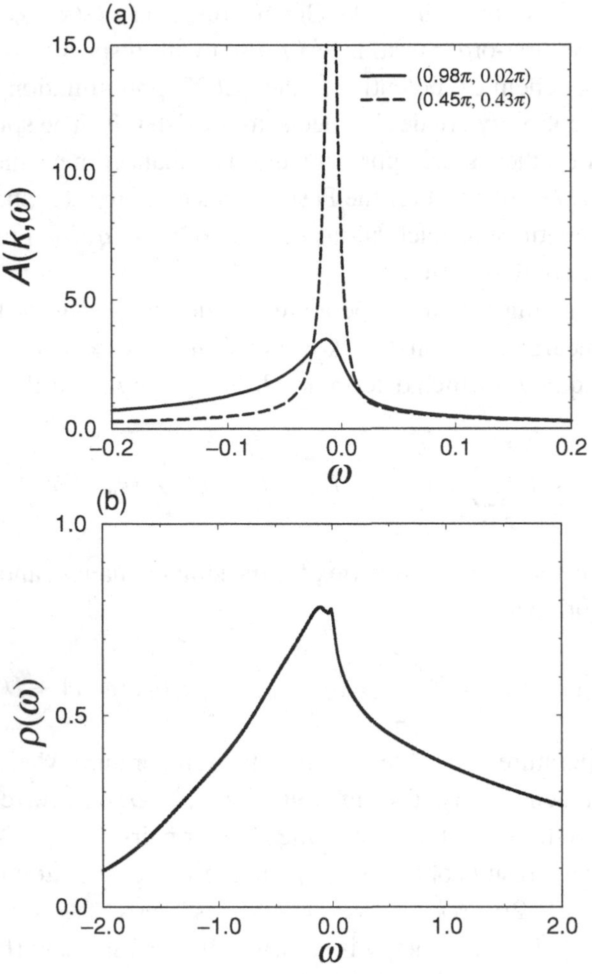

Fig. 9.19 (a) The single particle spectral weight and (b) the density of states obtained by FLEX. The parameters are the same as those in Fig. 9.18.

is inherent in the strong coupling superconductor in the quasi-two-dimensional systems. To simplify the explanation of the mechanism of the pseudogap we begin the discussion using the following attractive interaction model [16]:

$$\mathcal{H} = \sum_{k,\sigma} \varepsilon_k c_{k\sigma}{}^{\dagger} c_{k\sigma} + \sum_{k,k',q} V_{k-q/2,k'-q/2} c_{q-k',\downarrow}{}^{\dagger} c_{k',\uparrow}{}^{\dagger} c_{k,\uparrow} c_{q-k,\downarrow}, \tag{9.44}$$

where $V_{k,k'}$ is assumed to be the $d_{x^2-y^2}$ wave-pairing interaction:

$$V_{k,k'} = g\varphi_k \varphi_{k'}, \tag{9.45}$$

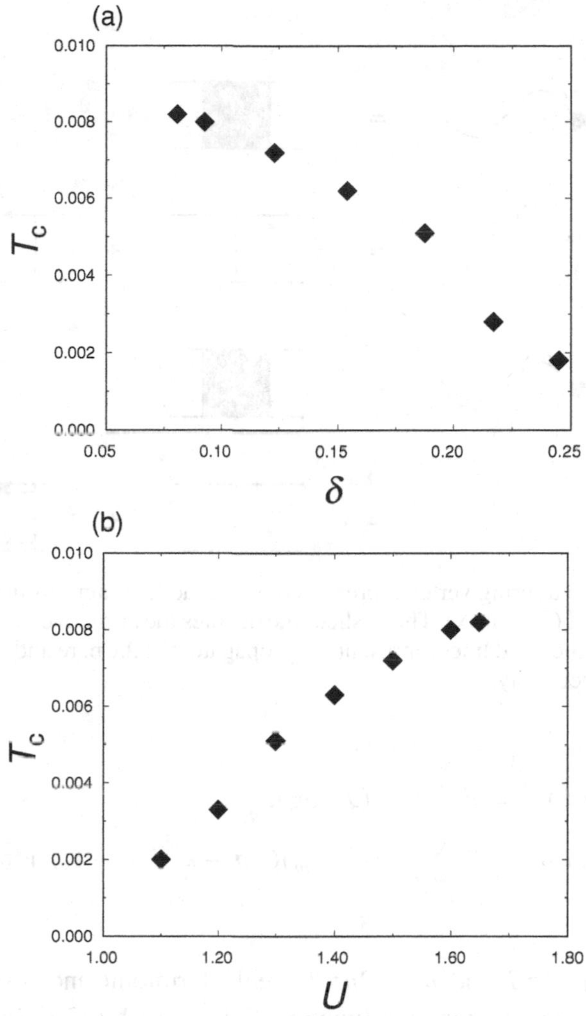

Fig. 9.20 The superconducting critical temperatures are shown as a function of δ and U, respectively. In (a) $U = 1.6$ and in (b) $\delta = 0.09$ are assumed.

$$\varphi_{\boldsymbol{k}} = \cos k_x - \cos k_y. \tag{9.46}$$

Here g is negative and $\varphi_{\boldsymbol{k}}$ is the $d_{x^2-y^2}$ wave-form factor.

To treat the superconducting fluctuation we adopt the T-matrix approximation. The self-energy of an electron is calculated as

$$\Sigma(\boldsymbol{k}, i\varepsilon_n) = T \sum_{\boldsymbol{q}, i\omega_n} T(\boldsymbol{q}, i\omega_n) G(\boldsymbol{q} - \boldsymbol{k}, i\omega_n - i\varepsilon_n), \tag{9.47}$$

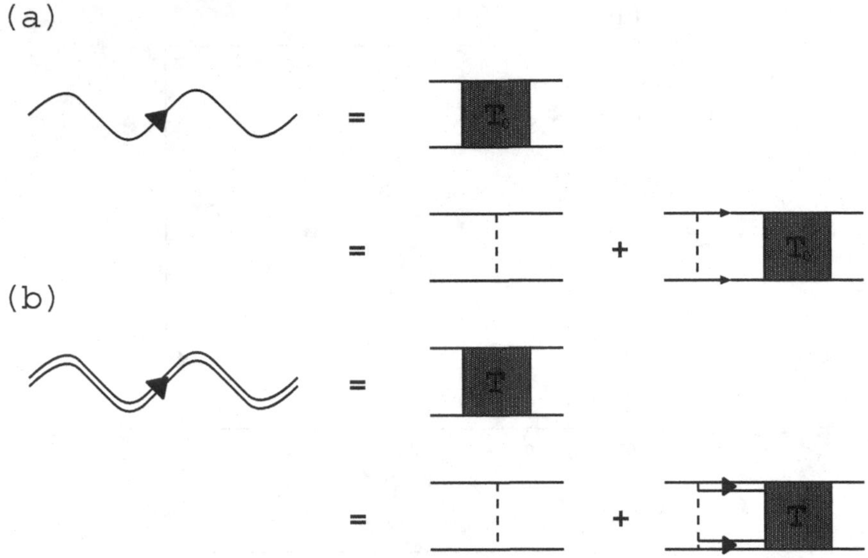

Fig. 9.21 The scattering vertex represented by the ladder diagrams in the particle–particle channel (T-matrix). The dashed line denotes the attractive interaction. The single and double solid lines represent the propagators of the bare and renormalized electrons, respectively.

$$T(\boldsymbol{q}, i\omega_n)^{-1} = g^{-1} + \chi_0(\boldsymbol{q}, i\omega_n), \tag{9.48}$$

$$\chi_0(\boldsymbol{q}, i\omega_n) = T \sum_{\boldsymbol{k}', \varepsilon_m} G(\boldsymbol{k}', i\varepsilon_m) G(\boldsymbol{q} - \boldsymbol{k}', i\omega_n - i\varepsilon_m) \varphi^2_{\boldsymbol{k}' - \boldsymbol{q}/2}. \tag{9.49}$$

Here $\varepsilon_m = (2m + 1)\pi T$ and $\omega_n = 2n\pi T$ are the fermionic and bosonic Matsubara frequencies, respectively. Green's function G is given by $G(\boldsymbol{k}, i\varepsilon_n) = (\omega - \varepsilon_{\boldsymbol{k}} - \Sigma(\boldsymbol{k}, i\varepsilon_n))^{-1}$.

The self-consistent T-matrix approximation is shown diagrammatically in Fig. 9.21. For simplicity we use first the one-loop approximation by replacing G with the unperturbed G_0. We show the self-energy part obtained by the approximation in Fig. 9.22. Surprisingly the real part of the self-energy possesses positive slope at the Fermi energy and the imaginary part of the self-energy shows a peak with negative sign. These behaviours, which are sharply in contrast to those of the Fermi liquid, originate from the resonance effect between the Cooper pair state and the quasi-particle state. The anomalous behaviour of the self-energy brings about the pseudogap in the energy spectrum of quasi-particles, as seen in Figs. 9.23 and 9.24.

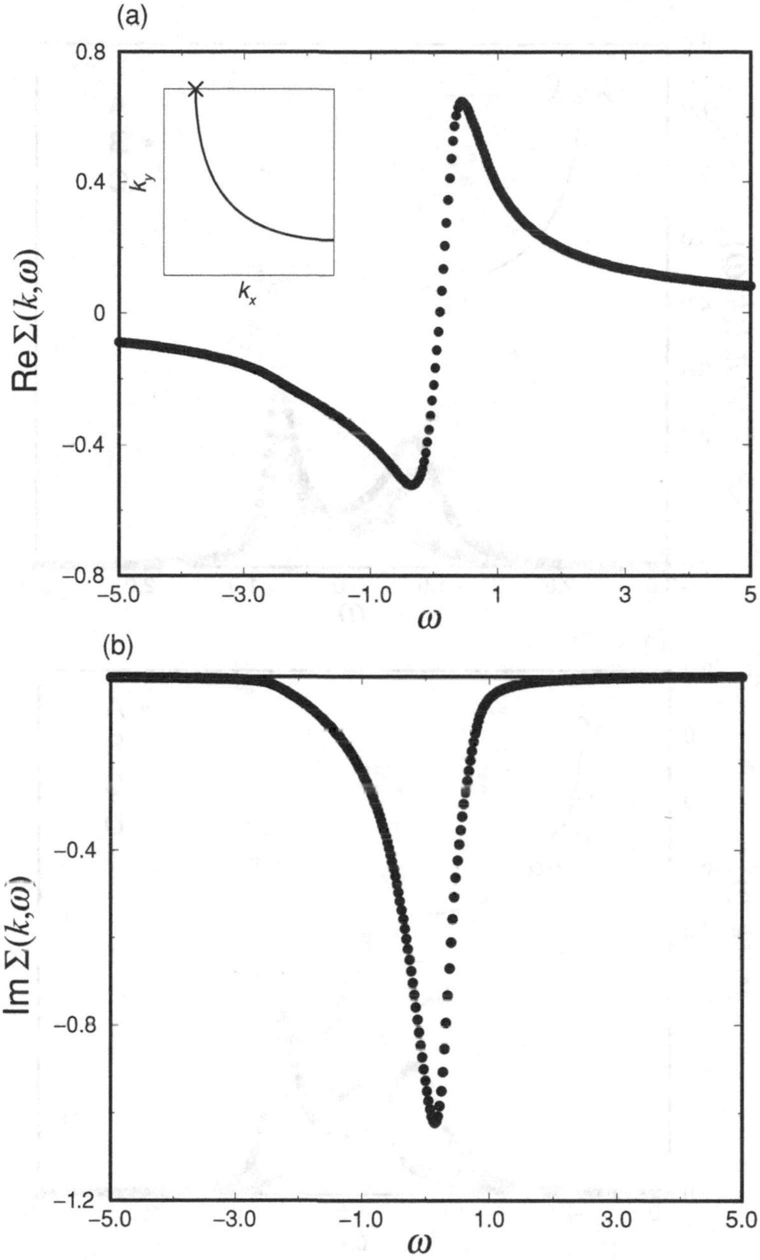

Fig. 9.22 The self-energy on the Fermi surface near $(0, \pi)$ obtained by the lowest order calculation. (a) The real part. (b) The imaginary part. Here, $\boldsymbol{k} = (0.589, \pi)$, $g = -1.0$ and $T = 0.21$. The $\boldsymbol{k}$-point is shown in the inset. $T_c = T_{MF} = 0.185$.

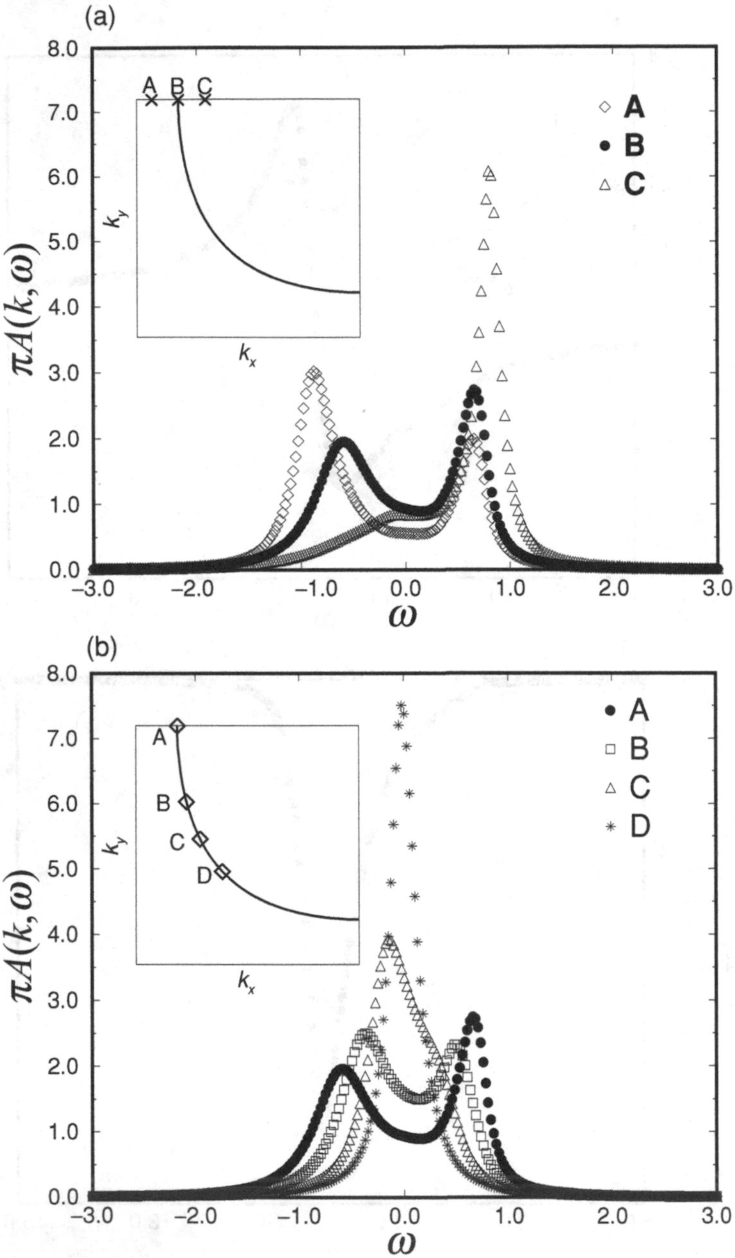

Fig. 9.23 The one-particle spectrum (a) across the Fermi surface near $(\pi, 0)$, (b) along the Fermi surface. The other parameters are the same as those in Fig. 9.22.

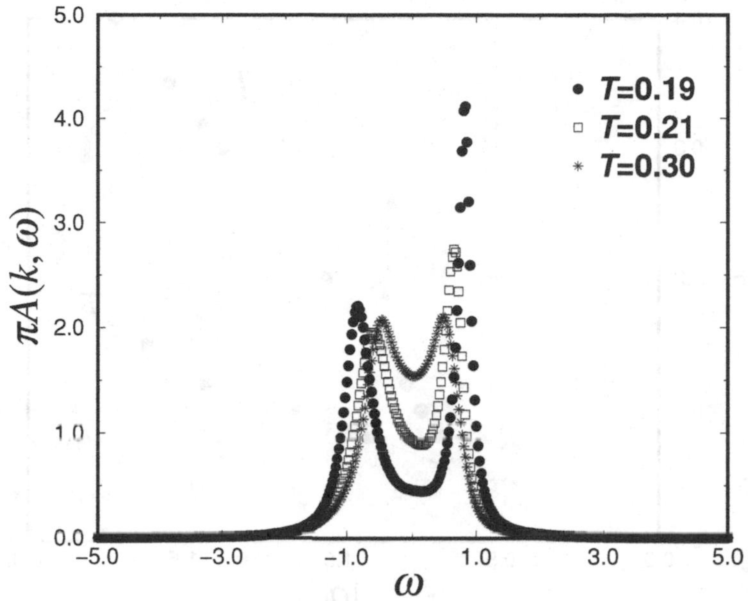

Fig. 9.24 The temperature dependence of a one-particle spectrum. The other parameters are the same as those in Fig. 9.22.

The self-consistent calculation has also been done. Owing to the superconducting fluctuations, the critical temperature T_c is much reduced from that obtained by the BCS mean field approximation, as shown in Fig. 9.25.

Now we extend the self-consistent T-matrix calculation to the superconducting state [17, 18]. The T-matrix is given by the following 2×2 matrix in the superconducting state:

$$T(\boldsymbol{q}, i\omega_n) = [g^{-1}\mathbf{1} + \chi(\boldsymbol{q}, i\omega_n)]^{-1}, \tag{9.50}$$

$$\chi(\boldsymbol{q}, i\omega_n) = \begin{pmatrix} K(\boldsymbol{q}, i\omega_n) & L(\boldsymbol{q}, i\omega_n) \\ L^*(\boldsymbol{q}, i\omega_n) & K(-\boldsymbol{q}, -i\omega_n) \end{pmatrix}, \tag{9.51}$$

where

$$K(\boldsymbol{q}, i\omega_n) = T \sum_{\boldsymbol{k}', \varepsilon_m} G(\boldsymbol{k}', i\varepsilon_m) G(\boldsymbol{q} - \boldsymbol{k}', i\omega_n - i\varepsilon_m) \varphi^2_{\boldsymbol{k}'-\boldsymbol{q}/2}, \tag{9.52}$$

$$L(\boldsymbol{q}, i\omega_n) = -T \sum_{\boldsymbol{k}', \varepsilon_m} F(\boldsymbol{k}', i\varepsilon_m) F(\boldsymbol{q} - \boldsymbol{k}', i\omega_n - i\varepsilon_m) \varphi^2_{\boldsymbol{k}'-\boldsymbol{q}/2}. \tag{9.53}$$

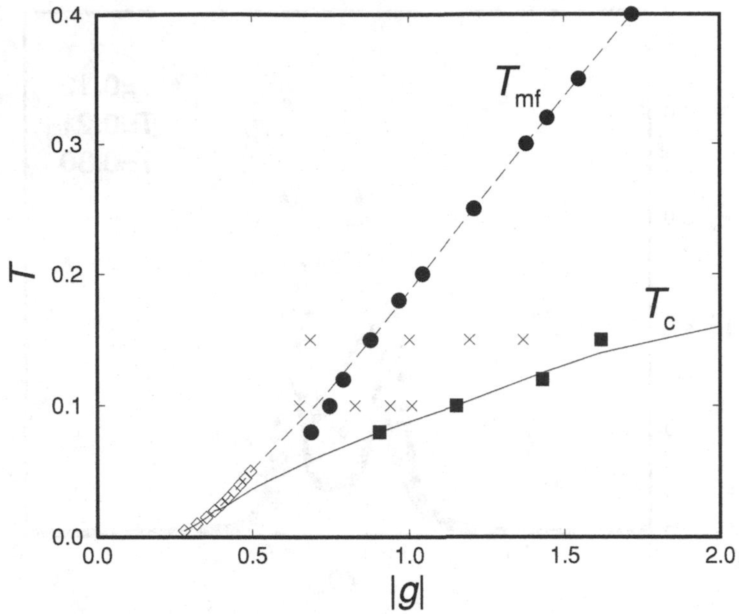

Fig. 9.25 The phase diagram obtained by the self-consistent T-matrix approximation. The critical temperature T_c is much reduced from T_{MF} given by the mean field theory by the superconducting fluctuations.

Here, $G(\boldsymbol{k}, i\varepsilon_m)$ and $F(\boldsymbol{k}, i\varepsilon_m)$ are the normal and anomalous Green's functions including the self-energy $\Sigma(\boldsymbol{k}, i\varepsilon_m)$, respectively:

$$G(\boldsymbol{k}, i\varepsilon_m) = \frac{i\varepsilon_m + \varepsilon_{\boldsymbol{k}} + \Sigma(\boldsymbol{k}, -i\varepsilon_m)}{[i\varepsilon_m - \varepsilon_{\boldsymbol{k}} - \Sigma(\boldsymbol{k}, i\varepsilon_m)][i\varepsilon_m + \varepsilon_{\boldsymbol{k}} + \Sigma(\boldsymbol{k}, -i\varepsilon_m)] - {\Delta_{\boldsymbol{k}}}^2}, \tag{9.54}$$

$$F(\boldsymbol{k}, i\varepsilon_m) = \frac{-\Delta_{\boldsymbol{k}}}{[i\varepsilon_m - \varepsilon_{\boldsymbol{k}} - \Sigma(\boldsymbol{k}, i\varepsilon_m)][i\varepsilon_m + \varepsilon_{\boldsymbol{k}} + \Sigma(\boldsymbol{k}, -i\varepsilon_m)] - {\Delta_{\boldsymbol{k}}}^2}, \tag{9.55}$$

where $\Delta_{\boldsymbol{k}} = \Delta\varphi_{\boldsymbol{k}}$ is the d-wave order parameter. The order parameter $\Delta_{\boldsymbol{k}}$ is determined by the gap equation

$$\Delta_{\boldsymbol{k}} = -gT \sum_{\boldsymbol{k}', i\varepsilon_m} F(\boldsymbol{k}', i\varepsilon_m)\varphi_{\boldsymbol{k}'}\varphi_{\boldsymbol{k}}. \tag{9.56}$$

The normal self-energy $\Sigma(\boldsymbol{k}, i\varepsilon_m)$ is given by the self-consistent T-matrix approximation

$$\Sigma(\boldsymbol{k}, i\varepsilon_m) = T \sum_{\boldsymbol{q}, i\omega_n} T_{11}(\boldsymbol{q}, i\omega_n) G(\boldsymbol{q} - \boldsymbol{k}, i\omega_n - i\varepsilon_m)\varphi^2_{\boldsymbol{k}-\boldsymbol{q}/2}. \tag{9.57}$$

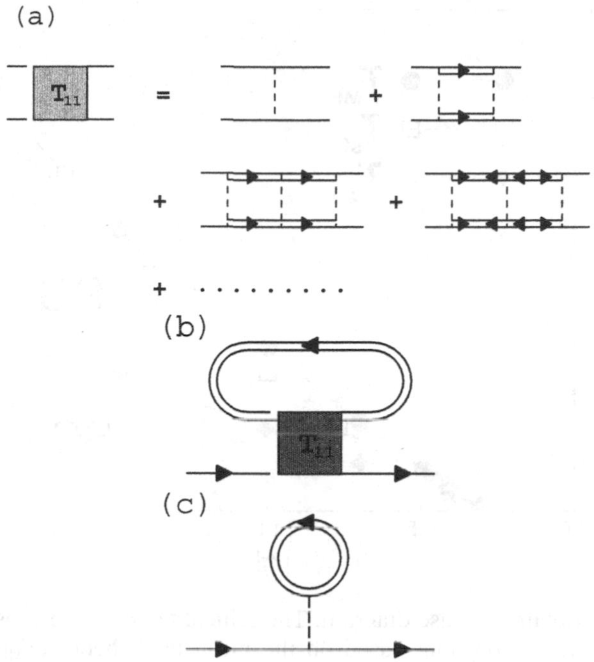

Fig. 9.26 (a) The diagonal component of the T-matrix in the superconducting state. The double solid lines denote the normal and anomalous Green's functions of electrons. (b) The normal self-energy calculated by the self-consistent T-matrix approximation. (c) The Hartree–Fock term which is excluded.

After carrying out the analytic continuation, the self-energy $\Sigma^{\mathrm{R}}(\boldsymbol{k}, \varepsilon)$, the order parameter $\Delta_{\boldsymbol{k}}$, the normal and anomalous Green's function $G^{\mathrm{R}}(\boldsymbol{k}, \varepsilon)$, $F^{\mathrm{R}}(\boldsymbol{k}, \varepsilon)$ and the 2×2 T-matrix are self-consistently determined for the real frequency. This procedure is shown diagrammatically in Fig. 9.26. The calculation is carried out both in the normal and the superconducting states.

The effects of the superconducting fluctuations are included in the self-energy. When we neglect the normal self-energy $\Sigma^{\mathrm{R}}(\boldsymbol{k}, \omega)$, this formalism reduces to the BCS mean field theory. In the normal state, the off-diagonal components $\Delta_{\boldsymbol{k}}$, $F(\boldsymbol{k}, i\varepsilon_n)$ and $L(\boldsymbol{q}, i\omega_n)$ vanish. In this case the above set of equations coincides with the self-consistent T-matrix calculation adopted in the normal state. The self-consistent T-matrix calculation gives a unified description for the pseudogap state, the superconducting state and their phase transition.

The pseudogap appears only in the quasi-two-dimensional system. In order to avoid the singularity arising from the two-dimensionality we maintain the denominator of the T-matrix as a small value $\alpha = 1 + gK(0, 0) - gL(0, 0) = 0.01$ in the superconducting state. The finite critical temperature is obtained by this procedure, which is justified in the weak three-dimensional coupling systems such as the

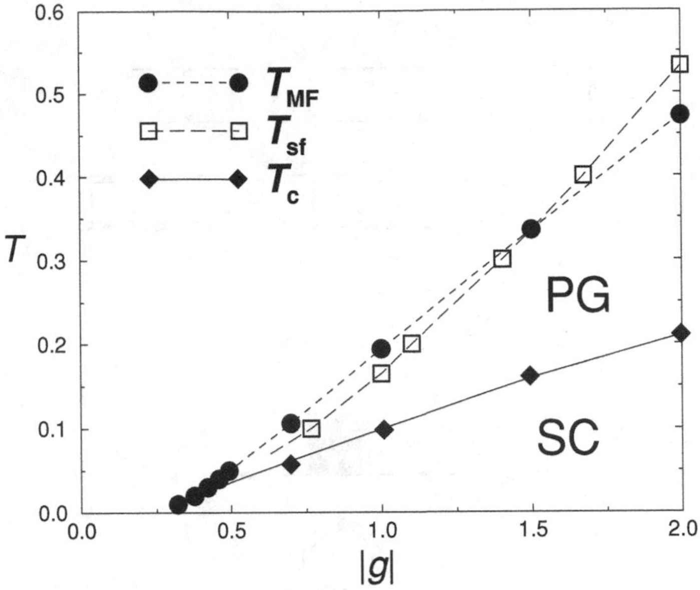

Fig. 9.27 The obtained phase diagram. The critical temperature T_c is suppressed by the fluctuations from that based on the mean field theory (T_{MF}). T_{sf} is the temperature where $1/|g| - \chi_0(\mathbf{0}, 0) = 0.1$.

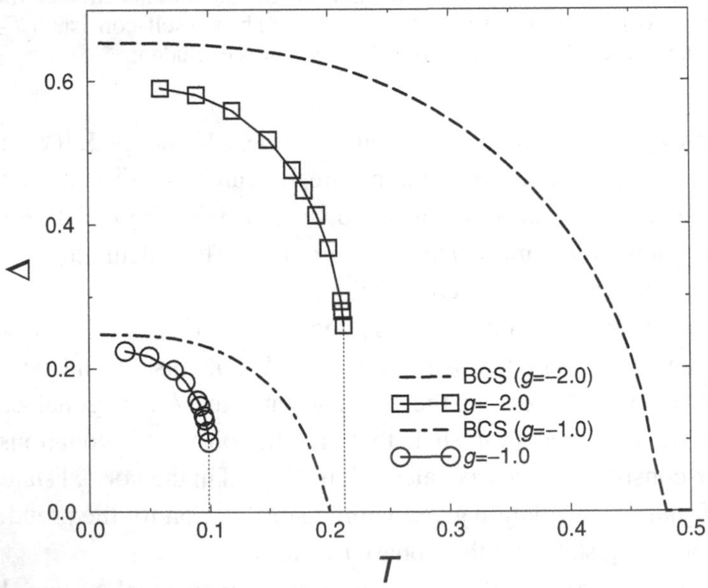

Fig. 9.28 The growth of the order parameter Δ. The open circles and open squares are the results for $g = -1.0$ and $g = -2.0$, respectively. The dash-dotted and broken lines show the results of the BCS theory for $g = -1.0$ and $g = -2.0$, respectively. The order parameter grows more rapidly than in the BCS theory.

high-T_c cuprates. The critical temperature T_c is reduced more as α decreases. The small finite value of α is considered as the three-dimensional coupling between layers in a real system. The choice of a small value of α makes no qualitative difference to the calculated results.

We show the phase diagram in Fig. 9.27. The suppression of T_c from T_{MF} becomes remarkable with increasing coupling constant $|g|$. The growth of the order parameter Δ is shown in Fig. 9.28. Once the superconducting order occurs, the effects of the fluctuations are drastically suppressed. There are two reasons. The amplitude mode is suppressed owing to the growth of the order parameter. Moreover, the weight of the phase mode shifts to high frequency, since the dissipation in the ordered state is reduced in the power law $\mathrm{Im}\, K(\boldsymbol{q}, \omega) \pm \mathrm{Im}\, L(\boldsymbol{q}, \omega) \propto \omega^4$, while it is exponentially reduced in the s-wave superconductor. As a result, the order parameter $\Delta_{\boldsymbol{k}} = \Delta\varphi_{\boldsymbol{k}}$ grows more rapidly than the result of the BCS theory. The rapid growth of the order parameter is also seen in the FLEX calculation for the superconducting state. The rapid growth below T_c in the FLEX calculation is caused by the suppression of the low-frequency spin fluctuations contributing to the depairing effect. Both effects exist in the high-T_c cuprates below T_c. As a result the value of $2\Delta/k_B T_c$ exceeds the BCS value.

The calculated single particle spectral weights at $\boldsymbol{k} = (\pi, 0.15\pi)$ and $\boldsymbol{k} = (0.5\pi, 0.25\pi)$ are shown in Figs. 9.29(a) and (b), respectively. The strong particle–hole asymmetry exists originally in the high-T_c cuprates and stabilizes the self-consistent solution, resulting in the pseudogap state. The bottom of the pseudogap is located apart from the Fermi level, and the gap structure is broad and asymmetric. On the other hand, below T_c the gap structure becomes clear and symmetric. The pseudogap is caused by the self-energy correction due to the superconducting fluctuations, while the superconducting gap is caused by the superconducting order. The effects of the self-energy correction on the spectrum are reduced in the superconducting state, since the fluctuations are suppressed there. Fig. 9.29 shows that the energy scale of the pseudogap and that of the superconducting gap are similar to each other, as well as their momentum dependence. The broad pseudogap, the sharp superconducting gap and the same energy scale are important properties observed by angle-resolved photoemission spectroscopy (ARPES).

We can see more clearly the transition from the pseudogap state to the superconducting state from the density of states (DOS) in Fig. 9.30. In the normal state the DOS shows a broad pseudogap and remains to some extent near the Fermi level. The bottom of the gap becomes deep as the temperature approaches T_c. Once the superconducting order occurs, the gap becomes deep and sharp with rapid growth of the order parameter. It should be noted that the energy scale of the pseudogap and that of the superconducting gap are almost the same. When the coupling constant $|g|$ increases, the energy scale of the pseudogap increases in accordance with

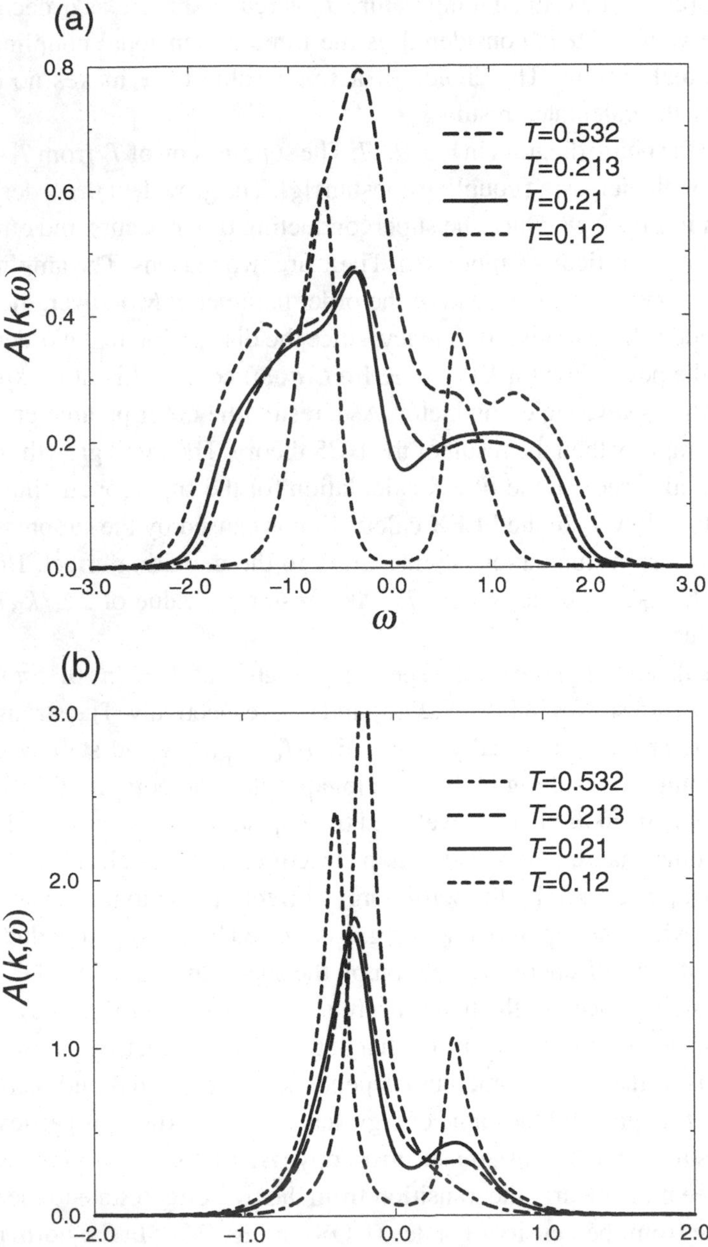

Fig. 9.29 The one-particle spectral weight for $g = -2$. (a) $\boldsymbol{k} = (\pi, 0.15\pi)$, (b) $\boldsymbol{k} = (0.5\pi, 0.25\pi)$, just below the Fermi level. $T = 0.213$ is just above T_c, $T = 0.21$ and $T = 0.12$ are below T_c.

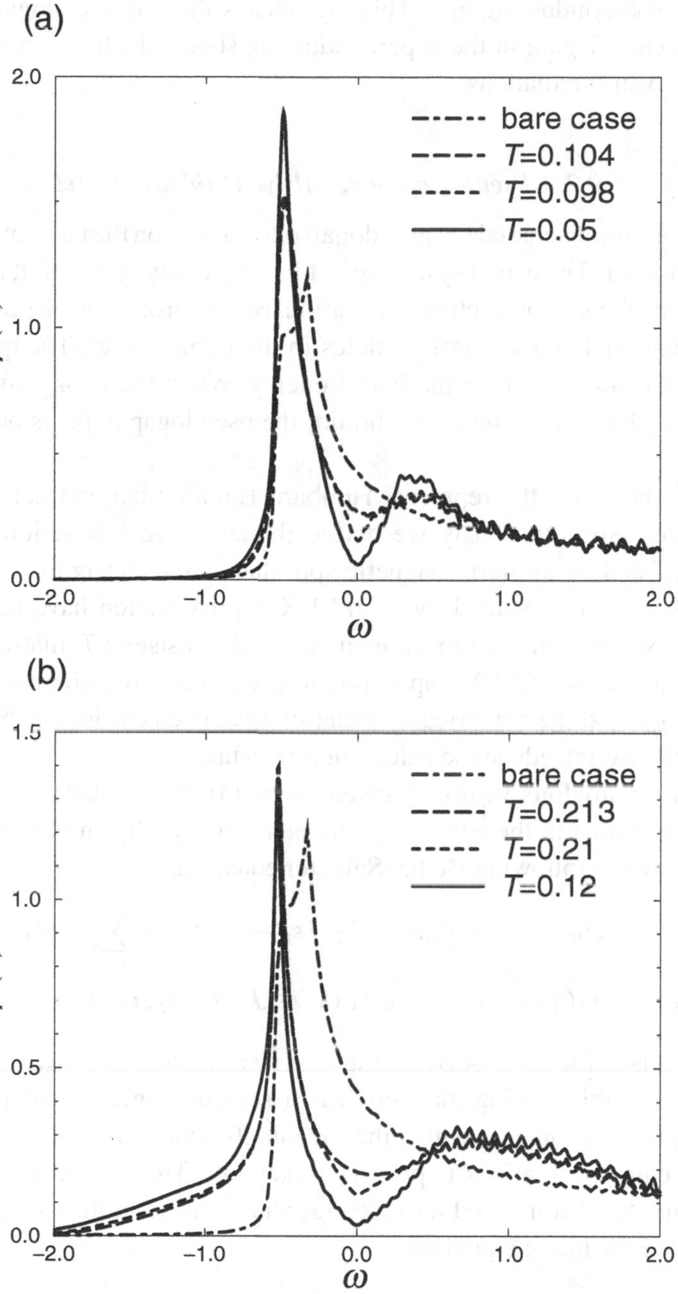

Fig. 9.30 The density of states for (a) $g = -1.0$ and (b) $g = -2.0$. $T = 0.104$ (a) and $T = 0.213$ (b) are above T_c. $T = 0.098$ and $T = 0.05$ (a) and $T = 0.21$ and $T = 0.12$ (b) are below T_c.

that of the superconducting gap. This fact means that the pseudogap state partly realizes the energy gain in the superconducting state, which is suppressed by the two-dimensional fluctuations.

9.2.4 Pseudogap in repulsive Hubbard model

Until now we have discussed the pseudogap phenomena on the basis of the attractive interaction model. The actual systems exist in the vicinity of the Mott transition and are affected by the strong electron correlation. By the strong correlation the electron system is renormalized to quasi-particles in the Fermi liquid. The quasi-particles compose a narrow band near the Fermi energy. When the strong attractive interaction is introduced into the Fermi liquid, the pseudogap appears near the Fermi energy.

Now we start with the repulsive Hubbard Hamiltonian and derive the quasi-particle states. Simultaneously we derive the attractive interaction for a $d_{x^2-y^2}$ channel mediated by antiferromagnetic spin fluctuations using the FLEX approximation. The results obtained by the FLEX approximation have been shown in the previous subsection. Moreover, using the self-consistent T-matrix approximation in addition to the FLEX approximation, we can derive the pseudogap state. This derivation and the numerical calculation have been carried out by Yanase and Yamada [19]. We introduce the calculation in detail.

Using the anomalous vertex V_{a} given by (9.43), the T-matrix is expressed by the ladder diagrams in the particle–particle channel as shown in Fig. 9.31, and is determined by the following Bethe–Salpeter equation:

$$T(\boldsymbol{k}_1, i\varepsilon_n{:}\boldsymbol{k}_2, i\varepsilon_m{:}\boldsymbol{q}, i\omega_n) = V_{\mathrm{a}}(\boldsymbol{k}_1 - \boldsymbol{k}_2, i\varepsilon_n - i\varepsilon_m) - T\sum_{\boldsymbol{k},\varepsilon_l} V_{\mathrm{a}}(\boldsymbol{k}_1 - \boldsymbol{k}_2, i\varepsilon_n - i\varepsilon_l)$$
$$\times G(\boldsymbol{k}, i\varepsilon_l)G(\boldsymbol{q} - \boldsymbol{k}, i\omega_n - i\varepsilon_l)T(\boldsymbol{k}, i\varepsilon_l{:}\boldsymbol{k}_2, i\varepsilon_m{:}\boldsymbol{q}, i\omega_n). \tag{9.58}$$

Generally, it is difficult to solve the above integral equation except for the case where the separable pairing interaction is assumed. Hence we adopt the following two approximations, by which the meaningful component as the $d_{x^2-y^2}$ wave superconducting fluctuations is properly taken out. The T-matrix at $\boldsymbol{q} = \omega_n = 0$ is approximately decomposed into the eigenfunctions with their respective eigenvalues of the Éliashberg equation:

$$T(\boldsymbol{k}_1, i\varepsilon_n{:}\boldsymbol{k}_2, i\varepsilon_m{:}\boldsymbol{q} = i\omega_n = 0) = \sum_{\alpha} \frac{g_\alpha \phi_\alpha(\boldsymbol{k}_1, i\varepsilon_n)\phi_\alpha{}^*(\boldsymbol{k}_2, i\varepsilon_m)}{1 - \lambda_\alpha}. \tag{9.59}$$

The eigenvalue λ_α and the eigenfunction $\phi_\alpha(\boldsymbol{k}, i\varepsilon_n)$ are derived from the Éliashberg equation (9.42). The index α denotes each mode included in the T-matrix. Now we

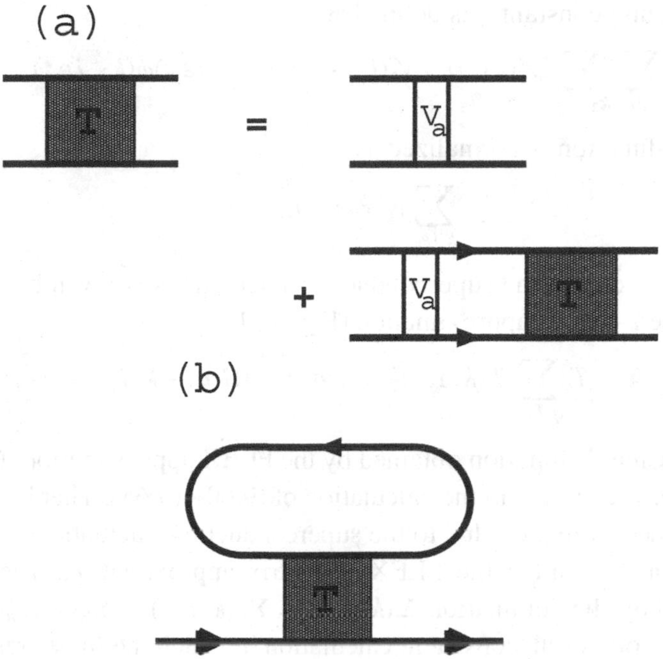

Fig. 9.31 (a) The T-matrix. (b) The self-energy due to the superconducting fluctuations.

take out the component with the maximum eigenvalue $\lambda_{\max}$ and the corresponding eigenfunction $\phi_{\max}$, which possesses the $d_{x^2-y^2}$ wave character. Since the superconducting transition is determined by the condition $\lambda_{\max} = 1$, the mode described by $\lambda_{\max}$ and $\phi_{\max}$ represents the $d_{x^2-y^2}$ wave superconducting fluctuations. The function $\phi_{\max}$ is the wave-function of the fluctuating Cooper pairs in the fluctuating regime. Hereafter we neglect the other modes, since they have no significant effect on the superconducting fluctuations. We omit the index 'max' for simplicity. Using the above approximation the T-matrix is expressed as

$$T(\boldsymbol{k}_1, i\omega_n{:}\boldsymbol{k}_2, i\omega_n{:}\boldsymbol{q}, i\omega_n) = \frac{g\lambda(\boldsymbol{q}, i\omega_n)\phi(\boldsymbol{k}_1, i\varepsilon_n{:}\boldsymbol{q}, i\omega_n)\phi^*(\boldsymbol{k}_2, i\varepsilon_n{:}\boldsymbol{q}, i\omega_n)}{1 - \lambda(\boldsymbol{q}, i\omega_n)}, \tag{9.60}$$

where

$$\lambda(\boldsymbol{q}, i\omega_n) = -T \sum_{\boldsymbol{k}, i\varepsilon_n} \sum_{\boldsymbol{p}, i\varepsilon_m} \phi^*(\boldsymbol{k}, i\varepsilon_n) V_{\mathrm{a}}(\boldsymbol{k} - \boldsymbol{p}, i\varepsilon_n - i\varepsilon_m) \times G(\boldsymbol{p}, i\varepsilon_m) G(\boldsymbol{q} - \boldsymbol{p}, i\omega_n - i\varepsilon_m)\phi(\boldsymbol{p}, i\varepsilon_m). \tag{9.61}$$

Here the coupling constant g is defined as

$$g = \sum_{\boldsymbol{k}_1, i\varepsilon_n} \sum_{\boldsymbol{k}_2, i\varepsilon_m} \phi^*(\boldsymbol{k}_1, i\varepsilon_n) V_{\mathrm{a}}(\boldsymbol{k}_1 - \boldsymbol{k}_2, i\varepsilon_n - i\varepsilon_m) \phi(\boldsymbol{k}_2, i\varepsilon_m), \tag{9.62}$$

and the wave-function is normalized as

$$\sum_{\boldsymbol{k}, i\varepsilon_n} |\phi(\boldsymbol{k}, i\varepsilon_n)|^2 = 1. \tag{9.63}$$

The self-energy due to the superconducting fluctuations is given by the one-loop diagram in the T-matrix approximation (Fig. 9.31):

$$\Sigma_{\mathrm{S}}(\boldsymbol{k}, i\varepsilon_n) = T \sum_{\boldsymbol{q}, i\omega_n} T(\boldsymbol{k}, i\varepsilon_n{:}\boldsymbol{k}, i\varepsilon_n{:}\boldsymbol{q}, i\omega_n) G(\boldsymbol{q} - \boldsymbol{k}, i\omega_n - i\varepsilon_n). \tag{9.64}$$

Here we use Green's function obtained by the FLEX approximation, $G_{\mathrm{F}}(\boldsymbol{k}, i\varepsilon_n) = (i\varepsilon_n - \varepsilon_{\boldsymbol{k}} - \Sigma_{\mathrm{F}}(\boldsymbol{k}, i\varepsilon_n))^{-1}$ in the calculation of (9.60)–(9.64). That is, we calculate the lowest-order correction due to the superconducting fluctuations for the FLEX approximation. We call it the FLEX+T-matrix approximation. The self-energy is determined by the summation $\Sigma(\boldsymbol{k}, i\varepsilon_n) = \Sigma_{\mathrm{F}}(\boldsymbol{k}, i\varepsilon_n) + \Sigma_{\mathrm{S}}(\boldsymbol{k}, i\varepsilon_n)$. Moreover, Yanase carried out a self-consistent calculation in which the fully dressed Green's function $G(\boldsymbol{k}, i\varepsilon_n) = (i\varepsilon_n - \varepsilon_{\boldsymbol{k}} - \Sigma_{\mathrm{F}}(\boldsymbol{k}, i\varepsilon_n) - \Sigma_{\mathrm{S}}(\boldsymbol{k}, i\varepsilon_n))^{-1}$ is used everywhere. As a result of self-consistency the effects of the superconducting fluctuations are reduced, but qualitatively similar results to the lowest order calculation are obtained.

In Fig. 9.32 we show the analytically continued self-energy $\Sigma^{\mathrm{R}}(\boldsymbol{k}, \omega)$ for the doping concentration corresponding to the under-doped cuprates. As shown in the previous section, the anomalous properties of the self-energy give rise to the pseudogap. The characteristics of the self-energy leading to the pseudogap are that, near the Fermi level $\omega = 0$, the real part has positive slope and the imaginary part has maximum absolute value, in contrast to the ordinary Fermi liquid. The large imaginary part reduces the single particle spectral weight near the Fermi level and gives rise to the pseudogap. In Fig. 9.32, the Fermi liquid behaviour is seen when we look at the large energy scale, $\omega \sim 0.5$. However, the anomalous behaviour leading to the pseudogap is clearly seen at a much smaller energy scale, $\omega \sim 0.05$. The anomalous behaviour vanishes around the cold spot $(\pi/2, \pi/2)$ owing to the $d_{x^2-y^2}$ wave symmetry of the fluctuating Cooper pairs.

It is an important point that the superconductivity and the pseudogap take place in renormalized quasi-particles that have small energy scale compared with the original bandwidth. That is, the pseudogap appears with much smaller energy scale than that of the electron systems. This is a newly observed physical phenomenon which is inherent in the high density electron system, since the strong coupling

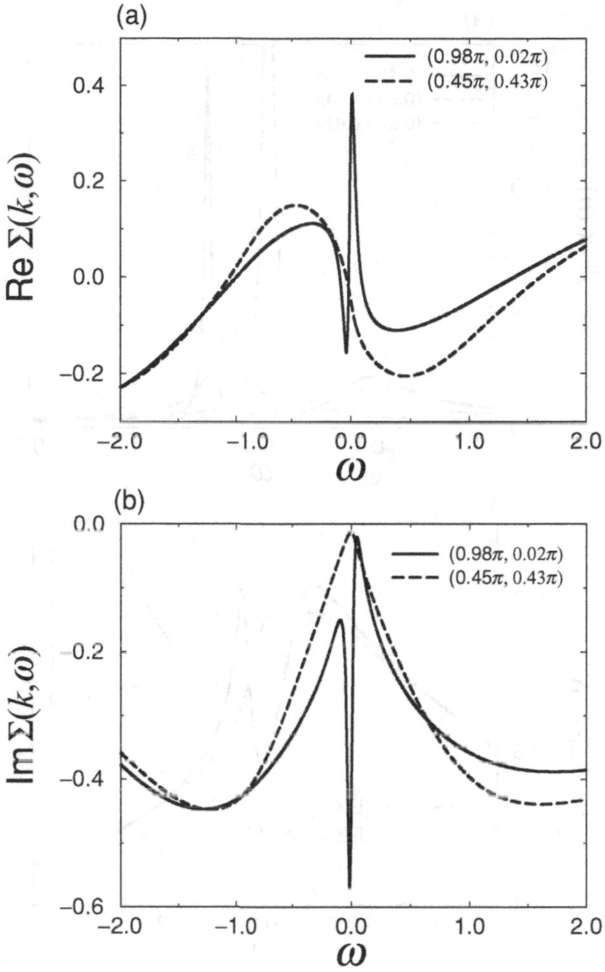

Fig. 9.32 The self-energy obtained by the FLEX+T-matrix approximation. (a) The real part. (b) The imaginary part. Here, $U = 1.6$, $\delta = 0.095$ and $T = 0.010$. The solid line and the broken line correspond to a hot spot $(\frac{63}{64}\pi, \frac{1}{64}\pi)$ and a cold spot $(\frac{29}{64}\pi, \frac{27}{64}\pi)$, respectively. Hereafter we put $t = 0.5$ and $t' = 0.25t$ in (9.4).

superconductors have been studied only for low density electron systems, as discussed by Legett and Nozières and Schmitt-Rink [20].

The results for the spectral weight obtained by the FLEX+T-matrix approximation are shown in Fig. 9.33. The pseudogap is clearly seen in the single particle spectral weight (Fig. 9.33(a)) and the DOS (Fig. 9.33(b)). Here it should be stressed that the pseudogap is derived from the self-energy correction due to the superconducting fluctuations, which are enhanced by the strong coupling superconductivity

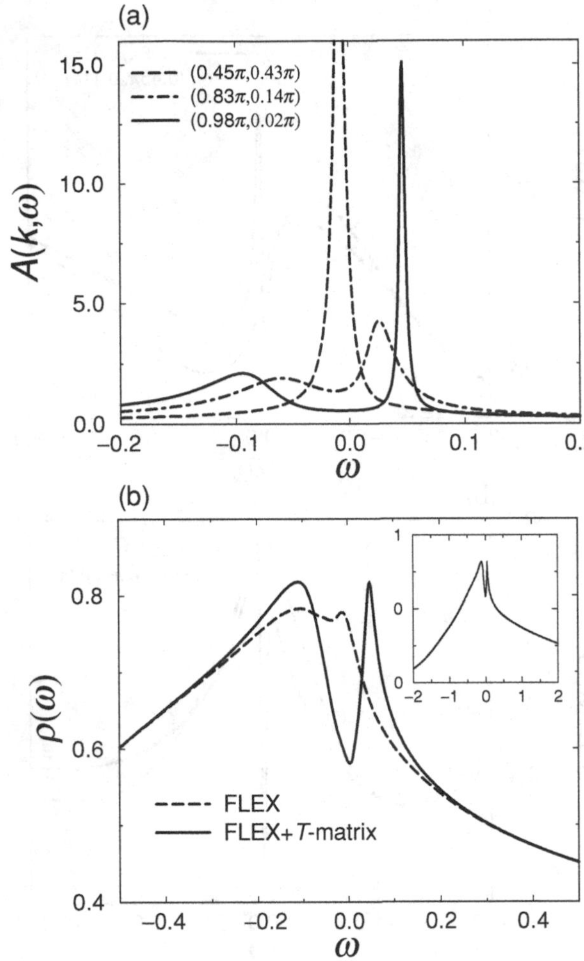

Fig. 9.33 (a) The single particle spectral weight obtained by the FLEX+T-matrix approximation. The broken, dash-dotted and solid lines correspond to cold spot $(0.45\pi, 0.43\pi)$, intermediate $(0.83\pi, 0.14\pi)$ and hot spot $(0.98\pi, 0.02\pi)$, respectively. (b) The density of states obtained by the FLEX (broken line) and the FLEX+T-matrix (solid line) approximations. The inset is the same result shown on a large scale. The other parameters are the same as those in Fig. 9.32.

and the quasi-two-dimensionality. Thus the scenario based on the resonance scattering mechanism is justified for the pseudogap phenomena.

The detailed results for the DOS are shown in Fig. 9.34, where the temperature $T = 1.25T_c$ is fixed to keep a fixed distance from the critical point. The doping dependence is shown in Fig. 9.34(a). The pseudogap becomes weak with increasing

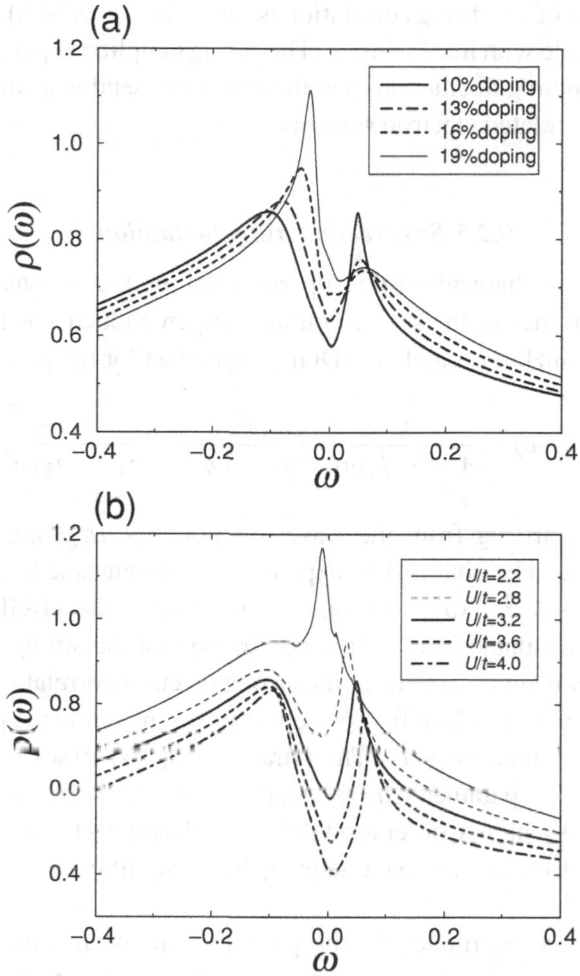

Fig. 9.34 The density of states obtained by the FLEX+T-matrix approximation. (a) The doping dependence and (b) the U dependence. In these figures, $T = 1.25T_c$.

hole doping. In the optimally doped region the gap is filled up and the DOS near the Fermi level increases. The effects of superconducting fluctuations almost disappear in the over-doped region $\delta > 0.2$. This is because both the critical temperature and the renormalization of the quasi-particles are reduced by the hole doping. Since the ratio T_c/ε_F decreases with increasing doping, the superconducting coupling $|g|$ is reduced and the effects of the superconducting fluctuations are reduced. It should be noted that the pseudogap becomes well-defined near the optimally doping $\delta \sim 0.15$. Thus the pairing scenario properly explains the pseudogap in the high-T_c cuprates, including their doping dependence.

The importance of the strong correlation is shown in Fig. 9.34(b). The pseudogap becomes remarkable with increasing U. The strong coupling superconductivity, the strong superconducting fluctuations and the resultant pseudogap are characteristics of the strongly correlated electron systems.

9.2.5 Superconducting fluctuations

Now we discuss the character of the superconducting fluctuations in order to emphasize the importance of the strong coupling superconductivity. Here we use the time-dependent Ginzburg–Landau (TDGL) expansion for the pair susceptibility:

$$t(\boldsymbol{q}, \omega) = \frac{g}{1 - \lambda(\boldsymbol{q}, \omega)} = \frac{g}{t_0 + b\boldsymbol{q}^2 - (a_1 + i a_2)\omega}, \tag{9.65}$$

where the factors arising from the wave-function $\phi(\boldsymbol{k}, i\omega_n)$ are neglected. The TDGL expansion can be obtained by expanding the eigenvalue function $\lambda(\boldsymbol{q}, \omega)$ as $1 - \lambda(\boldsymbol{q}, \omega) = t_0 + bq^2 - (a_1 + ia_2)\omega$. Now we discuss the TDGL parameters on the basis of the calculated results. It is confirmed that the strong coupling superconductivity is realized as a result of the strong electron correlation.

The parameter $t_0 = 1 - \lambda(\boldsymbol{0}, 0)$ represents the distance from the phase transition, and is sufficiently small near T_c. The parameter a_2 expresses the time scale of the fluctuations. The parameter a_1 is usually ignored within the weak coupling theory because it is a higher-order term than a_2 with respect to the superconducting coupling T_c/ε_F. However, the parameter a_1 has a significant effect in the strong coupling case.

The sign of a_1 is determined by the particle–hole asymmetry. The calculated results show that the sign of a_1 is negative in under-doped cuprates. This means a hole-like character of superconducting fluctuations. The parameter b represents the dispersion relation of the fluctuations and is related to the superconducting coherence length ξ_0 as $b \propto \xi_0{}^2$. The small b generally means strong superconducting fluctuations. The TDGL parameter b calculated by Yanase is shown in Fig. 9.35. We can see that the TDGL parameter b decreases with decreasing hole doping δ and/or with increasing U.

The TDGL parameter b is expressed by the Fermi liquid description within the weak coupling theory as $b = g_d \rho_d{}^*(0)\zeta(3)\bar{v}_F{}^{*2}/32\pi^2 T^2$, where g_d is the d-wave component of the residual interaction, $g_d = \sum_{\boldsymbol{k},\boldsymbol{k}'} \phi_d(\boldsymbol{k}) z_{\boldsymbol{k}} V_a(\boldsymbol{k} - \boldsymbol{k}') z_{\boldsymbol{k}'} \phi_d{}^*(\boldsymbol{k}')$. The effective density of states for quasi-particles $\rho_d{}^*(\varepsilon)$ is defined as $\rho_d{}^*(\varepsilon) = \sum_{\boldsymbol{k}} \delta(\varepsilon - \varepsilon_{\boldsymbol{k}}{}^*)|\phi_d(\boldsymbol{k})|^2$, which is enhanced by the renormalization. Here $\varepsilon_{\boldsymbol{k}}{}^*$ is the energy of a quasi-particle, $\varepsilon_{\boldsymbol{k}}{}^* = z_{\boldsymbol{k}}(\varepsilon_{\boldsymbol{k}} + \mathrm{Re}\,\Sigma^R(\boldsymbol{k}, 0))$. The $\bar{v}_F{}^*$ is the effective Fermi velocity for the d-wave symmetry and defined as $\bar{v}_F{}^{*2} = \int_F \rho_{k_F}{}^* v^*(k_F)^2 dk_F / \rho_d{}^*(0)$, where

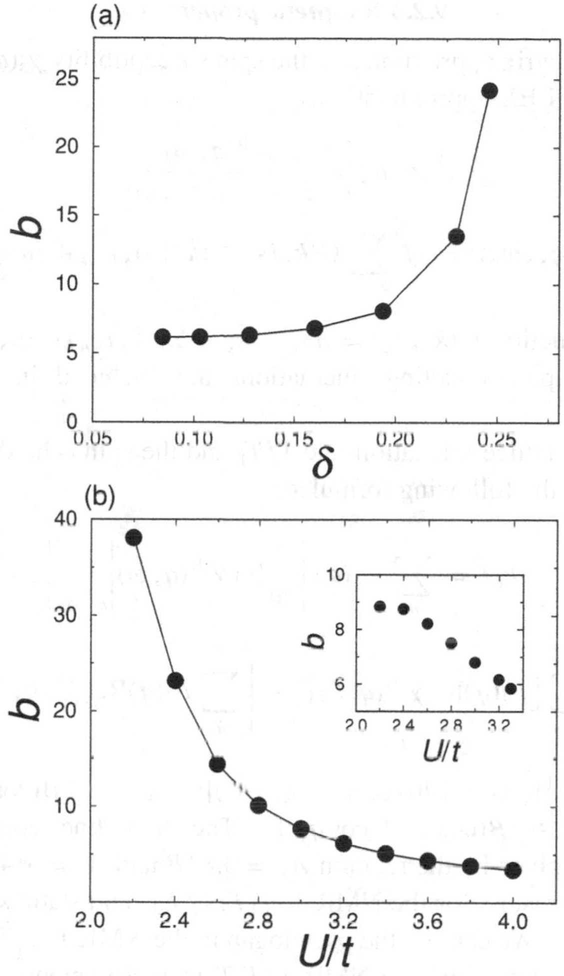

Fig. 9.35 The TDGL parameter b at T_c. (a) The doping dependence. Here, $U = 1.6$. (b) U dependence of b. Here $\delta = 0.09$. The inset shows b at $T = 0.0082$.

the integration is carried out on the Fermi surface and $\rho_{k_F}{}^* = |\phi_d(k_F)|^2/v^*(k_F)$. The velocity $v^*(k_F)$ is the absolute value of the velocity of the quasi-particle $v_k{}^* = d\varepsilon_k{}^*/d\boldsymbol{k}$ on the Fermi surface and is reduced by the renormalization, especially around $(\pi, 0)$. Since the effective Fermi velocity $\bar{v}_F{}^*$ is mainly determined by the region around $(\pi, 0)$, $\bar{v}_F{}^*$ is much reduced by the electron correlation.

The TDGL parameter b is proportional to $\bar{v}_F{}^{*2}/T^2$, which is proportional to the inverse square of the superconducting coupling $T_c/\varepsilon_F{}^*$. That is, the TDGL parameter b decreases with increasing superconducting coupling.

9.2.6 Magnetic properties

In the FLEX+T-matrix approximation, the spin susceptibility $\chi_s(\boldsymbol{q}, \omega)$ is obtained by extending the FLEX approximation:

$$\chi_s{}^{\mathrm{R}}(\boldsymbol{q}, \omega) = \frac{\chi_0{}^{\mathrm{R}}(\boldsymbol{q}, \omega)}{1 - U\chi_0{}^{\mathrm{R}}(\boldsymbol{q}, \omega)}, \tag{9.66}$$

$$\chi_0(\boldsymbol{q}, i\omega_n) = -T \sum_{\boldsymbol{k}, i\varepsilon_n} G(\boldsymbol{k}, i\varepsilon_n) G(\boldsymbol{k} + \boldsymbol{q}, i\varepsilon_n + i\omega_n), \tag{9.67}$$

where Green's function $G(\boldsymbol{k}, i\varepsilon_n) = (i\varepsilon_n - \varepsilon_{\boldsymbol{k}} - \Sigma_{\mathrm{F}}(\boldsymbol{k}, i\varepsilon_n) - \Sigma_{\mathrm{S}}(\boldsymbol{k}, i\varepsilon_n))^{-1}$. The effects of the superconducting fluctuations are included in the self-energy $\Sigma_{\mathrm{S}}(\boldsymbol{k}, i\varepsilon_n)$.

The NMR spin lattice relaxation rate $1/T_1$ and the spin echo decay rate $1/T_{2\mathrm{G}}$ are obtained from the following formulae:

$$1/T_1T = \sum_{\boldsymbol{q}} F_{\perp}(\boldsymbol{q}) \left[\frac{1}{\omega} \mathrm{Im}\, \chi_s{}^{\mathrm{R}}(\boldsymbol{q}, \omega) \bigg|_{\omega \to 0} \right], \tag{9.68}$$

$$1/T_{2\mathrm{G}}{}^2 = \sum_{\boldsymbol{q}} \left[F_{\parallel}(\boldsymbol{q}) \mathrm{Re}\, \chi_s{}^{\mathrm{R}}(\boldsymbol{q}, 0) \right]^2 - \left[\sum_{\boldsymbol{q}} F_{\parallel}(\boldsymbol{q}) \mathrm{Re}\, \chi_s{}^{\mathrm{R}}(\boldsymbol{q}, 0) \right]^2. \tag{9.69}$$

Here $F_{\perp}(\boldsymbol{q}) = \frac{1}{2}[\{A_1 + 2B(\cos q_x + \cos q_y)\}^2 + \{A_2 + 2B(\cos q_x + \cos q_y)\}^2]$ and $F_{\parallel}(\boldsymbol{q}) = \{A_2 + 2B(\cos q_x + \cos q_y)\}^2$. The hyperfine coupling constants A_1, A_2 and B are given by the relation $A_1 = 0.84B$ and $A_2 = -4B$.

The calculated results for the NMR $1/T_1T$, $1/T_{2\mathrm{G}}$ and static susceptibility are shown in Fig. 9.36. We can see the pseudogap in the NMR $1/T_1T$ in Fig. 9.36(a). In the FLEX approximation, the NMR $1/T_1T$ increases monotonously with decreasing temperature owing to the development of the spin fluctuations (see the inset to Fig. 9.36(a)). In the FLEX+T-matrix calculation, the NMR $1/T_1T$ increases with decreasing temperature from high temperature, shows a peak at T^* and decreases with decreasing temperature. This decrease above T_c is the well-known pseudogap in NMR measurements. This phenomenon is caused by the superconducting fluctuations. Since the DOS is reduced by the superconducting fluctuations, the low-frequency spin fluctuations are suppressed as a result. Since the NMR $1/T_1T$ measures the low-frequency component of the spin fluctuations, $1/T_1T$ decreases with approaching T_c. Thus the pseudogap observed in the NMR $1/T_1T$ takes place through the pseudogap in the single particle states.

The NMR $1/T_{2\mathrm{G}}$ also shows a pseudogap with the same onset temperature T^* as that in $1/T_1T$ (Fig. 9.36(b)). The NMR $1/T_{2\mathrm{G}}$ also decreases with approaching T_c. This is an effect of the superconducting fluctuations. However, the pseudogap

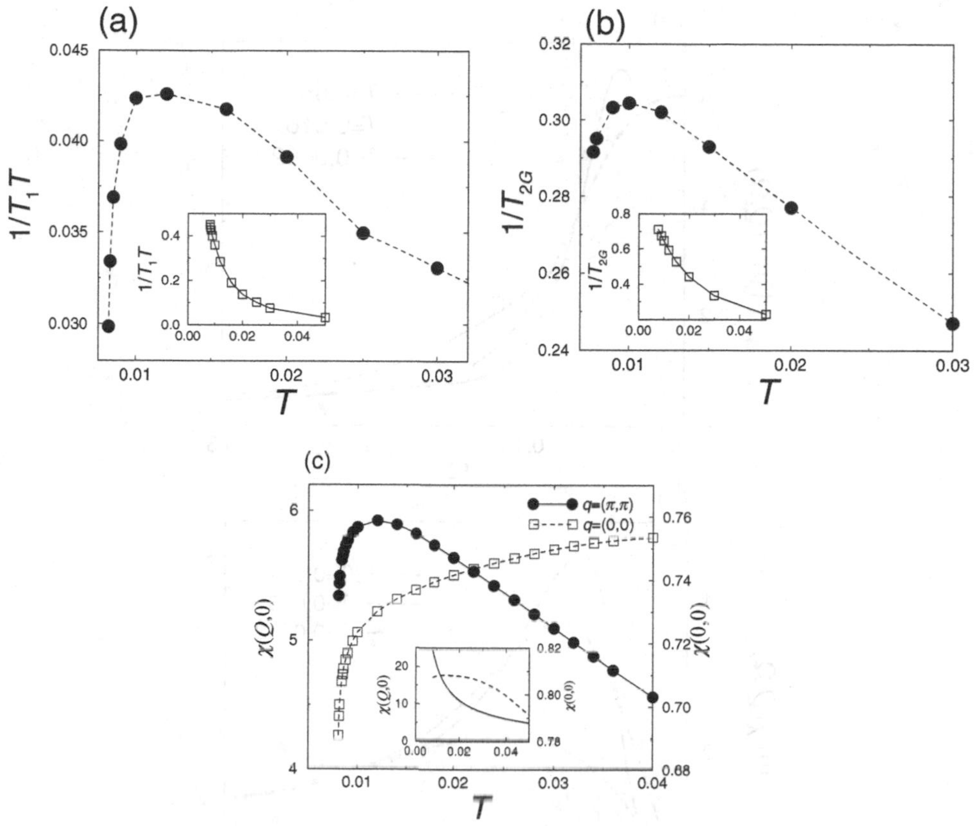

Fig. 9.36 (a) The NMR $1/T_1T$. (b) The NMR $1/T_{2G}$. (c) The static spin susceptibility, $\chi_s^R(\boldsymbol{q}, 0)$ at $\boldsymbol{q} = (0, 0)$ (open squares) and at $\boldsymbol{q} = (\pi, \pi)$ (closed circles). Here, the doping concentration is fixed to the under-doped region $\delta = 0.093$–0.103. These results are obtained by the FLEX+T-matrix approximation. The insets in (a), (b) and (c) show the results obtained by FLEX.

in the NMR $1/T_{2G}$ is weak compared with that in the NMR $1/T_1T$, since the NMR $1/T_{2G}$ measures the static spin susceptibility which reflects the total weight of the spin fluctuations. It should be noted that the pseudogap suppresses only the low-frequency component of the spin fluctuations, because the superconductivity has a smaller energy scale than that of the spin fluctuations.

Similar features of the NMR $1/T_1T$ and $1/T_{2G}$ have been observed in the superconducting state. The $1/T_{2G}$ remains even at low temperature, while the $1/T_1T$ rapidly decreases. Hence the above results for the pseudogap state are reasonable, since the pseudogap is a precursor of the d-wave superconductivity. While the many quantities show the pseudogap below the same onset temperature T^*, the uniform spin susceptibility $\chi_s{}^R(\mathbf{0}, 0)$ decreases from a much higher temperature than T^*, and the decrease in uniform susceptibility becomes more rapid near T^*. This rapid

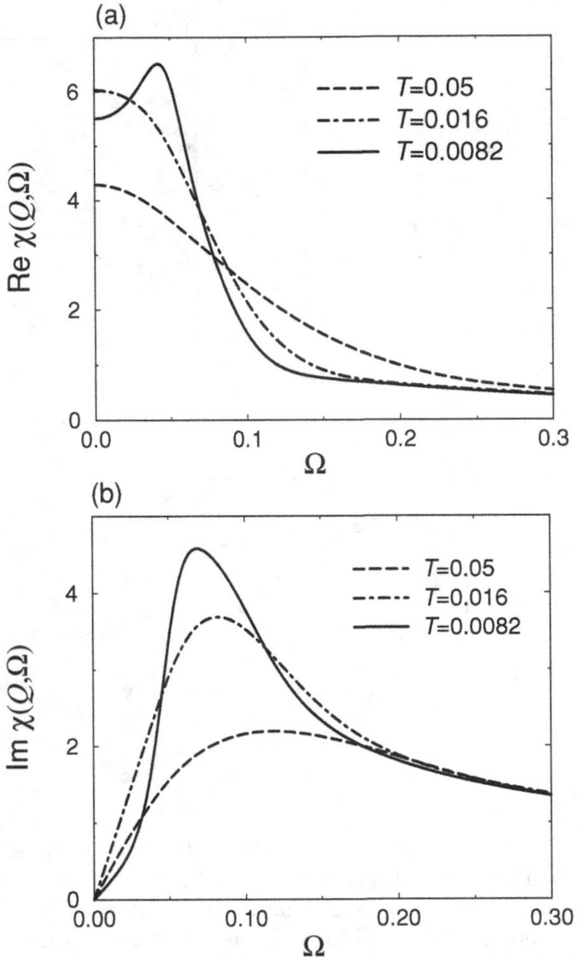

Fig. 9.37 The staggered spin susceptibility $\chi_s{}^R(\boldsymbol{Q}, \Omega)$ calculated by the FLEX+T-matrix approximation for under-doped cuprates. (a) The real part. (b) The imaginary part.

decrease is caused by superconducting fluctuations. Figure 9.36(c) shows the uniform susceptibility $\chi_s{}^R(\mathbf{0}, 0)$ and the staggered susceptibility $\chi_s{}^R(\boldsymbol{Q}, 0)$. The staggered susceptibility shows the pseudogap below the pseudogap onset temperature T^*. On the other hand, the uniform susceptibility $\chi_s{}^R(\mathbf{0}, 0)$ decreases with decreasing temperature from a much higher temperature, and decreases more rapidly below T^*. This is in good agreement with the NMR Knight shift measurements.

The decrease in uniform susceptibility is seen even in the FLEX approximation (see the inset in Fig. 9.36(c)). Hence the decrease in uniform susceptibility is not necessarily attributed to the superconducting fluctuations. However, the superconducting fluctuations significantly affect the uniform susceptibility and remarkably reduce it near the critical temperature T_c.

The frequency dependence of the spin susceptibility well describes the character of the pseudogap in magnetic properties. The results for the dynamical spin susceptibility $\chi_s{}^R(\boldsymbol{q}, \omega)$ at $\boldsymbol{q} = \boldsymbol{Q}$ are shown in Fig. 9.37. The real part is suppressed at low frequency in the pseudogap state ($T = 0.0082$ in Fig. 9.37(a)). Thus the magnetic ordering is suppressed by the superconducting fluctuations.

The imaginary part has been measured by inelastic neutron scattering and shows the pseudogap. The calculated results show that the imaginary part is remarkably suppressed at low frequency in the pseudogap state (Fig. 9.37(b)). This suppression is the pseudogap observed by the inelastic neutron scattering experiments. The pseudogap transfers the spectral weight of the spin fluctuations from the low-frequency region to the high-frequency region, and the total weight is not so reduced by the pseudogap. These features agree with the above explanation for the NMR $1/T_1T$ and $1/T_{2G}$. It should be noted that the d-wave pairing interaction is not so reduced by the pseudogap, because the pairing interaction due to the spin fluctuations originates from the relatively wide frequency region.

Now we discuss the commensurate and incommensurate structures of the spin fluctuations. We show the dynamical susceptibility Im $\chi_s{}^R(\boldsymbol{q}, \Omega)$ in Fig. 9.38. One general result is that the incommensurability increases with doping concentration δ. Another general result is that the superconducting fluctuations enhance the incommensurability.

9.2.7 *Self-consistent calculation*

Yanase carried out a self-consistent calculation including spin fluctuations, superconducting fluctuations and single-particle properties. Let us call it the SC-FLEX+T-matrix calculation. In the calculation, (9.38)–(9.43) and (9.60)–(9.64) are solved self-consistently, where the fully dressed Green's function $G(\boldsymbol{k}, i\varepsilon_n) = (i\varepsilon_n - \varepsilon_{\boldsymbol{k}} - \Sigma(\boldsymbol{k}, i\varepsilon_n))^{-1}$ is used. By the self-consistent calculation, the critical temperature T_c is reduced by the superconducting fluctuations. As mentioned before, there exists a singularity arising from the two-dimensionality. The singularity is actually removed by the weak coupling between layers. By taking the weak three-dimensionality into account, the critical temperature T_c is determined as the temperature at which $\lambda(\mathbf{0}, 0) = 0.98$. The method of determination makes no significant difference to the obtained results.

In order to study whether the feedback effects of the pseudogap on the spin fluctuations suppress the superconductivity itself, Yanase also carried out a modified FLEX (M-FLEX) calculation. In the M-FLEX calculation, the fully dressed Green's function is used only in (9.41), while in the other equations

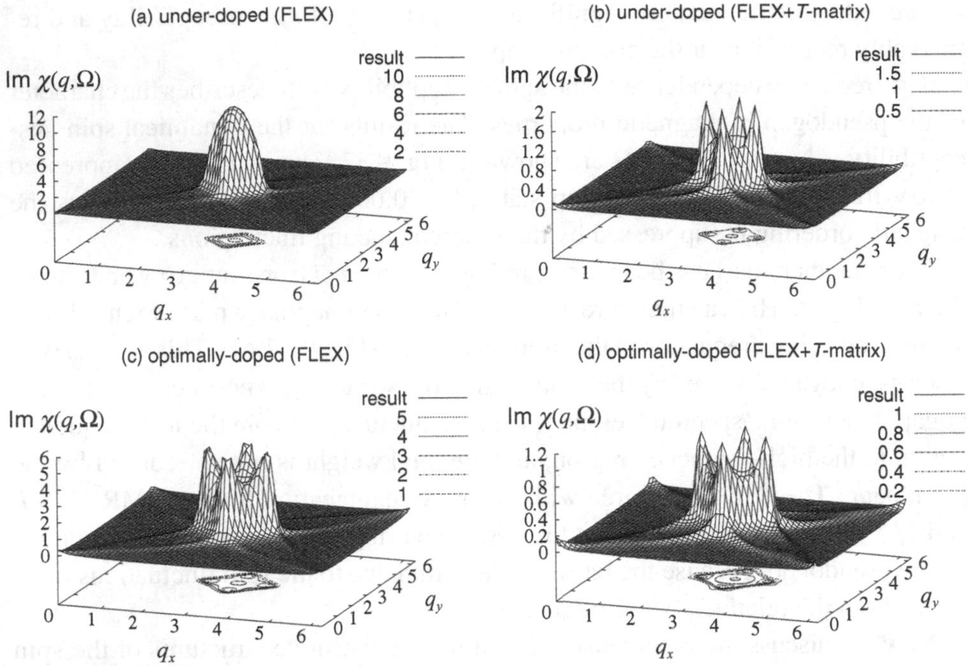

Fig. 9.38 The momentum dependence of the dynamical spin susceptibility $\mathrm{Im}\,\chi_s{}^R(\boldsymbol{q}, \Omega)$ at $\Omega = 0.01$. (a) The result for under-doped cuprates ($\delta = 0.095$ and $T = 0.01$), result obtained by the FLEX approximation. (b) Result obtained by the FLEX+T-matrix approximation. (c) The result for optimally doped cuprates ($\delta = 0.156$ and $T = 0.00789$. (d) Result obtained by the FLEX+T-matrix approximation.

$G^{\mathrm{F}}(\boldsymbol{k}, i\varepsilon_n) = (i\varepsilon_n - \varepsilon_{\boldsymbol{k}} - \Sigma_{\mathrm{F}}(\boldsymbol{k}, i\varepsilon_n))^{-1}$ is used. By this approximation the effects of the superconducting fluctuations on the spin fluctuations are included, although those on the single-particle properties are not included. Equations (9.38)–(9.43) and (9.60)–(9.64) given by the M-FLEX approximation are solved self-consistently. The results obtained by the FLEX, M-FLEX and SC-FLEX+T-matrix calculations are shown in Table 9.1. Here the critical temperatures T_{c} are determined by the condition $\lambda(\mathbf{0}, 0) = 0.98$ in the respective calculations for an equity.

The higher critical temperature $T_{\mathrm{c}} = 0.0098$ is obtained by the M-FLEX calculation, while $T_{\mathrm{c}} = 0.0084$ in the FLEX calculation. That is, the feedback effects enhance the superconductivity. This fact can be understood as follows. The spin fluctuations not only have a pairing effect but also a depairing effect. The former is from the relatively wide frequency region, and the latter from the low frequency component. The pseudogap strongly suppresses the low frequency components, while the total weight is not so reduced (Fig. 9.37). As a result, the depairing effect is more reduced than the pairing effect by the pseudogap.

Table 9.1 *The comparison among the FLEX, M-FLEX and SC-FLEX+T-matrix approximations. The critical temperature T_c, the effective pairing interaction $g\phi^2_{max}$, the damping at 'hot spot' γ_h and that at 'cold spot' γ_c at $T = T_c$ are shown. The parameters are $U = 1.6$ and $\delta = 0.083$–0.096. Here, the self-energy in the M-FLEX approximation is that due to the spin fluctuations $\Sigma^R_F(k, \omega)$*

	FLEX	M-FLEX	SC-FLEX+T-matrix
T_c	0.0084	0.0098	0.0031
$g\phi^2_{max}$	24.18	14.23	20.1529
γ_h	0.06970	0.02747	0.04277
γ_c	0.00995	0.00971	0.00298

Since the pseudogap itself in single-particle properties reduces the critical temperature, the lower critical temperature $T_c = 0.0031$ is obtained by the SC-FLEX+T-matrix calculation. As is seen from the fact that the pseudogap is easily caused by superconducting fluctuations, the depairing effect from superconducting fluctuations is more drastic than that from spin fluctuations.

Now we show the results obtained by the SC-FLEX+T-matrix calculation. Qualitatively similar results to the FLEX+T-matrix approximation are obtained, although the effects of superconducting fluctuations are reduced by the requirement of self-consistency. Since superconducting fluctuations suppress the antiferromagnetic order, we can treat a stronger electron–electron interaction U. The effect of superconducting fluctuations is clearly seen in the DOS shown in Fig. 9.39. The DOS near the Fermi level is reduced by superconducting fluctuations and the gap structure appears in the under-doped case. The pseudogap is suppressed by the hole doping. The pseudogap is thus properly obtained by the self-consistent calculation, including doping dependence.

Finally the phase diagram is shown in Fig. 9.40. The superconducting critical temperature T_c is determined by the self-consistent calculation and is suppressed by superconducting fluctuations. It should be noted that the suppression of T_c from the mean field (FLEX) value becomes remarkable with under-doping. An important result is that T_c has a maximum value at $\delta \sim 0.11$ and decreases with under-doping in the SC-FLEX+T-matrix calculation with $U = 2.4$, whereas T_c goes on increasing in the FLEX calculation. For $U = 1.6$, T_c does not decrease with under-doping even in the SC-FLEX+T-matrix calculation. The strong renormalization of the quasi-particles due to the electron correlation plays an essential role in describing the under-doped cuprates.

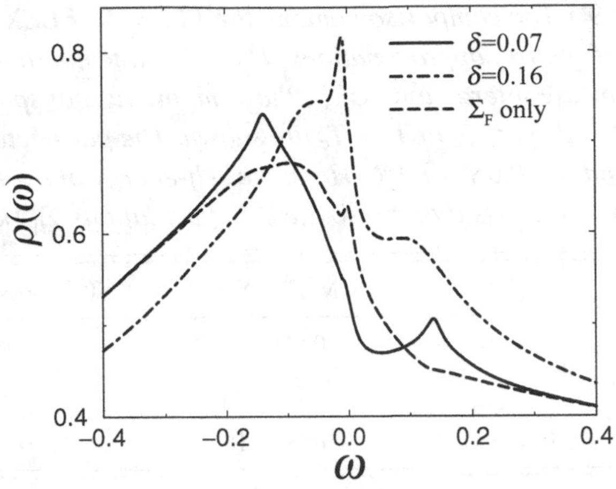

Fig. 9.39 The density of states obtained by the SC-FLEX+T-matrix approximation. The solid and dash-dotted lines show the under-doped case ($\delta = 0.073$ and $T = 0.004$) and the over-doped case ($\delta = 0.165$ and $T = 0.0035$), respectively. The long-dashed line shows the result for the under-doped case obtained by neglecting $\Sigma_S(\boldsymbol{k}, i\omega_n)$.

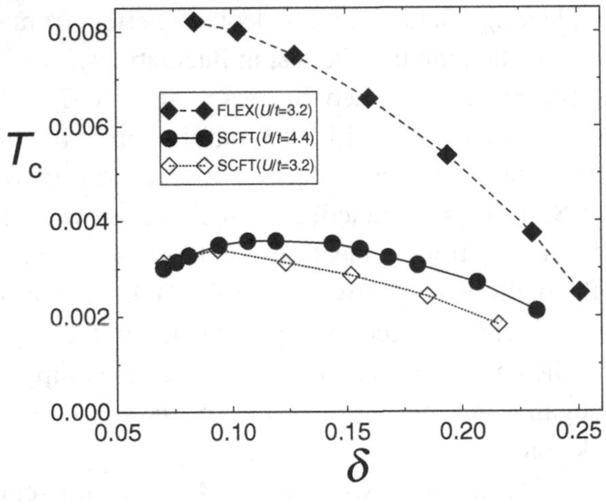

Fig. 9.40 The phase diagram obtained by the FLEX and SC-FLEX+T-matrix approximations. The closed circles show the results of the SC-FLEX+T-matrix approximation for $U = 2.4$. The closed and open diamonds show the results of the FLEX and SC-FLEX+T-matrix approximations, respectively, for $U = 1.6$.

9.3 Electron-doped cuprates

In this section let us apply the above calculation to the electron-doped cuprates. Here we describe the theoretical work carried out by Yanase and Yamada using the FLEX approximation [19]. There also exist electron-doped cuprate superconductors such as $Nd_{2-x}Ce_xO_{4-y}$ and $Pr_{2-x}Ce_xCuO_{4-y}$. Electron-hole symmetry is expected within the simple Hubbard model, including only nearest neighbour hopping. However, by the experiments for electron-doped cuprates, some different properties from hole-doped cuprates are pointed out. The antiferromagnetic ordered state is robust against carrier doping in the electron-doped cuprates compared with the hole-doped ones. A relatively low T_c is observed in the narrow doping range.

Now let us show that the theory used for hole-doped cuprates properly describes the essential properties of the electron-doped cuprates. The electron–hole asymmetry is taken into account by the next-nearest neighbour hopping terms t' in the Hubbard model. The main difference between the electron- and the hole-doped cuprates results from the shape of the Fermi surface and the distance from the Fermi level to the van Hove singularity. The Fermi surface of the electron-doped cuprates is given by lifting the chemical potential μ so as to make n larger than unity (see Fig. 9.4). As a result the Fermi level is lifted far from the van Hove singularity $(\pi, 0)$, and the DOS is rather low in the electron-doped cuprates. The low DOS means that the electron correlation is effectively weak. On the other hand, the nesting effect around $(\pi/2, \pi/2)$ is enhanced and the tendency towards antiferromagnetic order becomes robust. In the light of the d–p model, the carriers are confined to the Cu site in the electron-doped cases, while they are in the O sites for the hole-doped cases. This fact also makes robust the antiferromagnetic order. For the electron-doped systems $\delta = 1 - n$ is negative, n being the number of electrons per site. Owing to the above results, the spectrum of the spin fluctuations is confined to a rather small region of momentum space as shown in Fig. 9.41, and the attractive interaction mediated by the antiferromagnetic fluctuations is weak compared with that of the hole-doped cuprates. We therefore cannot expect the pseudogap phenomena for the electron-doped case, since the strong coupling superconductivity is not realized there. The critical temperature T_c is shown in Fig. 9.42 as a function of doping δ. With increasing electron doping, T_c decreases in accordance with experiment. As a result of the weak correlation in the electron-doped system, the width of the quasi-particle spectra shows T^2 and ω^2 dependence. Actually, Yanase obtained ω^2 dependence of $\mathrm{Im}\,\Sigma(\boldsymbol{k}, \omega)$.

9.3.1 Organic superconductor BEDT–TTF

The κ-type organic superconductor BEDT–TTF (κ-$(ET)_2X$) can be well described by the half-filled Hubbard Hamiltonian on a square lattice with nearest neighbour hopping integral t and next-nearest neighbour hopping integral t', as shown in Fig. 9.43.

The superconductivity of organic conductors has been argued by Kino and Kontani [21], Kondo and Moriya [22] and Schmalian [23] using the FLEX

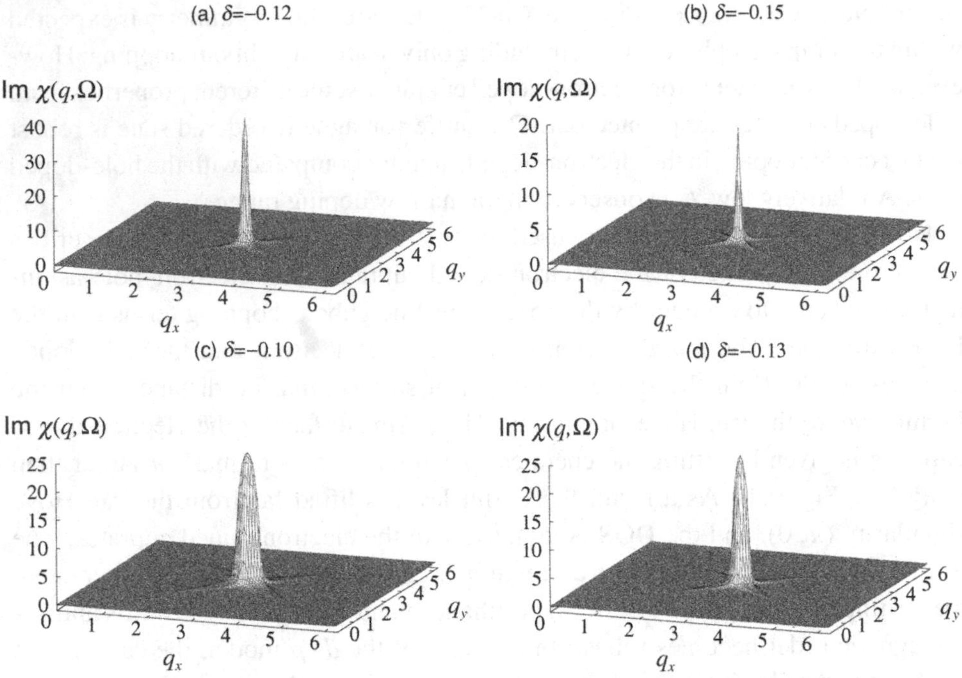

Fig. 9.41 The momentum dependence of the dynamical spin susceptibility. $\mathrm{Im}\,\chi_s{}^R(\boldsymbol{q}, \Omega)$ at $\Omega = 0.01$ calculated by the FLEX approximation. (a) $\delta = -0.123$ and (b) $\delta = -0.150$ at $t' = -0.25t$ and $T = 0.01$. (c) $\delta = -0.104$ and (d) $\delta = -0.130$ at $t' = -0.35t$ and $T = 0.005$.

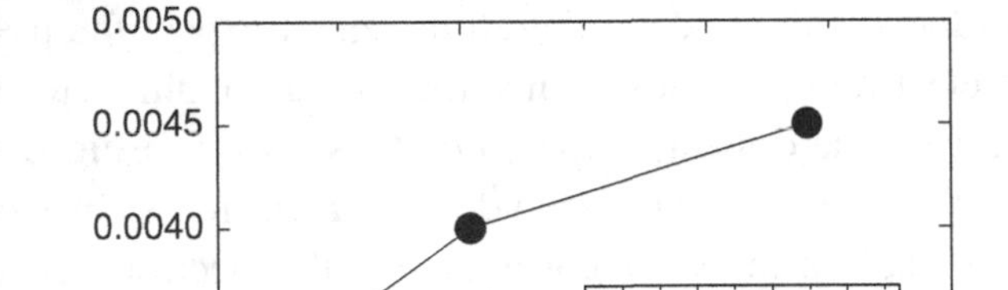

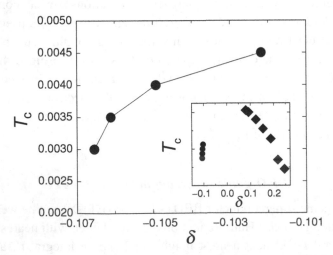

Fig. 9.42 The critical temperature T_c for electron-doped cuprates. The inset shows the phase diagram including both hole- and electron-doped cases.

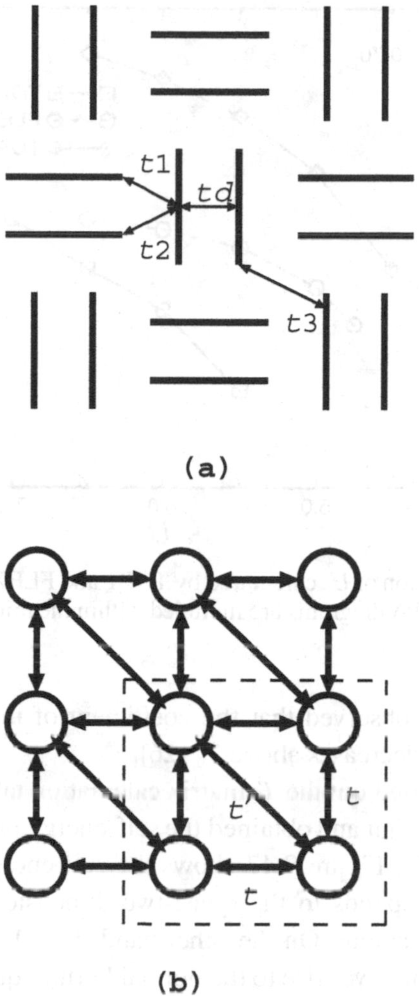

Fig. 9.43 The lattice structure of κ-$(ET)_2X$; t and t' are transfer integrals.

approximation. Jujo *et al.* studied superconductivity by the perturbation method with respect to the coulomb repulsion U [24].

The critical temperatures T_c obtained by the FLEX approximation and the third-order perturbation theory (TOPT) with respect to U are shown in Fig. 9.44 as functions of U. In Fig. 9.45(a), T_c is shown as a function of t' at fixed $U = 6.5$.

As shown in Fig. 9.45(b), the antiferromagnetic susceptibility is suppressed with increasing t' and the attractive interaction decreases, since the antiferromagnetic spin fluctuations are suppressed by approaching a triangular lattice.

The pseudogap phenomena are also observed in the κ-BEDT–TTF systems. The NMR relaxation rate $1/(T_1T)$ shows the pseudogap as shown in Fig. 9.46 [25].

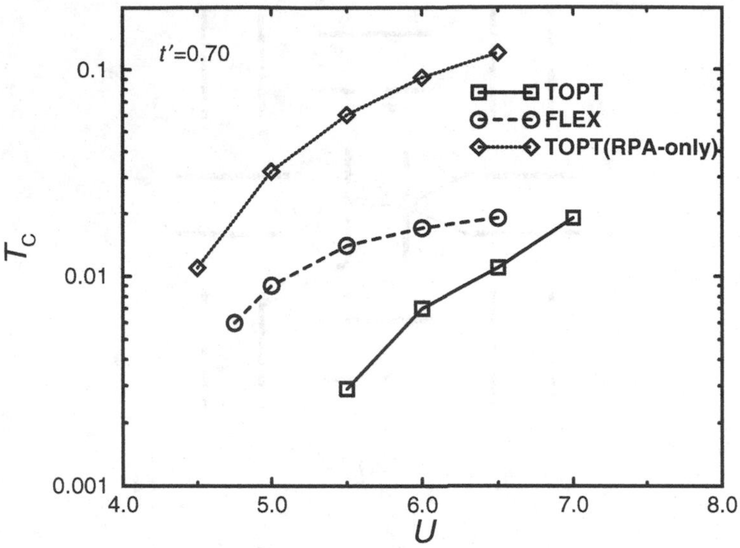

Fig. 9.44 T_c as a function of U, calculated by TOPT and FLEX. TOPT (RPA-only) means that only the RPA diagrams are included within the third-order perturbation theory.

Nakazawa and Kanoda observed that the coefficient of the T-linear term of the electronic specific heat decreases above T_c [26].

Jujo and Yamada carried out the T-matrix calculation taking the superconducting fluctuations into account and obtained the self-energy part due to the superconducting fluctuations [27]. Figure 9.47 shows the self-energy part. The interlayer coupling $t_z = 0.1$ corresponds to the quasi-two-dimensional case, in which we have the pseudogap behaviour. On the other hand, $t_z = 1.0$ corresponding to the three-dimensional system gives rise to the normal Fermi liquid behaviour. This fact confirms that the pseudogap is inherent in the two-dimensional superconducting fluctuations as well as a strong coupling superconductivity.

The Fermi surface and the spectral weight at point A are shown in Fig. 9.48. The pseudogap becomes clear with decreasing temperature. The pseudogap in the κ-BEDT–TTF system has been observed by Miyagawa *et al.* [25].

9.4 Triplet superconductor Sr_2RuO_4

The quasi-two-dimensional superconductor Sr_2RuO_4 which has the same crystal structure as the high-T_c cuprates was discovered in 1994 by Maeno *et al.* [8]. From a theoretical point of view Rice and Sigrist [9] pointed out the possibility of a spin-triplet pairing superconductivity in Sr_2RuO_4. The NMR Knight shift measurement

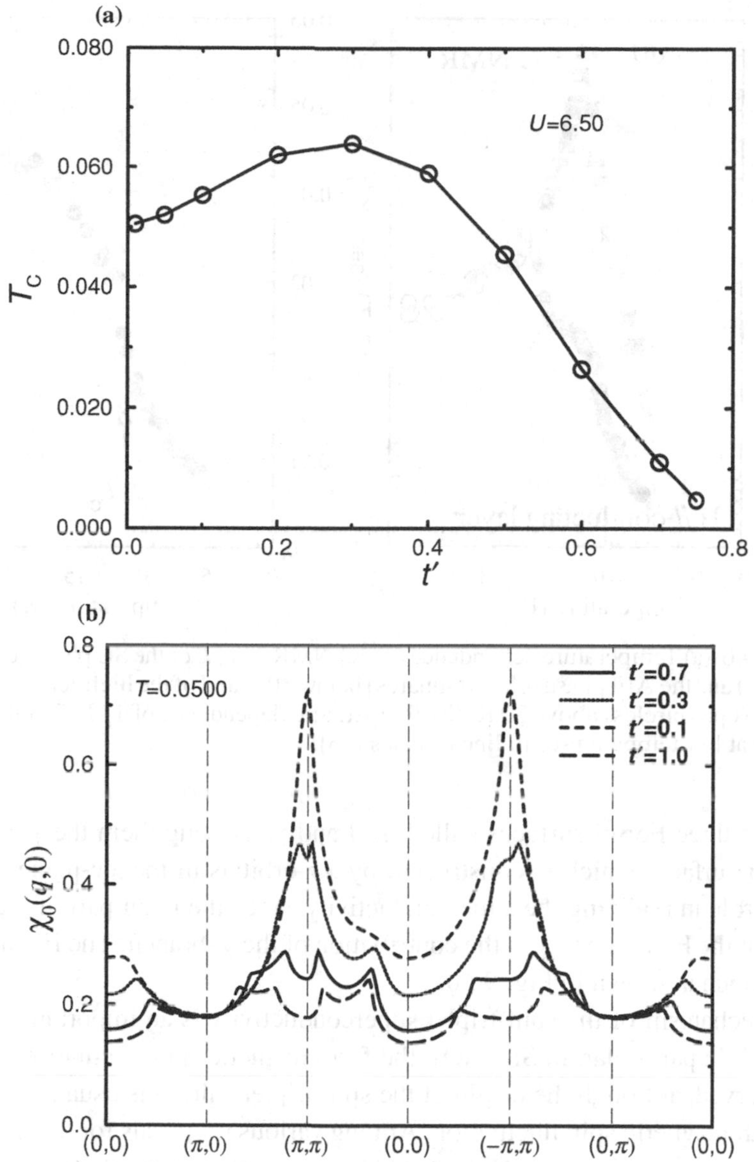

Fig. 9.45 (a) The t' dependence of T_c in TOPT, $U = 6.50$. (b) The momentum dependence of the static bare susceptibility $\chi_0(\boldsymbol{q}, 0)$ for various values of t'. Here, $T = 0.05$.

by Ishida *et al.* [10] showed that the spin-triplet pairing is realized in Sr_2RuO_4 because the Knight shift remains unchanged through the transition temperature $T_c = 1.5\,K$. The triplet superconductor Sr_2RuO_4 possesses two-dimensional RuO_2 networks in place of CuO_2 ones in the cuprates, as shown in Fig. 9.49. In Sr_2RuO_4

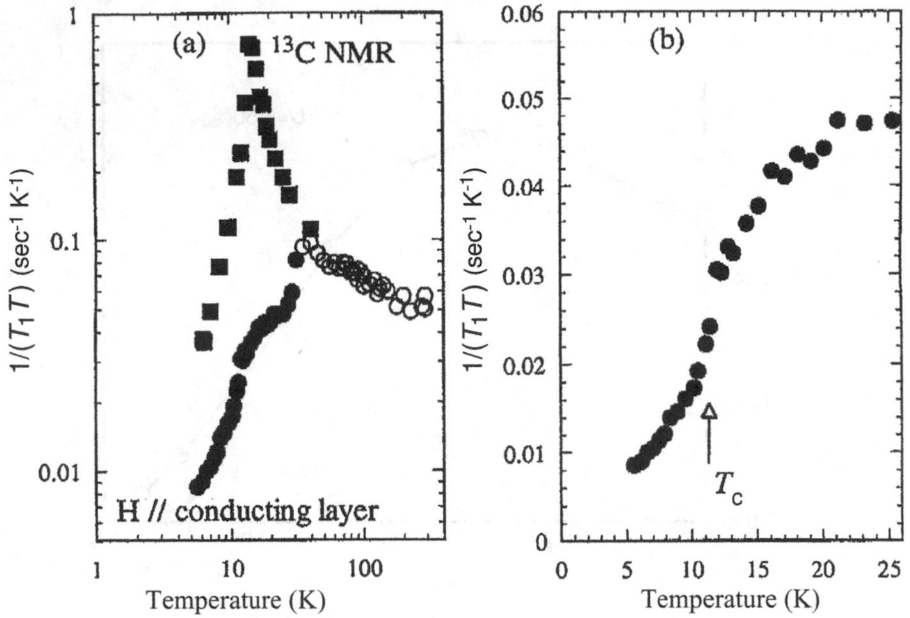

Fig. 9.46 (a) Temperature dependence of ^{13}C-NMR $1/T_1T$ of the SC phase (closed circles) and the AFI phase (closed squares) below 30 K and of the high-temperature phase (open circles) above 30 K. (b) Temperature dependence of $1/T_1T$ of the SC phase at low temperatures in linear scales [25].

there exist three Fermi surfaces called α, β and γ. Among them the γ-branch of the Fermi surface, which is constructed by xy-orbitals in the xy-plane, plays an essential role in realizing the superconductivity, since the main part of the density of states at the Fermi energy is the contribution of the γ-branch. The Fermi surface of a γ-branch is shown in Fig. 9.50.

The mechanism of the spin-triplet superconductivity is an important subject to be studied. In particular, in Sr_2RuO_4, the ferromagnetic spin fluctuations have not been observed, although the origin of the spin-triplet pairing is usually considered to be a paramagnetic spin fluctuation. Among various proposals for the triplet pairing of Sr_2RuO_4, Nomura and Yamada [11] carried out the perturbation calculation with respect to the on-site coulomb repulsion U for a two-dimensional Hubbard Hamiltonian and solved the Dyson–Gor'kov equation to successfully obtain the critical temperature for the spin-triplet superconducting state. In theory, the parameters related to the Hubbard model are adjusted to those for the main γ-branch of the Fermi surface in Sr_2RuO_4. The Hamiltonian is given by

$$\mathcal{H} = \sum_{\boldsymbol{k},\sigma} \varepsilon(\boldsymbol{k}) c_{\boldsymbol{k}\sigma}{}^{\dagger} c_{\boldsymbol{k}\sigma} + \frac{U}{2} \sum_{i} \sum_{\sigma \neq \sigma'} c_{i\sigma}{}^{\dagger} c_{i\sigma'}{}^{\dagger} c_{i\sigma'} c_{i\sigma}, \tag{9.70}$$

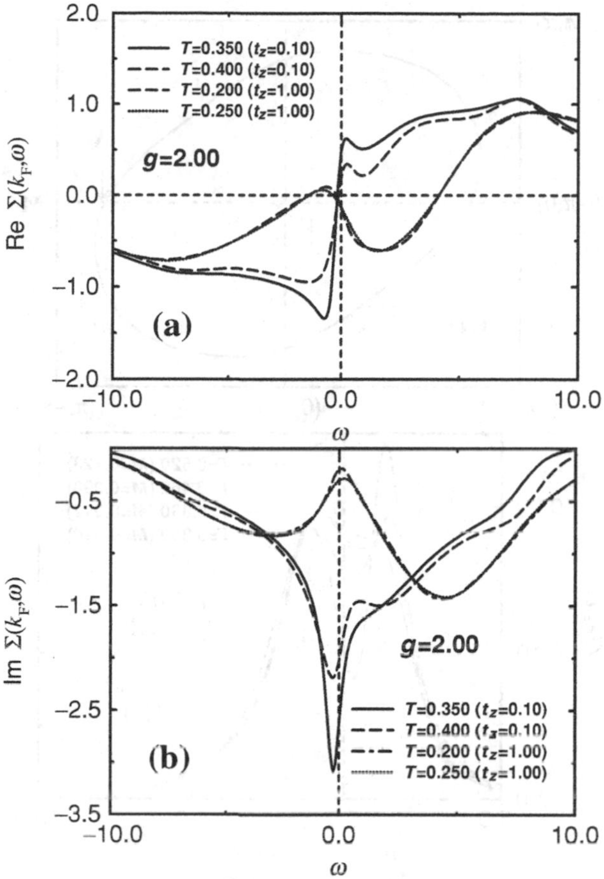

Fig. 9.47 The self-energy for $t_z = 0.1$ ($T_c = 0.337$) and $t_z = 1.0$ ($T_c = 0.195$). The coupling constant $|g| = 2$. (a) The real part. (b) The imaginary part.

where

$$\varepsilon(\boldsymbol{k}) = 2t_1(\cos k_x + \cos k_y) + 4t_2 \cos k_x \cos k_y. \tag{9.71}$$

The diagrams for the anomalous self-energy are shown up to third-order terms with respect to U in Fig. 9.51. The obtained critical temperature T_c for the spin-triplet state is shown in Fig. 9.52 as a function of U. The critical temperatures for the spin-triplet and the spin-singlet states are compared in Fig. 9.53 for the electron numbers $n = 0.630$ and 0.667, n being the electron number per spin. For $n = 0.667$, in good agreement with the actual electron number of the γ-branch, the spin-triplet states possess higher critical temperatures than those for singlet states. The t_2 dependence of T_c is shown in Fig. 9.54 for typical values of n. The critical temperature depends strongly on the filling number n. When the electron number per site, $2n$, is near

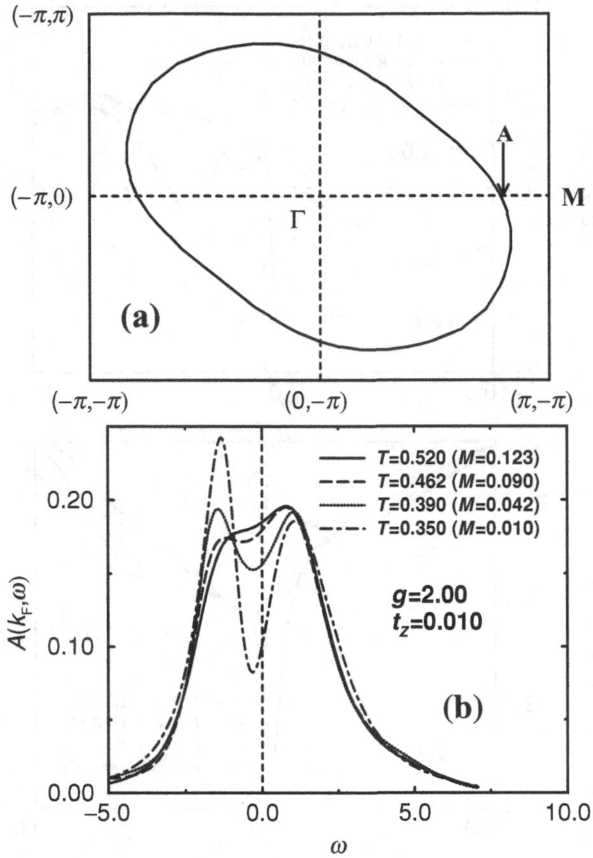

Fig. 9.48 (a) The Fermi surface for $t_z = 0$. (b) The one-particle spectrum at point A in (a) is shown for typical temperatures. M is the mass term which is a measure of the distance from T_c. The pseudogap becomes clear with decreasing T and M.

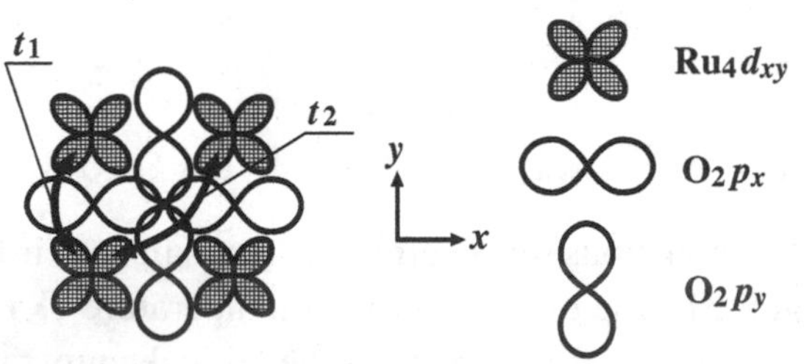

Fig. 9.49 The structure of the RuO_2 layer.

$2n \simeq 1$, the antiferromagnetic spin fluctuations are strong enough to realize d-wave superconductivity. The spin-triplet pairing is realized for electron filling n far from the half-filling number. This filling dependence is similar to that of the appearance of ferromagnetism [28].

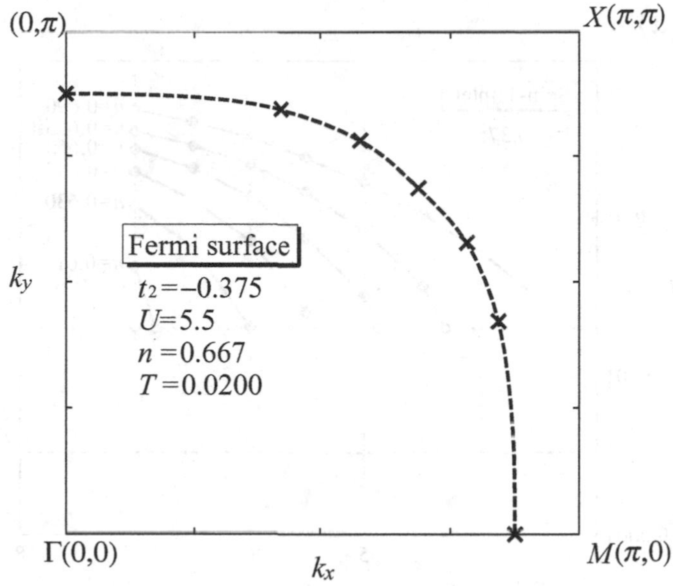

Fig. 9.50 The Fermi surface of the γ-branch.

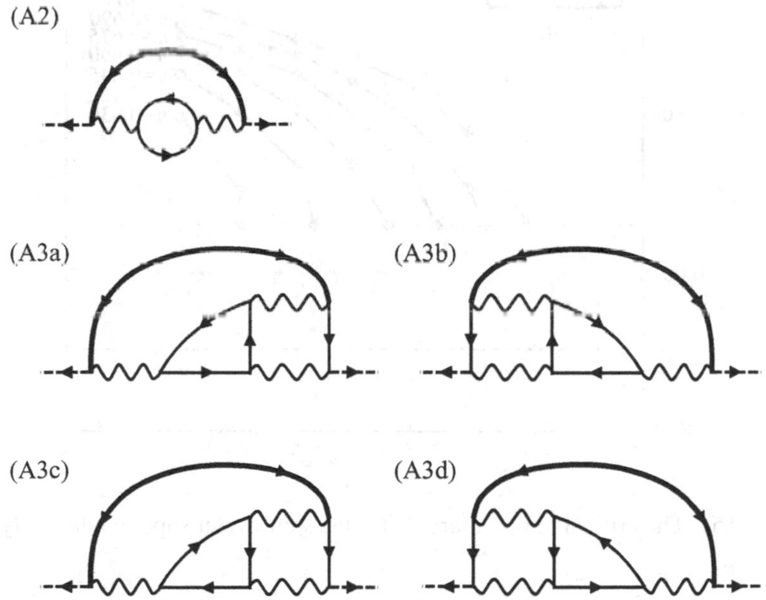

Fig. 9.51 Diagrams for the anomalous self-energy up to third-order terms.

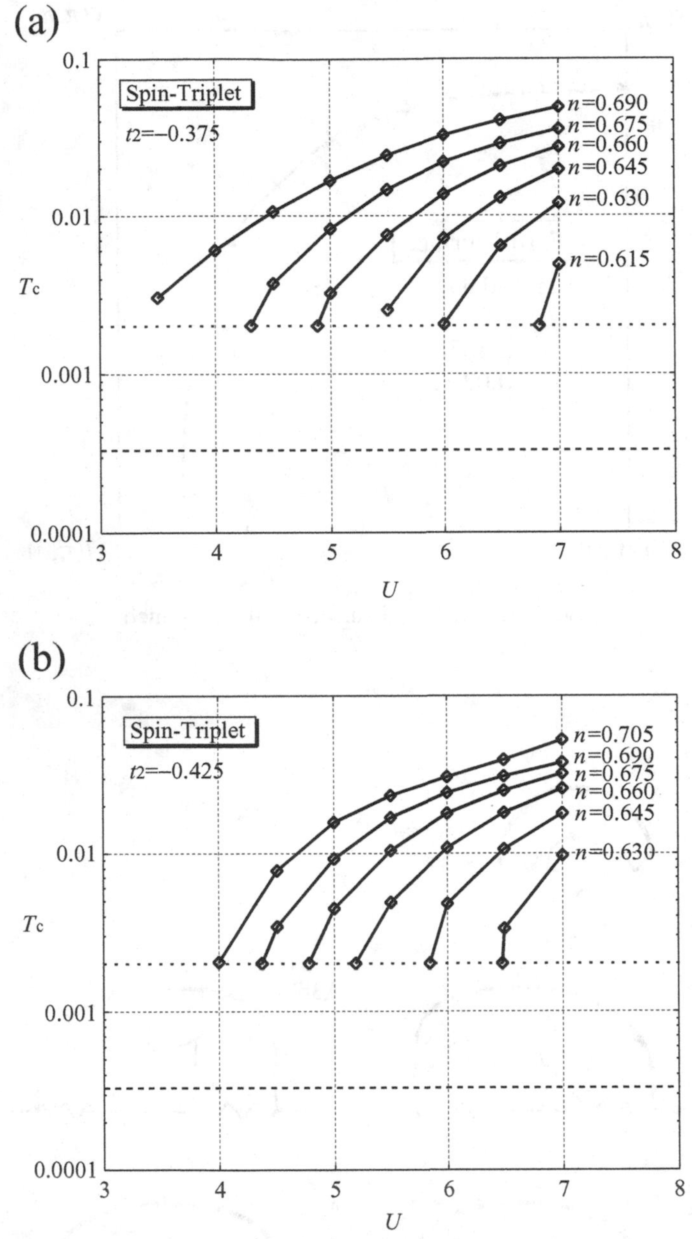

Fig. 9.52 The critical temperature T_c for the spin-triplet superconductivity.

In the perturbation calculation one of the third-order terms in the anomalous self-energy plays an essential role in realizing the spin-triplet superconducting state. It should be noted that the above calculation done by Nomura shows a similarity to the perturbation calculation for electron gas in the two-dimensional case by Chubakov

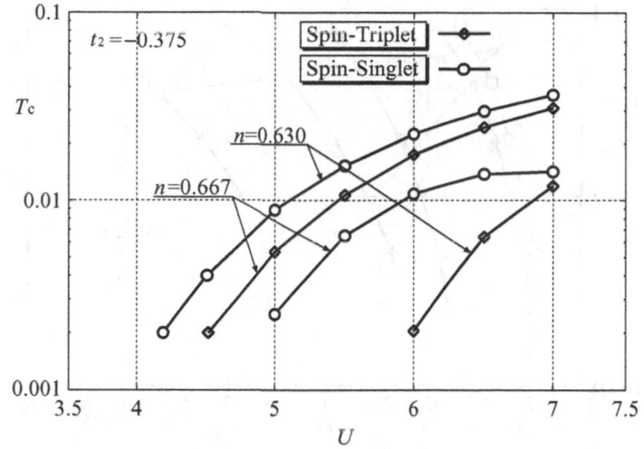

Fig. 9.53 The comparison of T_c between the spin-triplet and spin-singlet states.

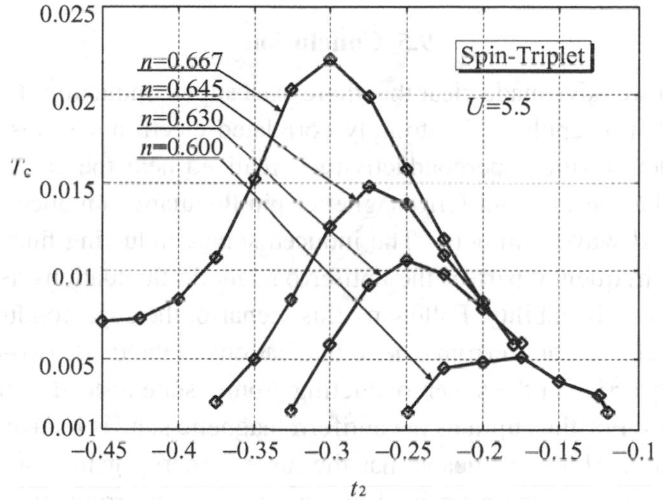

Fig. 9.54 The t_2 dependence of T_c for the spin-triplet superconducting state. Here, $U = 5.5$.

[29] and Feldman *et al.* [30], since in the latter calculation the same diagram as in Fig. 9.51 gives the essential contribution in realizing the spin-triplet p-state.

The single-band calculation based on the γ-band was extended to three bands to confirm the mechanism of the superconductivity. Moreover, by using the three-band calculation the temperature dependence of the specific heat below T_c is well reproduced by Nomura and Yamada [31].

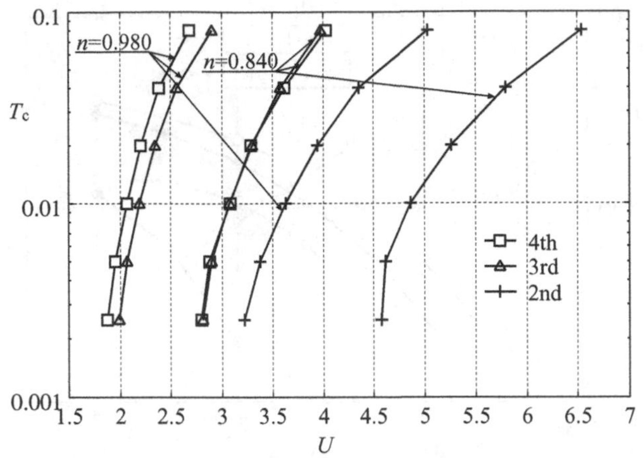

Fig. 9.55 The transition temperature T_c as a function of U for the nearly half-filled case. Here, $t = 1.0$, $t' = 0.1$ and $n = 0.98$. The pairing state is spin-singlet $d_{x^2-y^2}$ wave pairing.

9.5 Conclusion

In this chapter we have made clear that there exist superconducting states that arise from the coulomb repulsion in strongly correlated electron systems. In cuprates the spin-singlet pairing superconductivity is realized near the antiferromagnetic insulator. In this case the antiferromagnetic spin fluctuations induce attractive interaction with d-wave symmetry. The induced superconducting fluctuations suppress the low frequency part of the antiferromagnetic fluctuations as well as the antiferromagnetic instability. Following this scenario, the superconducting fluctuations induced by the antiferromagnetic fluctuations in the quasi-two-dimensional systems finally arrive at the superconducting ground state instead of the antiferromagnetic states. For the cuprates the antiferromagnetic state is realized only in the insulating states. This fact means that the kinetic energy gained in the metallic state is larger than the energy gained in the ordered antiferromagnetic state. The antiferromagnetic order makes the under-doped systems insulating by hindering the transfer motion of electrons and holes. The suppression of transfer motion is unfavourable in energy terms. This is why the superconducting state is realized in place of the antiferromagnetic states, even though there exist strong antiferromagnetic spin fluctuations.

In the weak electron correlation system the perturbation method with respect to the electron correlation U is justified. By this method we can derive the d-wave singlet pairing superconductivity for the systems from the overdoped cuprates to the optimally doped ones, and for the organic conductors. Moreover, we can derive the spin-triplet superconducting state for Sr_2RuO_4, using the same perturbation

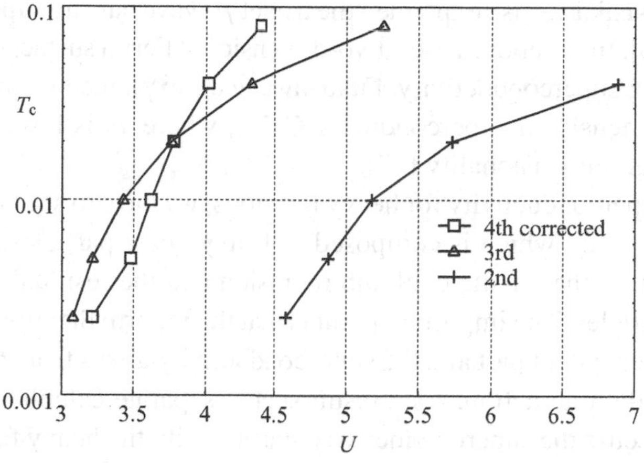

Fig. 9.56 The transition temperature T_c as a function of U for the triplet pairing case. Here $t = 1.0$, $t' = -0.375$ and $n = 1.334$. The pairing state is spin-triplet p-wave pairing state. The '4th corrected' means that the ladder diagrams are summed up to infinite order.

calculation. Thus we can conclude that the superconductivity in strongly correlated systems arises from the electron correlation itself. The symmetry of the pairing state is generally determined by the momentum dependence of interaction between quasi-particles. We can calculate the interaction term by using the perturbation theory with respect to the coulomb repulsion U.

The perturbation calculation for the vertex function of the Dyson–Gor'kov equation has been extended up to fourth-order terms by Nomura and Yamada [32]. They showed that the perturbation series converges smoothly for the d-wave pairing case and the momentum dependence of the pairing function is the same for all order terms. The fourth-order terms hardly change the critical temperature T_c, as shown in Fig. 9.55. On the other hand, the fourth-order terms for the p-wave pairing reduce the critical temperature T_c. As a result, the critical temperature T_c shows an oscillatory behaviour depending on the summed order. This behaviour arises from the contribution of particle–particle ladder diagrams. Nomura has showed that by summing up the ladder diagrams to infinite-order terms, we can recover the convergence of the perturbation calculation, as shown in Fig. 9.56. As a result, the conclusion obtained by the third-order perturbation theory holds also for the higher-order perturbation theory. From these calculations, we can conclude that the perturbation calculation gives reliable results for the superconductivity for both d- and p-wave pairing states.

Recently the perturbation calculation has been applied to the superconductivity in heavy fermions. H. Ikeda explained the d-wave superconductivity in $CeCu_2Si_2$ [33], Nisikawa *et al.* obtained the d-wave superconductivity in $CeCoIn_5$ [34] and

UPd_2Al_3 [35]. Nisikawa also explained the triplet p-wave pairing superconductivity for UNi_2Al_3 [36]. In the above cases, two-dimensional Fermi surfaces play essential roles in realizing superconductivity. Fukazawa *et al.* explained the d-wave pairing for the three-dimensional superconductor $CeIn_3$, where T_c is lower by one order owing to the high dimensionality [37].

To discuss superconductivity for heavy fermions, we have to start from the renormalized Hamiltonian, which is composed of heavy quasi-particles resulting from renormalization by the on-site coulomb repulsion and the residual interaction between quasi-particles. The important point is that the renormalization is done by the momentum-independent part and the superconductivity arises from the momentum dependence of the interaction. As a result such a separate calculation is justified. Thus we can discuss the superconductivity starting with the heavy fermion system. In this case both the kinetic energy and the interaction between quasi-particles are renormalized by the same factor z [34].

The superconductivity in strongly correlated electron systems was recently reviewed from a unified viewpoint in [38].

References

[1] Y. Kubo, Y. Shimakawa, T. Manoko and H. Igarashi, *Phys. Rev. B* **43** (1991) 7875.
[2] Y. Yanase and K. Yamada, *J. Phys. Soc. Jpn.* **68** (1999) 548.
[3] O. K. Anderson, A. I. Liechtenstein, O. Jepsen and F. Paulsen, *J. Phys. Chem. Solids* **56** (1995) 1573.
[4] H. Kohno and K. Yamada, *Prog. Theor. Phys.* **80** (1988) 623.
[5] H. Kontani, K. Kanki and K. Ueda, *Phys. Rev.* B **59** (1999) 14723; K. Kanki and H. Kontani, *J. Phys. Soc. Jpn.* **68** (1999) 1614.
[6] H. Kontani, *Phys. Rev. B* **64** (2001) 054413; *J. Phys. Soc. Jpn.* **70** (2001) 1873.
[7] H. Kohno and K. Yamada, *Prog. Theor. Phys.* **85** (1991) 13.
[8] Y. Maeno, H. Hashimoto, K. Yoshida, S. Nishizaki, T. Fujita, J. G. Bednorz and F. Lichtenberg, *Nature* **372** (1994) 532.
[9] T. M. Rice and M. Sigrist, *J. Phys. Condens. Matter* **7** (1995) L643.
[10] K. Ishida, H. Mukuda, Y. Kitaoka, A. Asayama, Z. Q. Mao, Y. Mori and Y. Maeno, *Nature* **396** (1998) 658.
[11] T. Nomura and K. Yamada, *J. Phys. Soc. Jpn.* **69** (2000) 3678.
[12] A. A. Abrikosov, L. P. Gor'kov and E. Dzyaloshinskii, *Method of Quantum Field Theory in Statistical Physics* (Prentice-Hall, 1963).
[13] N. E. Bickers, D. J. Scalapino and S. R. White, *Phys. Rev. Lett.* **62** (1989) 961; N. E. Bickers and D. J. Scalapino, *Ann. Phys.* (N. Y.) **193** (1989) 206.
[14] T. Hotta, *J. Phys. Soc. Jpn.* **63** (1994) 4126.
[15] Y. Yanase and K. Yamada, *J. Phys. Soc. Jpn.* **70** (2001) 1659.
[16] Y. Yanase and K. Yamada, *J. Phys. Soc. Jpn.* **68** (1999) 2999.
[17] T. Jujo, Y. Yanase and K. Yamada, *J. Phys. Soc. Jpn.* **69** (2000) 2240.
[18] Y. Yanase, T. Jujo and K. Yamada, *J. Phys. Soc. Jpn.* **69** (2000) 3664.
[19] Y. Yanase and K. Yamada, *J. Phys. Soc. Jpn.* **70** (2001) 1659.

[20] P. Nozières and S. Schmitt-Rink, *J. Low Temp. Phys.* **59** (1985) 195; A. J. Legett, *Modern Trends in the Theory of Condensed Matter*, eds. A. Pekalski and R. Przystawa (Springer-Verlag, 1980).
[21] H. Kino and H. Kontani, *J. Phys. Soc. Jpn.* **67** (1998) 3691.
[22] H. Kondo and T. Moriya, *J. Phys. Soc. Jpn.* **67** (1998) 3695.
[23] J. Schmalian, *Phys. Rev. Lett.* **81** (1998) 4232.
[24] T. Jujo, S. Koikegami and K. Yamada, *J. Phys. Soc. Jpn.* **68** (1999) 1331.
[25] H. Mayaffre, P. Wzietek, D. Jérome, C. Lenoir and P. Batail, *Phys. Rev. Lett.* **75** (1995) 4122; K. Kanoda, K. Miyagawa, A. Kawamoto and Y. Nakazawa, *Phys. Rev. B* **54** (1996) 76; K. Miyagawa, A. Kawamoto and K. Kanoda, *Phys. Rev. Lett.* **89** (2002) 017003.
[26] Y. Nakazawa and K. Kanoda, *Phys. Rev.* B **55** (1997) R8670.
[27] T. Jujo and K. Yamada, *J. Phys. Soc. Jpn.* **68** (1999) 2198.
[28] T. Nomura and K. Yamada, *J. Phys. Soc. Jpn.* **71** (2002) 404.
[29] A. V. Chubakov, *Phys. Rev. B* **48** (1993) 1097.
[30] J. Feldman, H. Knörrer, R. Sinclair and E. Trubowitz, *Helv. Phys. Acta* **70** (1997) 154.
[31] T. Nomura and K. Yamada, *J. Phys. Soc. Jpn.* **71** (2002) 1993.
[32] T. Nomura and K. Yamada, *J. Phys. Soc. Jpn.* **72** (2003) 2053.
[33] H. Ikeda, *J. Phys. Soc. Jpn.* **71** (2002) 1126.
[34] Y. Nisikawa, H. Ikeda and K. Yamada, *J. Phys. Soc. Jpn.* **71** (2002) 1140.
[35] Y. Nisikawa and K. Yamada, *J. Phys. Soc. Jpn.* **71** (2002) 237.
[36] Y. Nisikawa and K. Yamada, *J. Phys. Soc. Jpn.* **71** (2002) 2629.
[37] H. Fukazawa, H. Ikeda and K. Yamada, *J. Phys. Soc. Jpn.* **72** (2003) 884; H. Fukazawa and K. Yamada, *J. Phys. Soc. Jpn.* **72** (2003) 10.
[38] Y. Yanase, T. Jujo, T. Nomura, H. Ikeda, T. Hotta and K. Yamada, *Physics Report* **387** (2003) 1.

Appendix A. Feynman relation

We consider the following Hamiltonian:

$$\mathcal{H} = \mathcal{H}_0 + g\mathcal{H}_{\rm int}, \tag{A.1}$$

where $\mathcal{H}_{\rm int}$ is an interaction Hamiltonian and g is a coupling constant. The eigenvalue and the eigenfunction of $\mathcal{H}$ are written as $E_n(g)$ and $\varphi_n(g)$, respectively:

$$E_n(g) = \langle \varphi_n | \mathcal{H}_0 + g\mathcal{H}_{\rm int} | \varphi_n \rangle. \tag{A.2}$$

Differentiating this equation by g, we obtain

$$g\frac{\partial E_n(g)}{\partial g} = \langle \varphi_n(g) | g\mathcal{H}_{\rm int} | \varphi_n(g) \rangle + E_n(g)\frac{\partial}{\partial g}\langle \varphi_n | \varphi_n \rangle. \tag{A.3}$$

Here the second term vanishes because of $\langle \varphi_n | \varphi_n \rangle = 1$. Hence using the expectation value of $\mathcal{H}_{\rm int}$, we can obtain the eigenvalue $E_n(g)$ as

$$E_n(g) = E_n(0) + \int_0^g dg \frac{1}{g} \langle \varphi_n(g) | g\mathcal{H}_{\rm int} | \varphi_n(g) \rangle. \tag{A.4}$$

This relation holds for every eigenvalue. Therefore, if we consider the thermal average in place of the expectation value, the same relation holds for the free energy $F(g)$.

Appendix B. Second quantization

There are Fermi particles and Bose particles in nature. The former follow the Fermi–Dirac statistics and their wave-functions are antisymmetric with respect to their interchange. The latter follow the Bose–Einstein statistics and their wave-functions are symmetric with respect to their interchange. Let us assume a complete orthonormal set of one-particle wave-functions $\phi_i(\boldsymbol{r})$ $(i = 1, 2, \ldots)$, and put them in order. Then we represent a many-body wave-function Φ using the occupation number n_i, as

$$\Phi(n_1, n_2, \ldots, n_i, \ldots) = |n_1, n_2, \ldots, n_i, \ldots\rangle, \tag{B.1}$$

where the following relation holds:

$$\langle n_1{}', n_2{}', \ldots, n_i{}', \ldots |n_1, n_2, \ldots, n_i, \ldots\rangle = \delta_{n_1 n_1{}'}\delta_{n_2 n_2{}'} \ldots \delta_{n_i n_i{}'} \ldots \tag{B.2}$$

For this state Φ we define the annihilation and creation operators.
For Bose particles:

$$b_i\Phi(\ldots, n_i, \ldots) = \sqrt{n_i}\Phi(\ldots, n_i - 1, \ldots), \tag{B.3}$$

$$b_i{}^\dagger\Phi(\ldots, n_i, \ldots) = \sqrt{n_i + 1}\Phi(\ldots, n_i + 1, \ldots). \tag{B.4}$$

For a Fermi particle, using $A_i = \sum_{j<i} n_j$:

$$c_i\Phi(\ldots, n_i, \ldots) = \sqrt{n_i}(-1)^{A_i}\Phi(\ldots, n_i - 1, \ldots), \tag{B.5}$$

$$c_i{}^\dagger\Phi(\ldots, n_i, \ldots) = \sqrt{1 - n_i}(-1)^{A_i}\Phi(\ldots, n_i + 1, \ldots). \tag{B.6}$$

In the definition of the fermion operators the factor $(-1)^{A_i}$ is necessary, since the interchange between the Fermi particles gives a sign change.

Hereafter we write the creation and annihilation operators for both the Fermi particle and the Bose particle as $a_i{}^\dagger$ and a_i, respectively. The Bose (Fermi) particle satisfies the following commutation (anticommutation) relation:

$$[a_i, a_j]_\mp = [a_i{}^\dagger, a_j{}^\dagger]_\mp = 0, \tag{B.7}$$

$$[a_i, a_j{}^\dagger]_\mp = \delta_{ij}. \tag{B.8}$$

Using these operators, the one-particle operator $F_1 = \sum_i f_1(\boldsymbol{r}_i)$ is written as

$$F_1 = \sum_{i,j} \langle i | f_1 | j \rangle a_i{}^{\dagger} a_j, \tag{B.9}$$

$$\langle i | f_1 | j \rangle = \int \phi_i{}^*(\boldsymbol{r}) f_1(\boldsymbol{r}) \phi_j(\boldsymbol{r}) d\boldsymbol{r}. \tag{B.10}$$

The two-particle operator $F_2 = \frac{1}{2}\sum_{i \neq j} f_2(\boldsymbol{r}_i, \boldsymbol{r}_j)$ is written as

$$F_2 = \frac{1}{2} \sum_{\substack{ij \\ km}} \langle ij | f_2 | km \rangle a_i{}^{\dagger} a_j{}^{\dagger} a_m a_k, \tag{B.11}$$

$$\langle ij | f_2 | km \rangle = \int\int d\boldsymbol{r}_1 d\boldsymbol{r}_2 \phi_i{}^*(\boldsymbol{r}_1) \phi_j{}^*(\boldsymbol{r}_2) f_2(\boldsymbol{r}_1, \boldsymbol{r}_2) \phi_k(\boldsymbol{r}_1) \phi_m(\boldsymbol{r}_2). \tag{B.12}$$

The field operators $\psi(\boldsymbol{r})$ and $\psi^{\dagger}(\boldsymbol{r})$ are defined as

$$\psi(\boldsymbol{r}) = \sum_i \phi_i(\boldsymbol{r}) a_i, \tag{B.13}$$

$$\psi^{\dagger}(\boldsymbol{r}) = \sum_i \phi_i{}^*(\boldsymbol{r}) a_i{}^{\dagger}. \tag{B.14}$$

Here, depending on $a_i = b_i$ or $a_i = c_i$, $\psi(\boldsymbol{r})$ represents the Bose operator $\psi_{\mathrm{B}}(\boldsymbol{r})$ or the Fermi operator $\psi_{\mathrm{F}}(\boldsymbol{r})$. The field operators $\psi(\boldsymbol{r})$ and $\psi^{\dagger}(\boldsymbol{r})$ satisfy the following commutation relations:

$$[\psi(\boldsymbol{r}), \psi(\boldsymbol{r}')]_{\mp} = [\psi^{\dagger}(\boldsymbol{r}), \psi^{\dagger}(\boldsymbol{r}')]_{\mp} = 0, \tag{B.15}$$

$$[\psi(\boldsymbol{r}), \psi^{\dagger}(\boldsymbol{r}')]_{\mp} = \delta(\boldsymbol{r} - \boldsymbol{r}'), \tag{B.16}$$

where the sign of the commutation relation takes $+$ for the fermion and $-$ for the boson.

Appendix C. Interaction representation and thermal Green's function

When an electron interaction $\hat{\mathcal{H}}'$ is introduced into the system possessing the unperturbed Hamiltonian $\hat{\mathcal{H}}_0$, the total Hamiltonian $\mathcal{H}$ is given by

$$\hat{\mathcal{H}} = \hat{\mathcal{H}}_0 + \hat{\mathcal{H}}'. \tag{C.1}$$

Let us discuss how we calculate the physical quantities for this system. Assuming $\beta = 1/k_B T$, we define the operator $S(\tau)$ as $(k_B = 1)$

$$e^{-(\hat{\mathcal{H}}-\mu\hat{N})\tau} = e^{-(\hat{\mathcal{H}}_0-\mu\hat{N})\tau} S(\tau), \tag{C.2}$$

$$e^{(\hat{\mathcal{H}}-\mu\hat{N})\tau} = S^{-1}(\tau) e^{(\hat{\mathcal{H}}_0-\mu\hat{N})\tau}, \tag{C.3}$$

where μ is the chemical potential and $\hat{N}$ is the number operator. Using an operator A in Schrödinger's representation, we define the operator $A(\tau)$ in the interaction representation as

$$A(\tau) = e^{(\hat{\mathcal{H}}_0-\mu\hat{N})\tau} \hat{A} e^{-(\hat{\mathcal{H}}_0-\mu\hat{N})\tau}. \tag{C.4}$$

For example, when we take $\hat{\mathcal{H}}$ or $\hat{\mathcal{H}}'$ as A, we have

$$\hat{\mathcal{H}}(\tau) = e^{(\hat{\mathcal{H}}_0-\mu\hat{N})\tau} \hat{\mathcal{H}} e^{-(\hat{\mathcal{H}}_0-\mu\hat{N})\tau}, \tag{C.5}$$

$$\hat{\mathcal{H}}'(\tau) = e^{(\hat{\mathcal{H}}_0-\mu\hat{N})\tau} \hat{\mathcal{H}}' e^{-(\hat{\mathcal{H}}_0-\mu\hat{N})\tau}. \tag{C.6}$$

The creation and annihilation operators, $\psi^\dagger(\boldsymbol{r})$ and $\psi(\boldsymbol{r})$, are written in the interaction representation as

$$\psi^\dagger(\boldsymbol{r}, \tau) = e^{(\hat{\mathcal{H}}_0-\mu\hat{N})\tau} \psi^\dagger(\boldsymbol{r}) e^{-(\hat{\mathcal{H}}_0-\mu\hat{N})\tau}, \tag{C.7}$$

$$\psi(\boldsymbol{r}, \tau) = e^{(\mathcal{H}_0-\mu\hat{N})\tau} \psi(\boldsymbol{r}) e^{-(\hat{\mathcal{H}}_0-\mu\hat{N})\tau}. \tag{C.8}$$

The Hamiltonians $\hat{\mathcal{H}}(\tau)$ and $\hat{\mathcal{H}}'(\tau)$ in the interaction representation are given by replacing $\psi(\boldsymbol{r})$ and $\psi^\dagger(\boldsymbol{r})$, describing $\hat{\mathcal{H}}$ and $\hat{\mathcal{H}}'$ in the Heisenberg representation, with those in the interaction representation (C.7) and (C.8). When $\hat{\mathcal{H}}_0$ is a free particle Hamiltonian, since $\hat{\mathcal{H}}_0$ commutes with $\hat{N}$, we have simply

$$\hat{\mathcal{H}}_0(\tau) = \hat{\mathcal{H}}_0,$$
$$\hat{N}(\tau) = \hat{N}.$$

Differentiating (C.2) by τ and multiplying it by $e^{(\hat{\mathcal{H}}_0-\mu\hat{N})\tau}$ from the left side, we obtain the following equation for $S(\tau)$:

$$\frac{\partial S(\tau)}{\partial \tau} = -\hat{\mathcal{H}}'(\tau)S(\tau). \tag{C.9}$$

The solution of this equation which satisfies $S(0) = 1$ is given by

$$S(\tau) = T_\tau \exp\left\{-\int_0^\tau \hat{\mathcal{H}}'(\tau')d\tau'\right\}, \tag{C.10}$$

where T_τ is the ordering operator with respect to τ. We assume $\tau_1 > \tau_2$ and introduce $S(\tau_1, \tau_2)$ as

$$S(\tau_1, \tau_2) = T_\tau \exp\left\{-\int_{\tau_2}^{\tau_1} \hat{\mathcal{H}}'(\tau')d\tau'\right\}. \tag{C.11}$$

The $S(\tau)$ in (C.10) is written as $S(\tau, 0)$. When $\tau_1 > \tau_2 > \tau_3$, we have

$$S(\tau_1, \tau_3) = S(\tau_1, \tau_2)S(\tau_2, \tau_3). \tag{C.12}$$

For $\tau_1 > \tau_2$,

$$S(\tau_1, \tau_2) = S(\tau_1)S^{-1}(\tau_2). \tag{C.13}$$

From (C.2) the thermodynamic potential Ω is given by

$$e^{-\Omega/T} = \mathrm{Tr}\left\{e^{-(\hat{\mathcal{H}}-\mu\hat{N})/T}\right\} = \mathrm{Tr}\left\{e^{-(\hat{\mathcal{H}}_0-\mu\hat{N})/T}S(\beta)\right\}, \tag{C.14}$$

$$\Omega = \Omega_0 - T\ln\langle S(\beta)\rangle_0, \tag{C.15}$$

$$\Omega_0 = -T\ln\mathrm{Tr}\left\{e^{-(\hat{\mathcal{H}}_0-\mu\hat{N})/T}\right\}. \tag{C.16}$$

Here, $\langle\ \rangle_0$ means the thermal average in the unperturbed states.

The thermal Green's function $G_{\alpha\beta}(\boldsymbol{r}_1\tau_1;\boldsymbol{r}_2\tau_2)$ is defined by the operators $\psi_\alpha(\boldsymbol{r}_1)$ and $\psi_\beta{}^\dagger(\boldsymbol{r}_2)$ in the Schrödinger representation, as

$$G_{\alpha\beta}(\boldsymbol{r}_1\tau_1;\boldsymbol{r}_2\tau_2)$$
$$= \begin{cases} \mathrm{Tr}[e^{(\Omega+\mu\hat{N}-\hat{\mathcal{H}})/T}e^{(\hat{\mathcal{H}}-\mu\hat{N})(\tau_1-\tau_2)}\psi_\alpha(\boldsymbol{r}_1)e^{-(\hat{\mathcal{H}}-\mu\hat{N})(\tau_1-\tau_2)}\psi_\beta{}^\dagger(\boldsymbol{r}_2)] & (\tau_1 > \tau_2) \\ \pm\mathrm{Tr}[e^{(\Omega+\mu\hat{N}-\hat{\mathcal{H}})/T}e^{(\hat{\mathcal{H}}-\mu\hat{N})(\tau_1-\tau_2)}\psi_\beta{}^\dagger(\boldsymbol{r}_2)e^{-(\hat{\mathcal{H}}-\mu\hat{N})(\tau_1-\tau_2)}\psi_\alpha(\boldsymbol{r}_1)] & (\tau_1 < \tau_2). \end{cases} \tag{C.17}$$

The $\pm$ sign takes $+$ for fermions and $-$ for bosons. Putting $\tau_1 - \tau_2 = \tau$ and $\beta = 1/T$, we have the relation

$$G(\tau < 0) = \mp G(\tau + \beta). \tag{C.18}$$

Green's function (C.17) can also be represented by $\psi_\alpha(\boldsymbol{r}, \tau)$ and $\psi_\beta{}^\dagger(\boldsymbol{r}, \tau)$ in the Heisenberg representation as

$$\psi_\alpha(\boldsymbol{r}, \tau) = e^{(\hat{\mathcal{H}}-\mu\hat{N})\tau}\psi_\alpha(\boldsymbol{r})e^{-(\hat{\mathcal{H}}-\mu\hat{N})\tau}, \tag{C.19}$$

$$\psi_\beta{}^\dagger(\boldsymbol{r}, \tau) = e^{(\hat{\mathcal{H}}-\mu\hat{N})\tau}\psi_\beta{}^\dagger(\boldsymbol{r})e^{-(\hat{\mathcal{H}}-\mu\hat{N})\tau}, \tag{C.20}$$

$$\begin{aligned} G_{\alpha\beta}(\boldsymbol{r}_1\tau_1;\boldsymbol{r}_2\tau_2) &= -\mathrm{Tr}\{e^{(\Omega+\mu\hat{N}-\hat{\mathcal{H}})/T}T_\tau\psi_\alpha(\boldsymbol{r}_1, \tau_1)\psi_\beta{}^\dagger(\boldsymbol{r}_2, \tau_2)\} \\ &\equiv -\langle T_\tau\psi_\alpha(\boldsymbol{r}_1, \tau_1)\psi_\beta{}^\dagger(\boldsymbol{r}_2, \tau_2)\rangle. \end{aligned} \tag{C.21}$$

In the interaction representation, when $\tau_1 - \tau_2 = \tau > 0$, Green's function can be written by $\psi(\boldsymbol{r}_1, \tau_1)$ and $\psi^\dagger(\boldsymbol{r}_2, \tau_2)$ in the interaction representation as

$$\begin{aligned} G_{\alpha\beta}(\boldsymbol{r}_1\tau_1;\boldsymbol{r}_2\tau_2) &= -e^{\Omega/T}\,\mathrm{Tr}\{e^{-(\hat{\mathcal{H}}_0-\mu\hat{N})/T}S(\beta)S^{-1}(\tau_1)e^{(\hat{\mathcal{H}}_0-\mu\hat{N})\tau_1}\psi_\alpha(\boldsymbol{r}_1) \\ &\quad\times e^{-(\hat{\mathcal{H}}_0-\mu\hat{N})\tau_1}S(\tau_1)S^{-1}(\tau_2)e^{(\hat{\mathcal{H}}_0-\mu\hat{N})\tau_2}\psi_\beta{}^\dagger(\boldsymbol{r}_2)e^{-(\hat{\mathcal{H}}_0-\mu\hat{N})\tau_2}S(\tau_2)\} \\ &= -e^{\Omega/T}\,\mathrm{Tr}\{e^{-(\hat{\mathcal{H}}_0-\mu\hat{N})/T}S(\beta,\tau_1)\psi_\alpha(\boldsymbol{r}_1,\tau_1)S(\tau_1,\tau_2)\psi_\beta{}^\dagger(\boldsymbol{r}_2,\tau_2)S(\tau_2)\}. \end{aligned} \tag{C.22}$$

For $\tau_1 - \tau_2 = \tau < 0$:

$$G_{\alpha\beta}(\boldsymbol{r}_1\tau_1;\boldsymbol{r}_2\tau_2) = \pm e^{\Omega/T}\,\mathrm{Tr}\{e^{-(\hat{\mathcal{H}}_0-\mu\hat{N})/T}S(\beta,\tau_2)\psi_\beta{}^\dagger(\boldsymbol{r}_2,\tau_2)S(\tau_2,\tau_1)\psi_\alpha(\boldsymbol{r}_1,\tau_1)S(\tau_1)\}. \tag{C.23}$$

Equations (C.22) and (C.23) can be written in a formula as

$$G_{\alpha\beta}(\boldsymbol{r}_1\tau_1;\boldsymbol{r}_2\tau_2) = -e^{\Omega/T}\,\mathrm{Tr}\{e^{-(\hat{\mathcal{H}}_0-\mu\hat{N})/T}T_\tau[\psi_\alpha(\boldsymbol{r}_1,\tau_1)\psi_\beta{}^\dagger(\boldsymbol{r}_2,\tau_2)S(\beta)]\}. \tag{C.24}$$

Using (C.14), we have

$$\begin{aligned} G_{\alpha\beta}(\boldsymbol{r}_1\tau_1;\boldsymbol{r}_2\tau_2) &= -\frac{\mathrm{Tr}\{e^{-(\hat{\mathcal{H}}_0-\mu\hat{N})/T}T_\tau[\psi_\alpha(\boldsymbol{r}_1,\tau_1)\psi_\beta{}^\dagger(\boldsymbol{r}_2,\tau_2)S(\beta)]\}}{\mathrm{Tr}\{e^{-(\hat{\mathcal{H}}_0-\mu\hat{N})/T}S(\beta)\}} \\ &= -\frac{\langle T_\tau\psi_\alpha(\boldsymbol{r}_1,\tau_1)\psi_\beta{}^\dagger(\boldsymbol{r}_2,\tau_2)S\rangle_0}{\langle S\rangle_0}, \end{aligned} \tag{C.25}$$

where $\langle\cdots\rangle_0$ is

$$\langle\cdots\rangle_0 = \mathrm{Tr}\{e^{(\Omega_0+\mu\hat{N}-\hat{\mathcal{H}}_0)/T}\cdots\}, \tag{C.26}$$

and $S = S(\beta)$.

Now let us consider the Fourier transform of $G(\tau)$. Since $G(\tau)$ is the periodic function with period $2\beta = 2/T$,

$$G(\tau) = T\sum_l e^{-i\omega_l\tau}G(\omega_l), \tag{C.27}$$

$$G(\omega_l) = \frac{1}{2}\int_{-\beta}^{\beta} e^{i\omega_l\tau}G(\tau)d\tau, \tag{C.28}$$

$$\omega_l = l\pi T \qquad (l : \text{integer}). \tag{C.29}$$

Using (C.18), we rewrite the integration for $\tau < 0$ in (C.28) as

$$G(\omega_l) = \frac{1}{2}(1 \mp e^{-i\omega_l\beta})\int_0^\beta e^{i\omega_l\tau}G(\tau)d\tau. \tag{C.30}$$

Hence ω_l is confined to the following value:

$$\omega_l = \begin{cases} (2l+1)\pi T & \text{(fermions)} \\ 2l\pi T & \text{(bosons)} \end{cases} \tag{C.31}$$

For a free particle $\boldsymbol{k}$, Green's function is given by

$$G_{\boldsymbol{k}}{}^0(\tau) = -\langle T_\tau a_{\boldsymbol{k}}(\tau)a_{\boldsymbol{k}}{}^\dagger(0)\rangle, \tag{C.32}$$

$$a_{\boldsymbol{k}}(\tau) = e^{\tau(\hat{\mathcal{H}}_0-\mu N)}a_{\boldsymbol{k}}e^{-\tau(\hat{\mathcal{H}}_0-\mu\hat{N})} = e^{-(\varepsilon_{\boldsymbol{k}}-\mu)\tau}a_{\boldsymbol{k}}. \tag{C.33}$$

The Fourier transform of (C.32) is given by

$$G_k{}^0(\omega_l) = [i\omega_l + \mu - \varepsilon_k]^{-1}. \tag{C.34}$$

Now we consider the stationary property of the thermodynamic potential Ω in (C.15) with respect to the variation of the self-energy Σ. This property is important in calculating the electronic specific heat:

$$\begin{aligned}\langle S(\beta)\rangle_0 &= \left\langle T_\tau \exp\left\{-\int_0^\beta \mathcal{H}'(\tau')d\tau'\right\}\right\rangle_0 \\ &= 1 + \sum_{n=1}^{\infty} \frac{(-1)^n}{n!} \int_0^\beta \cdots \int_0^\beta d\tau_1 \cdots d\tau_n \langle T_\tau[\mathcal{H}'(\tau_1)\cdots\mathcal{H}'(\tau_n)]\rangle_0.\end{aligned} \tag{C.35}$$

From (C.35), Ω is given by

$$\begin{aligned}\Omega &= \Omega_0 - T\ln\langle S(\beta)\rangle_0 \\ &= \Omega_0 + T\sum_{n=1}^{\infty} \frac{(-1)^{n+1}}{n!} \int_0^\beta \cdots \int_0^\beta d\tau_1 \cdots d\tau_n \langle T_\tau[\mathcal{H}'(\tau_1)\cdots\mathcal{H}'(\tau_n)]\rangle_{\text{conn}} \\ &= \Omega_0 + \sum_{n=1}^{\infty} \Omega_n,\end{aligned} \tag{C.36}$$

where conn means taking only the connected diagrams and neglecting the disconnected ones. The term Ω_n is the contribution of the nth-order term to Ω, and is written as

$$\Omega_n = \frac{1}{2n} T \sum_{kl} \frac{1}{z_l - \varepsilon_k} \Sigma'_{kn}(z_l) \qquad (z_l = (2l+1)i\pi T), \tag{C.37}$$

where $\Sigma'_{kn}(z_l)$ includes the improper self-energy. Green's function $G_k(z_l)$ for the total Hamiltonian $\mathcal{H} = \mathcal{H}_0 + \mathcal{H}'$ is written as

$$\begin{aligned}G_k(z_l) &= G_k{}^0(z_l) + G_k{}^0(z_l)\Sigma_k(z_l)G_k(z_l) \\ &= G_k{}^0(z_l) + G_k{}^0(z_l)\Sigma'_k(z_l)G_k{}^0(z_l),\end{aligned} \tag{C.38}$$

where $G_k{}^0(z_l)$ is Green's function for $\mathcal{H}_0$ and $\Sigma_k(z_l)$ is the proper self-energy.

The factor $1/n$ in (C.37) makes our calculation difficult. To avoid the difficulty, we introduce a parameter λ representing the strength of the interaction $\mathcal{H}'$, and obtain

$$\frac{1}{n}\Sigma'_{kn}(z_l : \lambda) = \int_0^\lambda \Sigma'_{kn}(z_l : \lambda')d\lambda'/\lambda'. \tag{C.39}$$

Using (C.39), we obtain

$$\Omega = \Omega_0 + \frac{1}{2\beta}\sum_{kl}\int_0^\lambda G_k{}^0(z_l)\Sigma'_k(z_l : \lambda')d\lambda'/\lambda', \tag{C.40}$$

$$\Sigma'_k = \Sigma_k + \Sigma_k G_k{}^0(z_l)\Sigma_k + \cdots = \frac{(z_l - \varepsilon_k)\Sigma_k}{z_l - \varepsilon_k - \Sigma_k}. \tag{C.41}$$

From (C.40) and (C.41), Ω is given by

$$\Omega = \Omega_0 + \frac{1}{2\beta}\sum_{kl}\int_0^\lambda \frac{d\lambda'}{\lambda'}\Sigma_k(z_l : \lambda')G_k(z_l : \lambda'). \tag{C.42}$$

Thus the following relation is obtained:

$$\lambda\frac{\partial\Omega}{\partial\lambda} = \frac{1}{2\beta}\sum_{kl}\Sigma_k(z_l:\lambda)G_k(z_l:\lambda). \quad (C.43)$$

Now we consider the following equation:

$$Y = -\frac{1}{\beta}\sum_{kl} e^{z_l 0_+}\{\log(\varepsilon_k + \Sigma_k(z_l) - z_l) + G_k(z_l)\Sigma_k(z_l)\} + Y', \quad (C.44)$$

where Y' is the sum of all the diagrams in which $G_k{}^0(z_l)$ in the closed loop diagrams representing the terms of Ω are replaced by the full Green's function $G_k(z_l)$. Here $G_k(z_l)$ includes the self-energy correction. Considering Y as a function of $\Sigma_k(z_l)$, we obtain

$$\frac{\partial Y}{\partial\Sigma_k} = -\frac{1}{\beta}\sum_l \Sigma_k(z_l)[G_k(z_l)]^2 + \frac{\partial Y'}{\partial\Sigma_k}, \quad (C.45)$$

$$\frac{\partial Y'}{\partial\Sigma_k} = \frac{1}{\beta}\sum_n\sum_l\{G_k(z_l)\}^2\Sigma''_{kn}(z_l)$$

$$= \frac{1}{\beta}\sum_l\{G_k(z_l)\}^2\Sigma_k(z_l). \quad (C.46)$$

Here, $\Sigma''_{kn}(z_l)$ is the nth-order proper self-energy in which $G_k{}^0(z_l)$ is replaced by $G_k(z_l)$. From (C.45) and (C.46),

$$\frac{\partial Y}{\partial\Sigma_k(z_l)} = 0. \quad (C.47)$$

This result means that Y is invariant with respect to the first order of Σ_k. As a result, when we differentiate Y by λ with a fixed μ, we can neglect the change of Y through Σ_k and obtain the following result:

$$\lambda\frac{\partial Y}{\partial\lambda} = \frac{1}{2\beta}\sum_n\sum_{kl}G_k(z_l)\Sigma''_{kn}(z_l) = \frac{1}{2\beta}\sum_{kl}G_k(z_l)\Sigma_k(z_l) = \lambda\frac{\partial\Omega}{\partial\lambda}. \quad (C.48)$$

From this result and $Y(\lambda = 0) = \Omega_0$, we obtain $Y = \Omega$. The first-order correction of $Y' = \Omega'$ can be obtained by opening $2n$ electron lines in the closed loop diagrams, and the factor $1/2n$ in (C.37) is cancelled by the number $2n$ of opened diagrams. We obtain

$$\Omega' = \frac{1}{\beta}\sum_{kl}\frac{1}{z_l - \varepsilon_k - \Sigma_k(z_l)}\Sigma_k(z_l). \quad (C.49)$$

Hence, when we consider the first-order correction, (C.44) can be written as

$$\Omega = -\frac{1}{\beta}\sum_{kl}e^{z_l 0_+}\{\log[\varepsilon_k + \Sigma_k(z_l) - z_l]\}. \quad (C.50)$$

Now let us prove the following result:

$$J = \frac{1}{2\pi i}\int_C dz e^{z0_+}\left(\frac{\partial\Sigma_\sigma}{\partial z}\right)G_\sigma(z) = -\frac{1}{2\pi i}\int_C dz\Sigma_\sigma(z)\frac{\partial}{\partial z}G_\sigma(z) = 0, \quad (C.51)$$

where $T = 0$ and the integral path is taken along the imaginary axis. The integral J given by (C.51) is obtained by taking a derivative of $G_\sigma(z)$ in Ω' given by (C.49) with respect to z. In the case where the variable z is included in only the special $G_\sigma(z)$, $J = 0$ is easily shown by a partial integration. We therefore have to prove the relation when z is also included in the other Green's functions through energy conservation. For this purpose let us consider the following function:

$$\int_C dz_1 dz_2 dz_3 dz_4 \delta(z_1 + z_2 - z_3 - z_4) \left(\sum_{i=1}^{4}{}' \frac{\partial}{\partial z_i} \right) G_{\sigma_1}(z_1) G_{\sigma_2}(z_2) G_{\sigma_3}(z_3) G_{\sigma_4}(z_4), \tag{C.52}$$

where the sum $\sum'$ means taking the sum over the derivatives of the Green's functions possessing only σ spin. If the interaction is spin-independent, $\sigma_1 = \sigma_3 = \sigma$ and $\sigma_2 = \sigma_4 = \sigma'$. In this case, by partial integration, we obtain
For $\sigma' = \sigma$:

$$\left(\frac{\partial}{\partial z_1} + \frac{\partial}{\partial z_2} + \frac{\partial}{\partial z_3} + \frac{\partial}{\partial z_4} \right) \delta(z_1 + z_2 - z_3 - z_4) = 0. \tag{C.53}$$

For $\sigma' \neq \sigma$:

$$\left(\frac{\partial}{\partial z_1} + \frac{\partial}{\partial z_3} \right) \delta(z_1 + z_2 - z_3 - z_4) = 0. \tag{C.54}$$

Thus, $J = 0$ is obtained. The integral (C.51) can be rewritten as the integral along the real axis by analytic continuation. The result is given by

$$\int_{-\infty}^{\infty} d\omega f(\omega) \left(-\frac{1}{\pi} \mathrm{Im} \right) G_\sigma(\omega_+) \frac{\partial \Sigma_\sigma(\omega_+)}{\partial \omega} = 0, \tag{C.55}$$

where $f(\omega)$ is the Fermi distribution function.

Now let us prove the Luttinger theorem. As noted in Chapter 2, it is clear that the volume of the Fermi sphere is invariant with respect to the interaction from the principle of continuity. Generally the Fermi surface is transformed by the interaction. Therefore we need a proof showing that the volume surrounded by the Fermi surface is not changed by the introduction of electron interactions as far as the system remains in the Fermi liquid. The number of particles N is given by the thermal Green's function as

$$N - \sum_{k\sigma} \frac{1}{\beta} \sum_{l} e^{z_l 0_+} G_{k\sigma}(z_l). \tag{C.56}$$

Assuming $T = 0$ and changing the summation along the imaginary axis into the integral, from (C.51) we obtain

$$N = \sum_{k\sigma} \int \frac{dz}{2\pi i} e^{z0_+} \frac{\partial}{\partial z} \log[\varepsilon_k + \Sigma_k(z) - z]. \tag{C.57}$$

Then we transform the integral path along the imaginary axis into that along the real axis. Noting that Im $\Sigma_k(x)$ vanishes at $x = \mu$ and changes its sign, we obtain

$$N = \sum_{k\sigma} \frac{1}{\pi} \mathrm{Im} \log[\varepsilon_k + \Sigma_k(\mu) - \mu + i0_+] = \sum_{k\sigma} \theta(\mu - \varepsilon_k - \Sigma_k(\mu)). \tag{C.58}$$

When we define the Fermi surface by the following equation:

$$\mu - \varepsilon_k - \Sigma_k(\mu) = 0, \tag{C.59}$$

we obtain

$$N = \frac{2\Omega}{(2\pi)^3} V_{\rm FS} = \frac{2\Omega}{(2\pi)^3} \int d\boldsymbol{k} \theta(\mu - \varepsilon_{\boldsymbol{k}} - \Sigma_{\boldsymbol{k}}(\mu)). \tag{C.60}$$

The N in (C.60) is written by the volume ${V_{\rm FS}}^0$ for the non-interacting case as

$$N = \frac{2\Omega}{(2\pi)^3} {V_{\rm FS}}^0. \tag{C.61}$$

Thus, we obtain the relation

$$V_{\rm FS} = {V_{\rm FS}}^0. \tag{C.62}$$

The volume surrounded by the Fermi surface is invariant.

Appendix D. Linear response theory

Assuming a system in the thermal equibrium at $t = -\infty$, let us consider the grand canonical ensemble. Using the thermodynamic potential Ω, we write the density matrix ρ_0 as

$$\rho_0 = e^{-\beta\mathcal{H}_0}/\mathrm{Tr}\, e^{-\beta\mathcal{H}_0} = e^{\beta(\Omega-\mathcal{H}_0)}. \tag{D.1}$$

When a perturbation $\mathcal{H}'(t)$ due to an external field is applied to the system, we put

$$\mathcal{H} = \mathcal{H}_0 + \mathcal{H}'(t), \tag{D.2}$$

$$\rho = \rho_0 + \rho'(t), \tag{D.3}$$

and consider the first-order change of the system in $\mathcal{H}'(t)$. In the first-order term with respect to $\mathcal{H}'$, the following equation holds:

$$i\hbar\frac{\partial\rho'}{\partial t} = [\mathcal{H}_0, \rho'] + [\mathcal{H}', \rho_0]. \tag{D.4}$$

The solution of this equation is given by

$$\rho'(t) = -\frac{i}{\hbar}\int_{-\infty}^{t} dt' e^{-i\mathcal{H}_0(t-t')/\hbar}[\mathcal{H}'(t), \rho_0]e^{i\mathcal{H}_0(t-t')/\hbar}. \tag{D.5}$$

Using $\rho'(-\infty) = 0$, we can confirm that (D.5) satisfies (D.4) by differentiating (D.5) by t.

Assuming an operator A independent of t, we write the external perturbation $\mathcal{H}'(t)$ as

$$\mathcal{H}'(t) = -AF(t). \tag{D.6}$$

Later we assume $F(t) \sim e^{-i\omega t+\delta t}$, where δ is a positive infinitesimal. This time dependence represents A applied adiabatically from $t = -\infty$. Here we assume that owing to A the physical quantity B changes from its thermal average $B_0 = \mathrm{Tr}\, \rho_0 B = 0$ into

$$\langle B\rangle = \mathrm{Tr}\, \rho' B. \tag{D.7}$$

To obtain the linear response of B to A, we insert (D.5) into (D.7):

$$\begin{aligned}\langle B\rangle &= \frac{i}{\hbar}\int_{-\infty}^{t} F(t')\mathrm{Tr}\{e^{-i\mathcal{H}_0(t-t')/\hbar}[A, \rho_o]e^{i\mathcal{H}_0(t-t')/\hbar} B\}dt' \\ &= \frac{i}{\hbar}\int_{-\infty}^{t} e^{-i\omega t'+\delta t'}\mathrm{Tr}\{[A, \rho_0]B(t-t')\}dt'. \end{aligned} \tag{D.8}$$

Here if we put $t - t' = \tau$, we can see that $\langle B \rangle$ possesses the same time dependence $e^{-i\omega t+\delta t}$, and $\langle B \rangle$ can be written as

$$\langle B \rangle = e^{-i\omega t+\delta t}\chi_{BA}(\omega), \tag{D.9}$$

$$\begin{aligned}\chi_{BA}(\omega) &= \frac{i}{\hbar}\int_0^\infty e^{i\omega\tau-\delta\tau}\mathrm{Tr}[e^{-i\mathcal{H}_0\tau/\hbar}[A,\rho_0]e^{i\mathcal{H}_0\tau/\hbar}B]d\tau \\ &= -\frac{i}{\hbar}\int_0^\infty d\tau e^{i\omega\tau-\delta\tau}\langle[B(\tau),A]\rangle,\end{aligned} \tag{D.10}$$

$$B(\tau) = e^{i\mathcal{H}_0\tau/\hbar}Be^{-i\mathcal{H}_0\tau/\hbar}. \tag{D.11}$$

Equation (D.9) is the linear response to A.

Now let us rewrite (D.10) as follows. If we put

$$X(t) = \mathrm{Tr}(e^{-i\mathcal{H}_0 t/\hbar}[A,\rho_0]e^{i\mathcal{H}_0 t/\hbar}B), \tag{D.12}$$

(D.10) is written as

$$\chi_{BA}(\omega) = \frac{i}{\hbar}\int_0^\infty e^{i\omega t-\delta t}X(t)dt. \tag{D.13}$$

Using the partial integration for (D.13), we obtain

$$\chi_{BA}(\omega) = \frac{i}{\hbar}\left[-\frac{X(0)}{i\omega} - \int_0^\infty \frac{e^{i\omega t-\delta t}}{i\omega}\frac{dX(t)}{dt}dt\right]. \tag{D.14}$$

From

$$X(0) = -\int_0^\infty e^{-\delta t}\frac{dX(t)}{dt}dt, \tag{D.15}$$

we obtain

$$\chi_{BA}(\omega) = -\frac{i}{\hbar}\int_0^\infty \frac{e^{i\omega t}-1}{i\omega}e^{-\delta t}\frac{dX(t)}{dt}dt. \tag{D.16}$$

Let us calculate $dX(t)/dt$. As ρ_0 and $\exp[i\mathcal{H}_0 t/\hbar]$ commute each other,

$$X(t) = \mathrm{Tr}[e^{-i\mathcal{H}_0 t/\hbar}Ae^{i\mathcal{H}_0 t/\hbar}\rho_0 B - \rho_0 e^{-i\mathcal{H}_0 t/\hbar}Ae^{i\mathcal{H}_0 t/\hbar}B]. \tag{D.17}$$

Differentiating the first term by t and putting $\dot{A} \equiv i(\mathcal{H}_0 A - A\mathcal{H}_0)/\hbar$, we obtain

$$\frac{i}{\hbar}\mathrm{Tr}[e^{-i\mathcal{H}_0 t/\hbar}(A\mathcal{H}_0 - \mathcal{H}_0 A)e^{i\mathcal{H}_0 t/\hbar}\rho_0 B] = -\mathrm{Tr}[\dot{A}\rho_0 B(t)] = -\langle B(t)\dot{A}\rangle. \tag{D.18}$$

Adding the derivative of the second term in (D.17) by t, we obtain

$$\frac{dX(t)}{dt} = -\langle B(t)\dot{A}(0) - \dot{A}(0)B(t)\rangle, \tag{D.19}$$

where $\dot{A}(0) = \dot{A}$.

Introducing Green's function

$$K^{\mathrm{R}}(t) = -\frac{i}{\hbar}\theta(t)\langle B(t)\dot{A}(0) - \dot{A}(0)B(t)\rangle, \tag{D.20}$$

we can write (D.16) as

$$\chi_{BA}(\omega) = -\int_{-\infty}^{\infty} \frac{e^{i\omega t} - 1}{i\omega} K^{\mathrm{R}}(t)dt. \tag{D.21}$$

Using the Fourier transform $K^{\mathrm{R}}(\omega)$ of $K^{\mathrm{R}}(t)$, we obtain

$$\chi_{BA}(\omega) = -\frac{K^{\mathrm{R}}(\omega) - K^{\mathrm{R}}(0)}{i\omega}. \tag{D.22}$$

From this equation, we can calculate $\chi_{BA}(\omega)$.

For simplicity, here we put $T = 0$ and $F(t) = \delta(t)$. Then (D.8) becomes

$$\langle B \rangle \equiv \varphi_{BA}(t) = \begin{cases} 0 & (t < 0) \\ \frac{i}{\hbar}\langle 0|[A, B(t)]|0\rangle & (t > 0), \end{cases} \tag{D.23}$$

where $|0\rangle$ is the ground state. The function $\varphi_{BA}(t)$ is called the response function. Since A operates at $t = 0$, for $t < 0$, $\varphi_{BA}(t) = 0$. This result means causality. The general function $F(t)$ is written as

$$F(t) = \int_{-\infty}^{\infty} F(t')\delta(t - t')dt'. \tag{D.24}$$

Then $\langle B \rangle$ can be written as

$$\langle B \rangle = \int_{-\infty}^{\infty} F(t')\varphi_{BA}(t - t')dt'. \tag{D.25}$$

Next, let us consider the Fourier transform of the general form of $F(t)$

$$F(t) = \frac{1}{2\pi}\int_{-\infty}^{\infty} F(\omega)e^{-i\omega t + \delta t}d\omega. \tag{D.26}$$

Since $F(t)$ is real, $F(-\omega) = F^*(\omega)$. In this case $\chi_{BA}(\omega)$ in (D.9) is written as

$$\chi_{BA}(\omega) = \int_{0}^{\infty} e^{(i\omega - \delta)t'}\varphi_{BA}(t')dt'. \tag{D.27}$$

If we assume $\langle B \rangle$ is real, $\chi_{BA}(\omega) = \{\chi_{BA}(-\omega)\}^*$. For $F(t)$ in (D.26), $\langle B \rangle$ is given by

$$\langle B \rangle = \frac{1}{2\pi}\int_{-\infty}^{\infty} e^{-i\omega t + \delta t}F(\omega)\chi_{BA}(\omega)d\omega. \tag{D.28}$$

Using the complete set of eigenfunctions $|n\rangle$ and the eigenvalue E_n of $\mathcal{H}_0$, we put $E_n - E_0 = \omega_{n0}$. Writing $\langle 0|A|n\rangle$ as A_{0n}, we obtain

$$\varphi_{BA}(t) = \begin{cases} 0 & (t < 0) \\ \frac{i}{\hbar}\sum_n \{A_{0n}B_{n0}e^{i\omega_{n0}t} - B_{0n}A_{n0}e^{-i\omega_{n0}t}\} & (t > 0). \end{cases} \tag{D.29}$$

We can see that $\varphi_{BA}(t)$ is real from (D.29). Inserting this result into $\chi_{BA}(\omega)$, we obtain

$$\chi_{BA}(\omega) = \sum_n \left\{ \frac{B_{0n}A_{n0}}{\omega - \omega_{n0} + i\delta} - \frac{A_{0n}B_{n0}}{\omega + \omega_{n0} + i\delta} \right\}. \tag{D.30}$$

The $\chi_{BA}(\omega)$ has a singular point at $\omega = \pm\omega_{n0} - i\delta$. Since ω_{n0} is continuous, $\chi_{BA}(\omega)$ possesses a discontinuity when ω crosses the real axis. Thus, $\chi_{BA}(\omega)$ is analytic in the upper

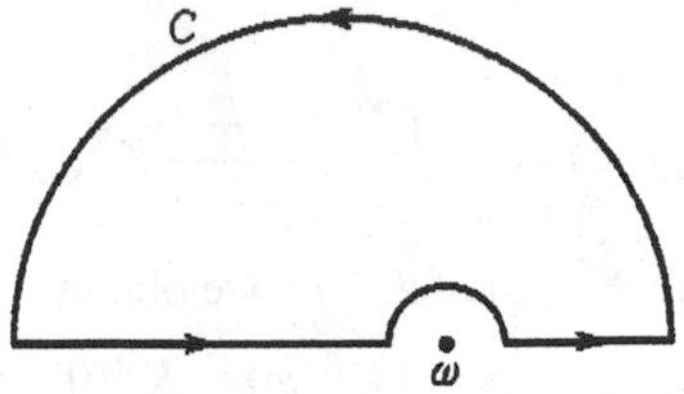

Fig. D.1 The integration path C.

half-plane. Now we consider the Fourier transform

$$\chi_{BA}(t) = \frac{1}{2\pi}\int_{-\infty}^{\infty} e^{-i\omega t+\delta t}\chi_{BA}(\omega)d\omega. \tag{D.31}$$

For $t < 0$, the integral path of ω can be chosen as that along the half-circle in the upper half-plane. By using this integral path, we can confirm that $\chi_{BA}(\omega)$ vanishes since it is analytic in the upper half-plane.

On the other hand, in the limit $\omega \to \infty$,

$$\chi_{BA}(\omega) \simeq \frac{1}{\omega}\langle 0|[B, A]|0\rangle. \tag{D.32}$$

If B and A commute each other, $\chi_{BA}(\omega)$ vanishes. In this case we calculate the next order terms, $1/\omega^2$:

$$\chi_{BA}(\omega) \simeq \frac{1}{\omega^2}\sum \omega_{n0}\{B_{0n}A_{n0} + A_{0n}B_{n0}\} = -\frac{1}{\omega^2}\langle 0|[[\mathcal{H}, A], B]|0\rangle. \tag{D.33}$$

In general, $\chi_{BA}(\omega)$ vanishes in the limit $\omega \to \infty$.

Here, we consider the following integral. The integral path C is shown in Fig. D.1. From the causality, $\chi_{BA}(\omega)$ is analytic in the upper half-plane the following integral vanishes:

$$\int_C \frac{\chi_{BA}(\omega')}{\omega' - \omega}d\omega' = 0, \tag{D.34}$$

where the contribution from the half-circle with infinite radius vanishes. Since the integral along the small half-circle gives $-i\pi\chi_{BA}(\omega)$, the following equation holds:

$$\int_{-\infty}^{\infty} \chi_{BA}(\omega')\mathrm{P}\frac{1}{\omega' - \omega}d\omega' = i\pi\chi_{BA}(\omega). \tag{D.35}$$

Here, if we put $\chi_{BA}(\omega) = \chi_{BA}'(\omega) + i\chi_{BA}''(\omega)$, where $\chi_{BA}'(\omega)$ and $\chi_{BA}''(\omega)$ are the real and imaginary parts, respectively, we obtain the equations

$$\chi_{BA}'(\omega) = \frac{1}{\pi}\int_{-\infty}^{\infty} \chi_{BA}''(\omega')\mathrm{P}\frac{1}{\omega' - \omega}d\omega', \tag{D.36}$$

$$\chi_{BA}''(\omega) = -\frac{1}{\pi}\int_{-\infty}^{\infty} \chi_{BA}'(\omega')\mathrm{P}\frac{1}{\omega' - \omega}d\omega'. \tag{D.37}$$

The above relations are called the Kramers–Kronig relation.

Moreover, the following equations hold as the sum rule:

$$\int_{-\infty}^{\infty} \chi_{BA}(\omega)d\omega = -i\pi \langle 0|[B, A]|0\rangle. \tag{D.38}$$

If $[B, A] = 0$, the following equation holds:

$$\int_{-\infty}^{\infty} \chi_{BA}(\omega)\omega d\omega = -i\pi \langle 0|[[\mathcal{H}_0, A], B]|0\rangle. \tag{D.39}$$

Appendix E. Transport equation derived by Éliashberg

We give a derivation of the transport equation in a degenerate Fermi system [1]. The response to an external field E is current density $j = ev$, and the time derivative of perturbation $\dot{A}$ is also given by j since $A = erE$ in (D.6).

From (D.22), the electric conductivity $\sigma_{\mu\nu}(\omega)$ is given by the correlation function between current fluctuations:

$$\sigma_{\mu,\nu}(\omega) = e^2 \sum_{kk',\sigma\sigma'} v_{k\mu} v_{k'\nu} \frac{1}{\omega} \mathrm{Im}\ K_{k\sigma,k'\sigma'}(\omega + i\delta). \tag{E.1}$$

Here $K_{k\sigma,k'\sigma'}(\omega + i\delta)$ is obtained by analytic continuation from a two-particle thermal Green's function $K_{k\sigma,k'\sigma'}(i\omega_m)$ in the upper half-plane, that is, we put $i\omega_m$ ($\omega_m > 0$) as $\omega + i\delta$ in the following:

$$K_{k\sigma,k'\sigma'}(i\omega_m) = \int_0^{1/T} d_\tau e^{i\omega_m \tau} \langle T_\tau a^\dagger{}_{k\sigma}(\tau) a_{k\sigma}(\tau) a^\dagger_{k'\sigma'} a_{k'\sigma'} \rangle, \tag{E.2}$$

$$a_{k\sigma}(\tau) = e^{(H-\mu N)\tau} a_{k\sigma} e^{-(H-\mu N)\tau}. \tag{E.3}$$

In our Hamiltonian, electron interactions exist among electrons, and give the self-energy and vertex correction in (E.2). There are two terms with and without vertex correction $\Gamma(\varepsilon, \varepsilon'; \omega)$, as shown in Fig. 7.5:

$$\begin{aligned} K_{k\sigma,k+q\sigma'}(i\omega_m) = &- T \sum_n G_{k+q}(\varepsilon_n + \omega_m) G_k(\varepsilon_n) \\ &- T^2 \sum_{n,n',k'} G_{k+q}(\varepsilon_n + \omega_m) G_k(\varepsilon_n) \Gamma_{kk';q}(\varepsilon_n, \varepsilon_{n'}; \omega_m) \\ &\times G_{k'+q}(\varepsilon_{n'} + \omega_m) G_{k'}(\varepsilon_{n'}). \end{aligned} \tag{E.4}$$

In order to carry out an analytic continuation in (E.4), we need to study the analytic properties of Γ. By using the Lehmann expansion of the two-particle Green's function $K(\varepsilon_n, \varepsilon'_n; \omega_m)$, Éliashberg showed that $K(\varepsilon, \varepsilon'; \omega)$ possesses singularities as a function of the complex variables ε, ε' and ω, when

$$\begin{aligned} &\text{(a)}\quad \mathrm{Im}\ \varepsilon = 0, \quad \mathrm{Im}(\varepsilon + \omega) = 0, \quad \mathrm{Im}\ \varepsilon' = 0, \quad \mathrm{Im}(\varepsilon' + \omega) = 0; \\ &\text{(b)}\quad \mathrm{Im}(\varepsilon + \varepsilon' + \omega) = 0; \\ &\text{(c)}\quad \mathrm{Im}\ \omega = 0, \quad \mathrm{Im}(\varepsilon - \varepsilon') = 0. \end{aligned} \tag{E.5}$$

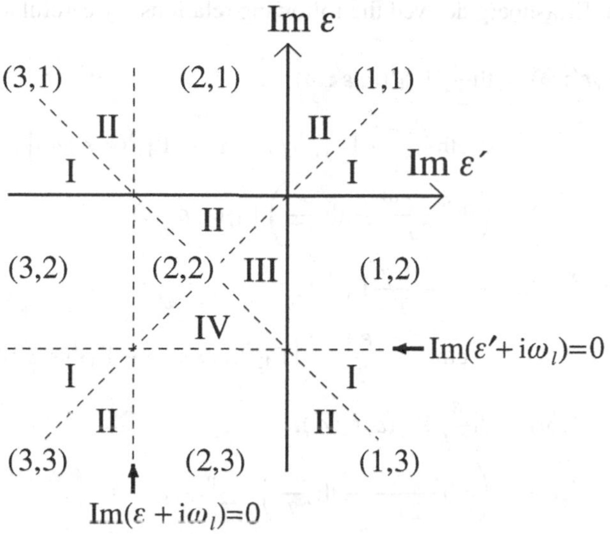

Fig. E.1 The definition of the region $[(l, m), N]$.

These singularities correspond to cuts parallel to the real axis in the complex planes of each argument. The whole space of the variables ε, ε' and ω is thus divided into several regions, in each of which Γ is an analytic function of any of its arguments, while the values of the other arguments are fixed.

We are interested in the properties of vertex $\Gamma(\varepsilon, \varepsilon'; \omega)$ as a function of ε and ε' for a fixed value of ω, with $\operatorname{Im}\omega > 0$. The analytic properties of Γ are shown in Fig. E.1, where singularities are plotted in the $\operatorname{Im}\varepsilon$, $\operatorname{Im}\varepsilon'$ plane. The lines drawn in the figure divide the plane into 16 regions, each of which corresponds to a function Γ which is analytic in that region for any of its arguments. Following Éliashberg, we obtain

$$K_{k\sigma,k'\sigma'}(\omega + i\delta) = -\frac{1}{4\pi i}\int_{-\infty}^{\infty} d\varepsilon \left[\text{th}\frac{\varepsilon}{2T} K_1(\varepsilon, \omega) + \left(\text{th}\frac{\varepsilon + \omega}{2T} - \text{th}\frac{\varepsilon}{2T} \right) \right.$$
$$\left. \times K_2(\varepsilon, \omega) - \text{th}\frac{\varepsilon + \omega}{2T} K_3(\varepsilon, \omega) \right], \qquad \text{(E.6)}$$

where

$$K_l(\varepsilon, \omega) = g_l(\varepsilon, \omega) + g_l(\varepsilon, \omega) \sum_{m=1}^{3} \frac{1}{4\pi i} \int_{-\infty}^{\infty} d\varepsilon' \mathcal{T}_{lm}(\varepsilon, \varepsilon' : \omega) g_m(\varepsilon', \omega). \qquad \text{(E.7)}$$

Here g_l is given by $(\omega > 0)$

$$g_1(\varepsilon, \omega) = G^{\mathrm{R}}(\varepsilon) G^{\mathrm{R}}(\varepsilon + \omega), \qquad \text{(E.8)}$$
$$g_2(\varepsilon, \omega) = G^{\mathrm{A}}(\varepsilon) G^{\mathrm{R}}(\varepsilon + \omega), \qquad \text{(E.9)}$$
$$g_3(\varepsilon, \omega) = G^{\mathrm{A}}(\varepsilon) G^{\mathrm{A}}(\varepsilon + \omega), \qquad \text{(E.10)}$$

where G^{R} and G^{A} are the retarded and advanced Green's functions, respectively. The quantities $\mathcal{T}_{lm}$ are connected with the functions Γ_{lm} which arise from the analytic continuation of the vertex part. Éliashberg derived the following relations, by careful analysis:

$$\mathcal{T}_{11}(\varepsilon, \varepsilon'; \omega) = \mathrm{th}\frac{\varepsilon'}{2T}\Gamma_{11}{}^{\mathrm{I}}(\varepsilon, \varepsilon'; \omega) + \mathrm{cth}\frac{\varepsilon' - \varepsilon}{2T}\left[\Gamma_{11}{}^{\mathrm{II}}(\varepsilon, \varepsilon'; \omega) - \Gamma_{11}{}^{\mathrm{I}}(\varepsilon, \varepsilon'; \omega)\right], \tag{E.11}$$

$$\mathcal{T}_{12}(\varepsilon, \varepsilon'; \omega) = \left(\mathrm{th}\frac{\varepsilon' + \omega}{2T} - \mathrm{th}\frac{\varepsilon'}{2T}\right)\Gamma_{12}(\varepsilon, \varepsilon'; \omega), \tag{E.12}$$

$$\mathcal{T}_{13}(\varepsilon, \varepsilon'; \omega) = -\,\mathrm{th}\frac{\varepsilon' + \omega}{2T}\Gamma_{13}{}^{\mathrm{I}}(\varepsilon, \varepsilon'; \omega) - \mathrm{cth}\,\frac{\varepsilon' + \varepsilon + \omega}{2T}\left[\Gamma_{13}{}^{\mathrm{II}}(\varepsilon, \varepsilon'; \omega) - \Gamma_{13}{}^{\mathrm{I}}(\varepsilon, \varepsilon'; \omega)\right], \tag{E.13}$$

$$\mathcal{T}_{21}(\varepsilon, \varepsilon'; \omega) = \mathrm{th}\frac{\varepsilon'}{2T}\Gamma_{21}(\varepsilon, \varepsilon'; \omega), \tag{E.14}$$

$$\begin{aligned}\mathcal{T}_{22}(\varepsilon, \varepsilon'; \omega) = {} & \left(\mathrm{cth}\frac{\varepsilon' - \varepsilon}{2T} - \mathrm{th}\frac{\varepsilon'}{2T}\right)\Gamma_{22}{}^{\mathrm{II}}(\varepsilon, \varepsilon'; \omega) \\ & + \left(\mathrm{th}\frac{\varepsilon' + \varepsilon + \omega}{2T} - \mathrm{cth}\frac{\varepsilon' - \varepsilon}{2T}\right)\Gamma_{22}{}^{\mathrm{III}}(\varepsilon, \varepsilon'; \omega) \\ & + \left(\mathrm{th}\frac{\varepsilon' + \omega}{2T} - \mathrm{cth}\frac{\varepsilon' + \varepsilon + \omega}{2T}\right)\Gamma_{22}{}^{\mathrm{IV}}(\varepsilon, \varepsilon'; \omega),\end{aligned} \tag{E.15}$$

$$\mathcal{T}_{23}(\varepsilon, \varepsilon'; \omega) = -\,\mathrm{th}\frac{\varepsilon' + \omega}{2T}\Gamma_{23}(\varepsilon, \varepsilon'; \omega), \tag{E.16}$$

$$\mathcal{T}_{31}(\varepsilon, \varepsilon'; \omega) = \mathrm{th}\frac{\varepsilon'}{2T}\Gamma_{31}{}^{\mathrm{I}}(\varepsilon, \varepsilon'; \omega) + \mathrm{cth}\frac{\varepsilon' + \varepsilon + \omega}{2T}\left[\Gamma_{31}{}^{\mathrm{II}}(\varepsilon, \varepsilon'; \omega) - \Gamma_{31}{}^{\mathrm{I}}(\varepsilon, \varepsilon'; \omega)\right], \tag{E.17}$$

$$\mathcal{T}_{32}(\varepsilon, \varepsilon'; \omega) = \left(\mathrm{th}\frac{\varepsilon' + \omega}{2T} - \mathrm{th}\frac{\varepsilon'}{2T}\right)\Gamma_{32}(\varepsilon, \varepsilon'; \omega), \tag{E.18}$$

$$\mathcal{T}_{33}(\varepsilon, \varepsilon'; \omega) = -\,\mathrm{th}\frac{\varepsilon' + \omega}{2T}\Gamma_{33}{}^{\mathrm{I}}(\varepsilon, \varepsilon'; \omega) - \mathrm{cth}\frac{\varepsilon' - \varepsilon}{2T}\left[\Gamma_{33}{}^{\mathrm{II}}(\varepsilon, \varepsilon'; \omega) - \Gamma_{33}{}^{\mathrm{II}}(\varepsilon, \varepsilon'; \omega)\right]. \tag{E.19}$$

When we approximate the Green's functions by poles near the Fermi surface, we obtain

$$g_1(\varepsilon, \omega) \simeq \left[G^{\mathrm{R}}(\varepsilon)\right]^2 = z_k{}^2(\varepsilon - \varepsilon_k^* + i\delta)^{-2}, \tag{E.20}$$

$$g_2(\varepsilon, \omega) = 2\pi i\frac{z_k{}^2\delta(\varepsilon - \varepsilon_k^*)}{\omega + 2i\gamma_k^* - \boldsymbol{v}\cdot\boldsymbol{q}}, \tag{E.21}$$

$$g_3(\varepsilon, \omega) = g_1(\varepsilon, \omega)^*. \tag{E.22}$$

Here we have assumed $\omega \ll T$ and $vq \ll T$. Among the quantities g_l, only the function $g_2(\boldsymbol{k}\varepsilon : \boldsymbol{q}\omega) = G^{\mathrm{R}}_{\boldsymbol{k}+\boldsymbol{q}}(\varepsilon + \omega)G^{\mathrm{A}}_{\boldsymbol{k}}(\varepsilon)$ depends appreciably on ω and $\boldsymbol{q}$.

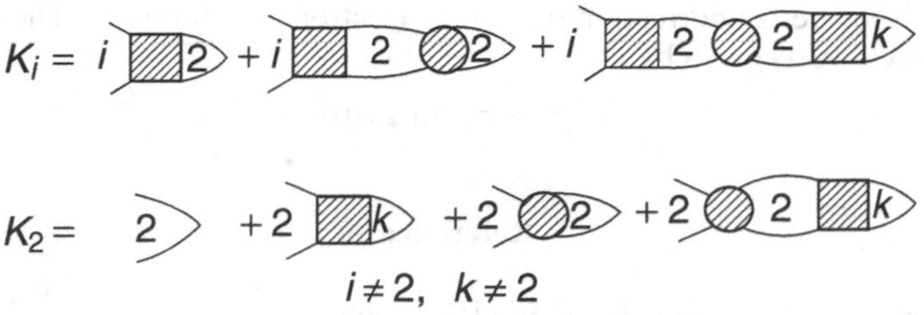

Fig. E.2 Diagrams for K_i and K_2.

We consider now the properties of the quantities $\mathcal{T}_{lm}$. We introduce the irreducible parts $\mathcal{T}^{(1)}_{lm}$ obtained as a result of the analytic continuation from all diagrams $\Gamma^{(1)}(\varepsilon_n, \varepsilon'_n; \omega_m)$ which do not contain a pair of lines of the type $G(\varepsilon_n + \omega_m)G(\varepsilon_n)$. We can easily verify that $\mathcal{T}_{lm}$ satisfies the equation, $k = (\boldsymbol{k}, \varepsilon)$:

$$\mathcal{T}_{lm}(k, k'; q) = \mathcal{T}^{(1)}_{lm}(k, k'; q) + \frac{1}{2i(2\pi)^4} \int d^4k d^4k' \mathcal{T}^{(1)}_{lj}(k, k''; q) g_j(k'', q) \mathcal{T}_{jm}(k'', k'; q). \qquad \text{(E.23)}$$

This equation means that $\mathcal{T}_{lm}$ can be written as a sum of diagrams containing different numbers of irreducible parts $\mathcal{T}^{(1)}$, which we depict by shaded rectangles and which are joined by pairs of lines g_l that we call section l. We have shown that among the three functions g_i only g_2 depends appreciably on ω and $\boldsymbol{q}$ when ω and $\boldsymbol{q}$ are small. We introduce for each function $\mathcal{T}_{lm}$ the totality of diagrams $\mathcal{T}^{(0)}_{lm}$ which does not contain the section 2. We shall then obtain one equation for $\mathcal{T}_{22}$:

$$\mathcal{T}_{22}(k, k'; q) = \mathcal{T}^{(0)}_{22}(k, k'; q) + \frac{1}{2i(2\pi)^4} \int d^4k \mathcal{T}^{(0)}_{22}(k, k''; q) g_2(k'', q) \mathcal{T}_{22}(k'', k'; q). \qquad \text{(E.24)}$$

We show now that the conductivity $\sigma_{\mu\nu}$ can be expressed in terms of the single function $\mathcal{T}_{22}$ only, while the other $\mathcal{T}_{lm}$ determine the values of renormalization constants. Bearing in mind the case $\omega \ll T$, $qv \ll T$, we retain the dependence on ω and $\boldsymbol{q}$ only in g_2 and $\mathcal{T}_{l2}$. From (E.1) and (E.6), we need only be interested in those diagrams $K_1(\varepsilon, \omega)$, $K_3(\varepsilon, \omega)$ and $K_2(\varepsilon, \omega)$ which contain at least one section 2.

All those diagrams for K_1, K_3 and $K_2(\varepsilon, \omega)$ are illustrated in Fig. E.2, in which the rectangles correspond to the quantities $\mathcal{T}^{(0)}$ which do not contain section 2, and a circle represents $\mathcal{T}_{22}$. Substituting the expressions for K_i corresponding to these diagrams into (E.6), and applying (E.21), we obtain

$$\sigma_{\mu\nu}(\omega) = \frac{ie^2}{2\Omega} \left[\sum_k v^*_{k\mu} \frac{1}{2T} \frac{\cosh^{-2}[(\varepsilon^*_k - \mu)/2T]}{\omega - \boldsymbol{v} \cdot \boldsymbol{q} + 2i\gamma^*_k} v^*_{k\nu} + \frac{1}{2} \sum_{kk'} z_k v^*_{k\mu} \frac{1}{2T} \frac{\cosh^{-2}[(\varepsilon^*_k - \mu)/2T] \mathcal{T}_{22}(k, k'; \omega)}{(\omega - \boldsymbol{v} \cdot \boldsymbol{q} + 2i\gamma^*_k)(\omega - \boldsymbol{v} \cdot \boldsymbol{q} + 2i\gamma^*_{k'})} z_{k'} v^*_{k\nu} \right], \qquad \text{(E.25)}$$

where ε^*_k, v^*_k and z_k are the energy, velocity of a quasi-particle and renormalization factor.

The term $\mathcal{T}_{22}$ is the vertex correction due to the electron correlation U. The damping rate of a quasi-particle γ_k^* is given by

$$\gamma_k^* = -z_k \text{ Im } \Sigma_k(0). \tag{E.26}$$

Reference

[1] G. M. Éliashberg, *Sov. Phys. JETP* **14** (1962) 886.

Index